2nd International Conference on Chemical Engineering (ICCE 2016)

IOP Conference Series: Materials Science and Engineering
Volume 162

Bandung, Indonesia
26 – 27 October 2016

Editor:

Katherine Kho

ISBN: 978-1-5108-3525-2
ISSN: 1757-8981

Printed from e-media with permission by:

Curran Associates, Inc.
57 Morehouse Lane
Red Hook, NY 12571

proceedings.com

Some format issues inherent in the e-media version may also appear in this print version.

Printed by Curran Associates, Inc. (2017)

Additional copies of this publication are available from:

Curran Associates, Inc.
57 Morehouse Lane
Red Hook, NY 12571 USA
Phone: 845-758-0400
Fax: 845-758-2633
Email: curran@proceedings.com
Web: www.proceedings.com

Table of contents

Volume 162

**Second International Conference on Chemical Engineering (ICCE) UNPAR
26–27 October 2016, Bandung, Indonesia**

**Accepted papers received: 17 November 2016
Published online: 12 December 2016**

Preface

011001
OPEN ACCESS
Second International Conference on Chemical Engineering (ICCE) UNPAR

Katherine Kho

011002
OPEN ACCESS
Peer review statement

Invited lectures

012001
OPEN ACCESS
Product engineering and sustainability

L P B M Janssen and C H C Janssen.....1

Separation Process

012002
OPEN ACCESS
Solar powered desalination system using Fresnel lens

M T B F Sales.....5

012003
OPEN ACCESS
Design of protonation constant measurement apparatus for carbon dioxide capturing solvents

S Ma'mun, E Amelia, V Rahmat, D R Alwani and D Kurniawan.....11

012004
OPEN ACCESS
Adsorption laboratory experiment for undergraduate chemical engineering: Introducing kinetic, equilibrium and thermodynamic concepts

S Muryanto and S Djatmiko Hadi.....17

012005
OPEN ACCESS
Evaluation of phase separator number in hydrodesulfurization (HDS) unit

A D Jayanti and A Indarto.....24

012006
OPEN ACCESS
Continuous absorption of CO_2 in packed column using MDEA solution for biomethane preparation

A Mindaryani, W Budhijanto and S S Ningrum.....30

012007
OPEN ACCESS
Influence of mid-point temperature of heavy hydrocarbons separator to the liquefaction process for small LNG plant

H A Hakim and A Indarto.....36

012008
OPEN ACCESS
Comparison on pore development of activated carbon produced by chemical and physical activation from palm empty fruit bunch

A Hidayat and B Sutrisno.....43

New and Renewable Energy

012009
OPEN ACCESS
Synthesis of renewable diesel through hydrodeoxygenation reaction from nyamplung oil (*Calophyllum Inophyllum* oil) using NiMo/Z and NiMo/C catalysts with rapid heating and cooling method

B H Susanto, M B Prakasa and M H Shahab.....51

012010
OPEN ACCESS
Hydrogen production by plasma electrolysis reactor of KOH-ethanol solution

N Saksono, T Batubara and S Bismo.....58

012011
OPEN ACCESS
Improvement of bio-oil yield and quality in co-pyrolysis of corncobs and high density polyethylene in a fixed bed reactor at low heating rate

D Supramono and S Lusiani.....64

012012
OPEN ACCESS
Temperature effect towards methane gas production and performances of anaerobic fixed bed reactors

H Budiastuti, D Widyabudiningsih and D R D Kurnia.....76

012013
OPEN ACCESS
Performance of continuous stirred tank reactor (CSTR) on fermentative biohydrogen production from melon waste

K Cahyari, Sarto, S Syamsiah and A Prasetya.....82

012014
OPEN ACCESS
Upgrading of bio-oil from the pyrolysis of biomass over the rice husk ash catalysts

B Sutrisno and A Hidayat.....88

Process System Engineering

012015
OPEN ACCESS
Vacuum infusion equipment design and the influence of reinforcement layers addition to the resin infusion time

A H Saputra and G Setyarso.....95

012016
OPEN ACCESS
Review of neural network modelling of cracking process

M N Rosli and N Aziz.....101

012017
OPEN ACCESS
Simulation of ethane steam cracking with severity evaluation

M N Rosli and N Aziz.....107

012018
OPEN ACCESS
Evaluation of shrinking core model in leaching process of Pomalaa nickel laterite using citric acid as leachant at atmospheric conditions

K C Wanta, I Perdana and H T B M Petrus.....113

012019
OPEN ACCESS
Kinetics study of carbon raiser on the reduction of nickel laterite from Pomalaa, Southeast Sulawesi

H T B M Petrus, A R Rhamdani, A D P Putera, I W Warmada, A T Yuliansyad and I Perdana.....118

012020
OPEN ACCESS
Effect of catalyst concentration and reaction time on the extraction of glucomannan from porang (*Amorphophallus oncophyllus*) flour via acid hydrolysis

A C Kumoro, T H A Yuganta, R Ratnawati and D S Retnowati.....125

Chemical Reaction Engineering

012021
OPEN ACCESS
The capacity of modified nickel catalysts derived from discharged catalyst of fertilizer plants for NOx treatment

T M P Ha, N T Luong and P N Le.....132

012022
OPEN ACCESS
UV/Fenton photo-oxidation of Drimarene Dark Red (DDR) containing textile-dye wastewater

T Hudaya, J Anthonios and E Septianto.....141

012023
OPEN ACCESS
Effect of initiator concentration to low-density polyethylene production in a tubular reactor

A Azmi and N Aziz.....147

012024
OPEN ACCESS
Transesterification of palm oil using sodium silicate base catalyst from geothermal sludge

I Perdana, N Nugrahanti, Sofiyah and I M Bendiyasa.....154

012025
OPEN ACCESS
Indonesian low rank coal oxidation: The effect of H_2O_2 concentration and oxidation temperature

S S Rahayu, F Findiati and F Aprilia.....161

012026
OPEN ACCESS
Application of a hybrid Electrocoagulation-Fenton process in yarn dye wastewater: Kinetic study

L. Riadi, L. Sapei, T. Lidiawati and Y.E. Agustin.....166

Fine Chemicals and Pharmaceuticals

012027
OPEN ACCESS
Pulverization of coffee silverskin extract as a source of antioxidant

S Tan, S P Kusumocahyo and D I Widiputri.....171

012028
OPEN ACCESS
Extraction and identification of bioactive compounds from agarwood leaves

N Y Lee, M A C Yunus, Z Idham, M S H Ruslan, A H A Aziz and N Irwansyah.....177

012029
OPEN ACCESS
Natural red dyes extraction on roselle petals

H M Inggrid, Jaka and H Santoso.....183

Supercritical Technology

012030
OPEN ACCESS
Supercritical carbon dioxide (SC-CO_2) extraction of essential oil from *Swietenia mahagoni* seeds

N S M Norodin, L M Salleh, Hartati and N M Mustafa.....190

012031
OPEN ACCESS
Optimization of supercritical carbon dioxide extraction of Piper Betel Linn leaves oil and total phenolic content

A H A Aziz, M A C Yunus, N H Arsad, N Y Lee, Z Idham and A Q A Razak.....197

012032
OPEN ACCESS
The effect of filler addition and oven temperature to the antioxidant quality in the drying of Physalis angulata leaf extract obtained by subcritical water extraction

R F Susanti and Desy Natalia.....204

Nanotechnology, Polymer, and Membrane Processes

012033
OPEN ACCESS
Preparation of silver nanoparticles loaded graphene oxide nanosheets for antibacterial activity

T T T Vi and S J Lue.....211

012034
OPEN ACCESS
The effect of activated carbon support surface modification on characteristics of carbon nanospheres prepared by deposition precipitation of Fe-catalyst

H Kristianto, A A Arie, R F Susanti, M Halim and J K Lee.....216

012035
OPEN ACCESS
Separating xylose from glucose using spiral wound nanofiltration membrane: Effect of cross-flow parameters on sugar rejection

N F M Roli, H W Yussof, M N A Seman, S M Saufi and A W Mohammad.....222

012036
OPEN ACCESS
Optimization of process condition of nanosilica production by hydrothermal method

N Qisti, N S Indrasti and Suprihatin.....228

012037
OPEN ACCESS
Synthesis and characterization of polyester thin film composite membrane via interfacial polymerization: Fouling behaviour of uncharged solute

K H Mah, H W Yussof, M N A Seman and A W Mohammad.....237

IOP Conf. Series: Materials Science and Engineering **162** (2017) 011001 doi:10.1088/1757-899X/162/1/011001

Preface

The International Conference on Chemical Engineering (ICCE) 2016 was held in Bandung, Indonesia from October 26-27, 2016. ICCE 2016 was organized by the Department of Chemical Engineering, Parahyangan Catholic University in cooperation with the University of New South Wales (UNSW) Australia and Nano Center Indonesia This conference addressed the important role of innovation in chemical process and product design. The objective of this conference was to highlight trends and research focuses in various areas of chemical engineering disciplines, especially when related to the innovation in chemical engineering fundamental concepts and applications.

The proceedings are a compilation of the accepted papers and represent important results of the conference. This book is divided into seven specific topics in the area of chemical engineering: Separation Process, New and Renewable Energy, Process System Engineering, Chemical Reaction Engineering, Fine Chemicals and Pharmaceuticals, Supercritical Technology and Nanotechnology, Polymer and Membranes.

We would like to acknowledge all people who were involved in ICCE 2016. Each individual and institution's support was very meaningful for the success of this conference. Especially, we would like to thank the organizing committee for their valuable advice in the organization and helpful peer review of the papers.

We sincerely hope that ICCE 2016 would serve as an ideal meeting point for academia, industrial practitioners, and authorities from around the globe, where every participant can share and at the same time update their most recent ideas and discoveries.

Katherine Kho, PhD. Parahyangan Catholic University

Peer review statement

All papers published in this volume of *IOP Conference Series: Materials Science and Engineering* have been peer reviewed through processes administered by the proceedings Editors. Reviews were conducted by expert referees to the professional and scientific standards expected of a proceedings journal published by IOP Publishing.

IOP Conf. Series: Materials Science and Engineering **162** (2017) 012001 doi:10.1088/1757-899X/162/1/012001

Product engineering and sustainability

L P B M Janssen[1,a*] and C H C Janssen [2,b]

[1] ex. Department of Chemical Engineering, University of Groningen, Groningen, the Netherlands

[2] ex. Facultad de Quimica, Universidad Nacional Autonomos Mexico Mexico City, Mexico

E-mail : [a]l.p.b.m.janssen@rug.nl, [b]engineerjanssen@gmail.com

Abstract. Chemical engineering has made a substantial contribution to the improvement of the environment during the last decades. Many processes have become more sustainable and harmful waste streams are minimised. However, considerable improvement of sustainability can still be obtained in product engineering and design. Especially the aspects that are important at the end of the life cycle of the product can be improved considerably. A priority list for the design of more sustainable products is presented and illustrated with examples of dailychemical engineering practice.

1. Introduction

During the last decades the awareness of sustainability has induced several changes in chemical engineering practice. In recent years, however, the emphasis in sustainable chemical engineering has been lying mainly in the field of process improvement. Judicious optimization of individual process steps and improved process integration have led to significant minimisation or even to the complete prevention of undesired waste streams in process industry. Significant further contributions from chemical engineering to a more sustainable world can also be achieved in the area of product design and product engineering.

In order to design for optimal product sustainability it is important to consider the total life cycle of a product and to distinguish clearly between the three life cycle steps of a product: production, utilisation and disposal. The requirements for production and utilisation are generally well considered in the concepts of basic product engineering and they are also directly related to marketing and product economics[1]. However, the end of use conditions still need more emphasis in order to come to a real 'green' product design.

These end of use requirements contradict often to the properties that are desired during the actual utilisation of the product. For example, in biodegradable packaging hydrophobicity is important for good product stability but for a well-controlled and swift degradation after use a certain degree of hydrophilicity of the product is indispensable. A complicating factor in the acceptance of a sustainable design is the fact that higher quality during the utilisation step can directly result in a higher price where the willingness the pay more for the sustainability of a products is only very moderate. As a rule of thumb it is often assumed that the customer is not willing to pay more than 10% extra for a more sustainable product.

Sustainability considerations are currently not taken into account sufficiently. The focus seems to be on the development phase of a product. First we discuss shortly this development phase. Consequently, we discuss some sustainability considerations. Finally a set of illustrations is given how the sustainability of a product can be improved so that its impact at the end of use is less severe to the environment.

2. The development phase.

During the design of a product two distinct steps can be distinguished (figure 1). In the first step, the diverging step, as many new ideas as possible are generated while in the second step, the converging step, the new ideas are tested and evaluated. The objective of the diverging step is to obtain as many new alternatives for the product and process as possible without too much concern about any possible constraints It is clear that in this brain storming step of the product development not too much consideration is generally given to the sustainability of the product and the process.

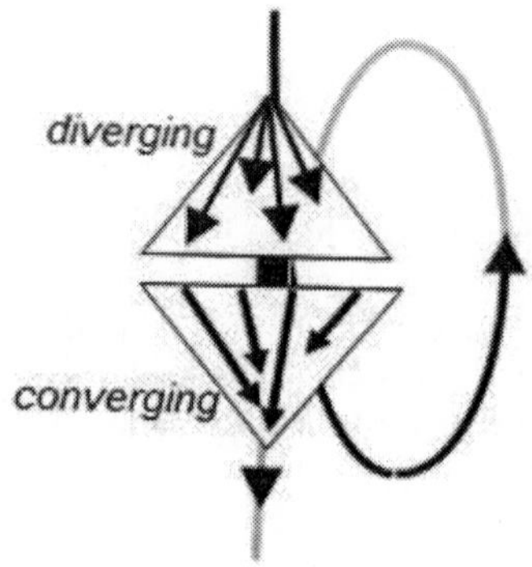

Figure 1: Two steps in product design: idea generation (diverging) and evaluation (converging)

This will mainly occur in the second step, the evaluation phase of the product. Here the process converges by critically scrapping many alternatives to come to the final design. In this step emphasis lies on the interactions between product, process, marketing and sustainability. In this step special attention must be given to the following considerations:

- existing know-how and experience,
- considering financial and time constraints
- environmental impact of the product.

In a normal development path the converging and diverging step will be passed in an alternating way many times. However, the more the final product concept is approached, the more important the converging step becomes and the more emphasis has to be given to the sustainability of the final product.

3. Sustainability considerations.

In order to decrease the impact of a product at the end its use, more attention needs to be paid to the disposal and the recycling in the development phase of the product. When considering the end of use alternatives for a general product several different alternatives can be distinguished. These alternatives can be ordered according to their sustainability hierarchy or environmental preference. Two additional effects should be taken into consideration when designing a product :

- environmentally undesired components should be avoided (like Volatile Organic Solvents (VOS) and certain heavy metals)
- preferably very limited amounts of components with a limited availability should be used (like for instance Lithium or Lanthanum)

Although the relative importance of the different environmental considerations strongly depends on the product considered they can be put in a generalized order of preference:

- avoid the use of hazardous components
- make the product re-usable
- avoid depleting feed stocks
- design the product for easy 'back to feedstock'
- re-use the product as filler
- re-use the recycle for lower quality products
- facilitate clean pyrolysis
- avoid hazardous smoke at incineration
- avoid tailings in case of land fill

4. Illustrations

Examples of the list above can be found in many all-day products. This list is not extensive but serves as an illustration of the possibilities and challenges in sustainable product engineering.

1. Hazardous components should already in the design phase of the product be avoided as much as possible. VOS can sometimes be substituted by more benign solvents like ionic liquids[2,3] or supercritical carbon dioxide[4]. A very well-known example of VOS substitution is water based paint, where the irreversible solidification of the binder occurs by a chemical reaction after the water has evaporated. Another possibility to avoid hazardous materials is the use of modern organic pigments that make it possible to eliminate heavy metals for colored products.
2. Re-usable products, like deposited packaging are preferred if possible. Although this seems to be an obvious conclusion it can only be applied in a very limited way. Too often plastic bottles are used for storage of other materials like gasoline or organic solvents. These components can diffuse into the bottle material and therefore form severe restrictions to re-use. Special care has to be taken with food packaging because these pollutions can form a serious health hazard on re-use.
3. Avoiding depleting feed stocks is especially important if the materials are scarce and have a limited availability. Lithium and Lanthanium are examples of elements that are widely used in electronics and for batteries and indispensable for the present electric car technology. The total minable amount of these elements is limited which implies that prices and availability can change quickly. Moreover most of these elements are concentrated in limited areas in the world and therefore subject to geopolitical tensions. The threat of a delivery stop of Li for electrical cars by P.R. China to the Japanese car industry has recently increased the research in other types of batteries. But also if the base materials are not scarse the use of renewable resources is important for avoiding depletion. Several new bioplastic products have been introduced in de last decade using common materials like celluloses and starches[5]. Also the production of fuels and platform chemicals from biomass contributes to the avoiding of depletion[6].
4. Re-use the product as a filler is only scarcely applied in plastics and rubber industry but it forms an economically attractive way of re-using old materials. Co-injection molding techniques allow to form an object which inside consists of reused polymers while the outside consists of virgin material[7]. Another example of re-use as filler can be found in the tire industry where the objective is to substitute a few percent of the virgin rubber in car tires by partly devulcanized rubber from old tires. Research of this application looks promising.
5. Back to feed stock can form an alternative for the reuse of plastic products. Because of this possibility thermoset materials should be avoided in the design as much as possible. When using thermoplastics, the returned products can be regranulized and the polymer can be devolatized to remove traces of impurities. Successively the granules can be extruded or injection molded to be formed into new products. Care has to be taken that no mixing of different types of polymers occurs. Due to the bad miscibility of (high molecular weight) polymers in general, traces of impurities will result in very poor product quality. During the design of a product it is preferential to use only one type of polymer or to have easily separable units each consisting of one single material.
6. Re-use for lower quality can be an alternative for the use of the recycle for the same product. It is particularly attractive if the possibility to purify the recycle stream is limited like in food packaging. A rather recent new technique is the use of devulcanized tire rubber to reinforce bitumen roof coverings[8]. This technique can also be used instead of back to feed stock. A typical other example is the production of large sheets of PET (poly ethylene terephthalate) for agricultural use. Plants are protected against winter conditions and against

IOP Conf. Series: Materials Science and Engineering **162** (2017) 012001 doi:10.1088/1757-899X/162/1/012001

pests by these sheets made of PET bottles.

7. Pyrolysis of waste streams is often considered if the alternatives above are technically or economically not feasible. Many of our waste streams still possess a large amount of useful components that can be obtained by cracking or pyrolysis. Relatively clean biomass can be used to obtain syngas. Another product from the pyrolysis of organic waste is biochar that can be used as fertilizers. Cracking of plastic waste streams can produced an oily substance with a slightly higher certain number than conventional diesel oil[9]. The pyrolysis of tires is reported to produce methane gas, combustible fluids and carbon black.
8. Incineration can occur in a controlled way in large incineration ovens or uncontrolled by small local fires. Controlled incineration can sometimes be used to obtain part of the formation energy that is still available in the waste stream and it is sometimes called thermal recycling. If the product is likely to be burned in an uncontrolled way (open fires at a country side) it is important that during the design of the product such materials are chosen, that the smoke from the fires does not contain harmful components.
9. Land fill as a way of disposal should in general be considered to be a last resort. If land fill is likely, the product should not contain components that can be leached out easily and come into contact with the soil and the water. This can result in serious pollution of the drinking water[10].

5. Conclusions

After the development in process engineering product engineering forms an area with interesting new possibilities in chemical engineering. Whereas many efforts have been successful to make processes more environmental friendly, improving the sustainability of products still forms a challenge. Although different products require different measures to improve their sustainability it is possible to present a generalized list of considerations and to order these measures in an order of urgency or desirability.

6. References

[1] Wesseling J A, Kiil S and Vigild M E 2007 Design and Development, John Wiley & Sons, Hoboken

[2] Janssen C H C, Kroon M C, Metz S, Spronsen J van and Witkamp G J 2010 *J Chem Eng Data*, **55** 3391

[3] Janssen C H C., Macías-Ruvalcaba N A., Aguilar-Martínez M and Kobrak M 2015 *International Reviews in Physical Chemistry* **34**(4) 591

[4] Muljana H, Picchioni F, Heeres H J and Janssen L P B M 2009 *Carbohydrate Polym* **78** 511

[5] Janssen L P B M and Moscicki L 2009 Thermoplastic Starch, Wiley VHC Weinheim

[6] Girisuta B, Janssen L P B M and Heeres H J 2007 *Ind Eng Chem Res.* **46** 1969

[7] Gomes M, Martino D, Pontes A J and Viana J C 2011 *Polym Eng Sci* **51** 2398

[8] Sutanto P, Picchioni F, Janssen L P B M 2006 *J Appl Pol Sci* **102** 5028

[9] Sharuddin S D A, Abnisa F, Daud W M A W and Aroun M K 2016 *Energy Conversion and Management* **115** 308

[10] Kohar I and Janssen L P B M 2008 Small scale water purification, Royal Netherlands Academy of Sciences the Hague

IOP Conf. Series: Materials Science and Engineering **162** (2017) 012002 doi:10.1088/1757-899X/162/1/012002

Solar powered desalination system using Fresnel lens

M T B F Sales, MEE, PhDTM
Chemical Engineering Department, University of St. La Salle, Bacolod City, Philippines

E-mail : tashie.sales@usls.edu.ph

Abstract. The Philippines is surrounded by coastal areas and these areas can be a potential source for potable water. This study aims to design and construct a solar powered desalination system using Fresnel lens. The experimental study was conducted using polluted salt water for the sample and desalination was carried out using the designed system. The desalination system was composed of the solar concentrator, solar still and the condenser system. The Fresnel lens was made of acrylic plastic and was an effective solar concentrator. Solar stills made of dark colored glass bottles were effective in absorbing the solar energy. The condenser system made of polybutylene and polystyrene were effective in condensing the vapor at ambient temperature. The shortest time of vaporization of the salt water was at 293 sec and the optimum angle of position of the lens was 36.42^0. The amount of condensate collected was directly proportional to the amount of salt water in the solar still. The highest mean efficiency of the designed set-up was 34.82%. The water produced by the solar powered desalination system using Fresnel lens passed the standards set by WHO (World Health Organization) for drinking water.

1. Introduction

The Philippines is an archipelago composed of about 7,107 islands and 18,000 km of coast line. Only 14% of these islands are inhabited but about 60% of the population live in coastal areas. These coastal areas provide food, livelihood and economic support to the people who live there but sometimes these areas can also be a threat to the populace when natural disasters occur. Natural disasters like typhoons, flash floods and drought can affect their day to day living when mass shortages of food, clean water and shelter happens.

Clean water is very important because the human body is composed principally of water which comprises about 60% of the body weight. The human body is dependent on water because it carries out the toxins from the vital organs, transfers the nutrients to the cells and provides a moist environment to the ear, nose and throat tissues. Humans need to drink an average of 2.5 liters of potable water per day in order to prevent dehydration [1].

Bacolod City is located in the northwestern coast of the Negros Island Region. BACIWA (Bacolod City Water District) is an agency that is mandated to maintain and operate the water supply and distribution systems for domestic, industrial, municipal and agricultural uses within the district of Bacolod. It has to supply about 1.25 million liters of water per day to the consumers in order to meet the average daily requirement of human consumption excluding other water requirements like cooking, bathing and washing. Presently, the Bacolod district is suffering from the effects of El Niño and BACIWA claims that there will be a water crisis by 2025 if other alternatives for water resources are not found [2]. Unfortunately, BACIWA can only serve about 42% of the population within the service area and about 28% within the service area cannot be supplied with water [3]. As a result, many people still do not have access to adequate supply of potable water because there is insufficient supply of potable water. There is a need to look for other sources of potable water.

The sun's energy is a renewable energy source and this is free. Bacolod City receives on an average of 300 W/m^2 of incoming solar radiation [4]. This energy can be used in many ways and they offer minimal environmental problems. This renewable energy resource can be used to power a household desalination system using Fresnel lens.

Solar powered desalination technology using Fresnel lens is not yet developed and used in the Philippines. The plan to use the solar energy from the sun to power the desalination system is not new but the design of using Fresnel lens to speed up the evaporation rate has not been tested in the Philippines.

Harnessing the sun's energy to power the desalination system using Fresnel lens can be another alternative source for potable water; therefore this study was conducted in order to determine whether a solar powered desalination system using Fresnel lens can produce potable water.

2. Materials and Methods

The study was divided into two phases. The first phase included the design and construction of the desalination system. The second phase of the study included the pilot testing of the solar powered desalination system and the testing of the parameters of the product from the solar powered desalination system as required by World Health Organization for drinking water quality.

The first phase of the research design involved the design and construction of the desalination system. The design set-up of the desalination system included the solar concentrator, the solar still, the condenser tubes, and the condensate collector. The illustration for the design set-up of the desalination system is shown below.

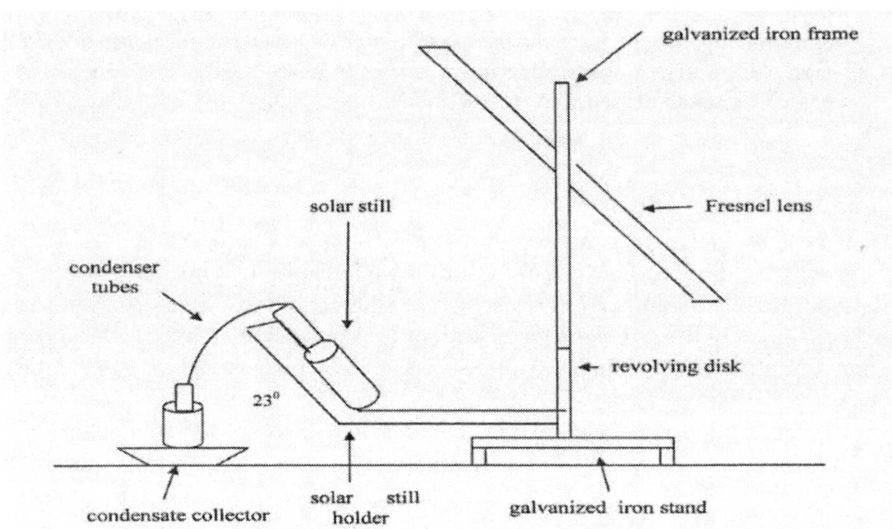

Figure 1. Design set-up of the desalination system

The solar concentrator was composed of the Fresnel lens and its support. Solar concentrator used was Fresnel lens that was made of acrylic plastic and had the following dimensions : 1067 mm high and 584 mm wide. The framed Fresnel lens was hinged to two galvanized iron pipes 857 mm high and spaced 792 mm apart. The hinges allowed the lens to rotate 360^0 in vertical motion. The galvanized iron stands were attached to a revolving disk made of galvanized iron pipe that rotated the panel in any direction and was about 190 mm high. The revolving disk was attached to two galvanized iron pipe supports spaced 792 mm apart. Galvanized iron was chosen because the stand had to be stable enough to support the lens even during windy conditions. See figure below.

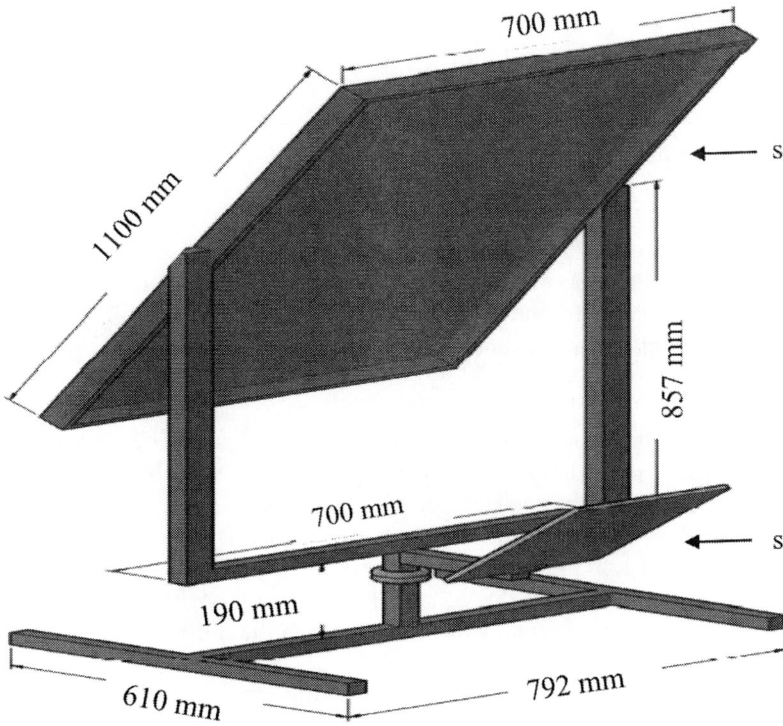

Figure 2. Design of solar concentrator

The solar still that held the salt water were 1000 ml capacity dark colored glass bottles. Dark colored glass bottles have 92% heat absorption property [5]. Solar still holder was made of galvanized iron No. 22 with a gage thickness of about 0.79 mm. Condenser system was composed of the condenser tubes and the condensate collector. Condenser tubes were made of 12.7 mm diameter hose made of polybutylene. The condensate collector was made of 500 ml glass bottles enclosed in a polystyrene container measuring 240 mm x 300 mm x 330 mm. The polystyrene was 12.7 mm thick.

The second phase of the study included the pilot testing of the desalination system and the testing of the parameters. The design set-up was tested by adjusting the angle of the lens with respect to the altitude of the solar noon measured in degrees. The altitude of the solar noon was taken from the daily monitoring of the website of PAGASA and DOST. The angle of the lens was measured relative to the ground. The position of the sun varies every hour so three different angles were used in the experiment taken every 20 minutes per hour to determine the angle and the time at which vaporization starts. Testing time was only from 10 am to 2 pm. No definite angle was used because actual data was taken during the experiment. The angle of the lens and the time of the day with the shortest time of vaporization was deduced to be the optimum position for the design.

The desalination of the salt water was based on the optimum design set-up. The salt water from Purok Bolinao of Barangay Banago coastal area was tested for the following parameters before desalination.

Three different volumes of salt water were passed through the desalination system using the optimum position. The volumes were 300 ml, 500 ml and 600 ml based on the 1000 ml capacity of the still. Time of vaporization and the amount of condensate collected was determined for each volume of salt water that was tested.

The efficiency of the method was measured in terms of the energy required for the vaporization versus the energy of the sun's radiation. Equation (1) was used to determine the efficiency:

$$\text{Efficiency} = \frac{\text{Energy required for the vaporization of distillate recovered}}{\text{Energy in the sun's radiation that falls on the still}} \times 100\% \qquad (1)$$

The efficiency was calculated for the different amounts of salt water used based on the energy required for the vaporization of distillate recovered per energy in the sun's radiation that falls on the still. The formula of List was used to solve for the energy in the sun's radiation and software developed by David Lawrence was used to calculate for the conditions required by List's formula [6].

3. Results and Discussions

For the design set-up of the desalination system, the angle of inclination of the solar still with respect to the ground was maintained at 23^0 all throughout the experiment. Using this angle to distill the salt water proves that the sun's position does not vary much in the Philippines [7].

Results of the study showed that the mean angle of the lens was at 39.5^0.Vaporization time occurred faster between 11:40 am to 12:40 pm. Angle of the position of the lens also did not vary much at this period because Bacolod is situated on the equator so the sun's position does not vary much in this location. The shortest time of vaporization was at 293 sec so the optimum angle of position of the lens was 36.42^0.

Results of the condensate collected indicates that the amount of condensate collected was directly proportional to the amount of salt water in the solar still. Only about 25% of the salt water in the solar still evaporated and condensed. Evaporation is dependent on the solar energy emitted by the sun and this solar energy was affected because the experiment was performed during the rainy months. Furthermore, the sun moves farther from the equator during the last half of the year so the solar energy was not intense enough to increase the amount of vaporization. The figure below shows the relationship between the amount of condensate collected of the designed set-up and the amount of saltwater in the solar still.

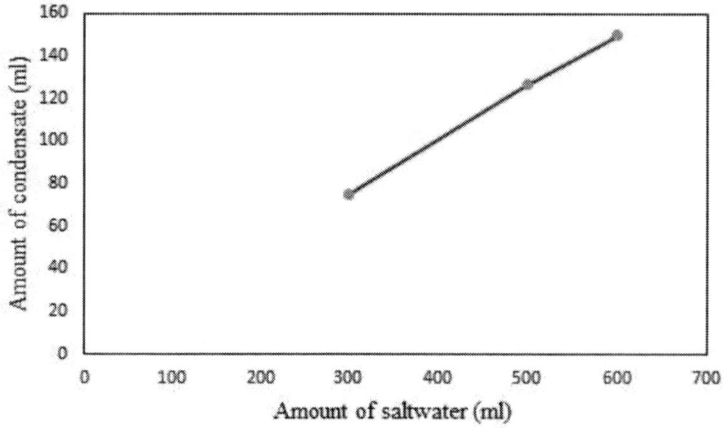

Figure 3. Amount of saltwater vs amount of condensate

The graph above shows that the amount of condensate collected is directly proportional to the amount of salt water in the solar still. This means that more condensate can be collected if the amount of salt water in the solar still was increased. The reason behind this is that there is less air space above the liquid when the amount of salt water in the solar still is increased. Less air space indicates that fewer vapor molecules would come into contact with the air molecules thereby increasing the amount of liquid to vaporize. When vapor molecules become entrapped with the gases in the air space, their kinetic energies decreases leading to condensation back into the solar still.

The salt water was taken from the shores of Purok Bolinao in Barangay 1 of Banago. The sample salt water was slightly cloudy in color and presence of sediments were evident.

Table 1 shows the results of the parameters of the salt water that were tested:

Table 1. Characteristics of the salt water

Parameter	Methodology	Results	WHO Standards
A. Physical			
Color	colorimetric	126 PtCo units	5PtCo units
turbidity	nephelometric	44.7 NTU	5 NTU
B. Chemical			
chloride	titrimetric	2615 ppm	250 ppm
Iron	spectrophotometric	0.37 ppm Fe	1 ppm Fe
manganese	spectrophotometric	0.80 ppm Mn	0.5 ppm Mn
pH	electrode membrane	7.21	6.5 – 8.5
sulfate	spectrophotometric	370 ppm	250 ppm
total dissolved solids	electrode membrane	664 g/L	500 g/L
C. Bacteriological			
Most Probable Number	Multiple Tube Fermentation Technique	21.6 x 10^4 MPN/ 100 ml sample	0- 50 MPN/100 ml sample

Results of Table 1 shows that concentrations for color, turbidity, chloride, manganese, sulfate, total dissolved solids and most probable number for the salt water were higher than the standard values set by World Health Organization for drinking water. High concentrations for these parameters indicate the presence of pollutants in the salt water. The high level of MPN is due to the fact that people living near the coastal area of Purok Bolinao do not have access to proper sanitary disposal systems so direct dumping of waste water is directly done into the seawater. The existing characteristics of the salt water in Purok Bolinao implies that it was not potable since it did not pass the standards for drinking water set by WHO.

The condensate collected was observed to be colorless and very clear. Table 2 shows the results of the parameters of the condensate that were tested.

Table 2. Characteristics of the condensate

Parameter	Methodology	Results	WHO Standards
A. Physical			
Color	colorimetric	9 PtCo units	5PtCo units
Turbidity	nephelometric	5.35 NTU	5 NTU
B. Chemical			
Chloride	titrimetric	30 ppm	250 ppm
Iron	spectrophotometric	0.03 ppm Fe	1 ppm Fe
manganese	spectrophotometric	0.4 ppm Mn	0.5 ppm Mn
pH	electrode membrane	7.11	6.5 – 8.5
sulfate	spectrophotometric	2 ppm	250 ppm
total dissolved solids	electrode membrane	7 g/L	500 g/L
C. Bacteriological			
Most Probable Number	Multiple Tube Fermentation Technique	< 1	0 – 50 MPN/100 ml sample

IOP Conf. Series: Materials Science and Engineering **162** (2017) 012002 doi:10.1088/1757-899X/162/1/012002

Results of Table 2 shows that all parameters passed the standard value set by WHO standards for drinking water except for color and turbidity. The condensate is slightly turbid because its value is 7% higher than the accepted value but the value for color is higher than the accepted value. Color is an aesthetic parameter which means that it is an indication of the presence of a coloring matter in the water. Since the bacteriological parameter passed the test, the presence of color may be due to matter of chemical origin. The presence of this chemical was not tested since the parameters set by WHO were just limited to the parameters mentioned in Table 1.

The characteristics of the condensate produced by the designed set-up using the Fresnel lens implies that it is potable since it passed the standards for drinking water except for the aesthetic parameter. Table 3 shows the mean efficiency of the designed desalination set-up using different amounts of saltwater.

Table 3. Efficiency of the system

Amount of Saltwater [ml]	Efficiency [%]
300	17.18
500	29.44
600	34.82

Resultsof Table 3 shows that the greatest amount of salt water had the highest efficiency for the designed desalination set-up. This was due to the fact that the greatest amount of salt water in the experiment had the greater heat requirement. The amount of heat required to vaporize the salt water was dependent on the amount of condensate recovered. The energy required to vaporize the salt water increased because it was directly proportional to the amount of salt water used in the operation.

The low efficiency of the designed set-up implies that the solar powered desalination system was dependent on the solar energy.

4. Conclusions

The design set-up for the solar powered desalination system using Fresnel lens can desalinate saltwater. Fresnel lens made of acrylic plastic can be used as a solar concentrator for a solar powered desalination system. Solar stills inclined at 23^0 with respect to the ground can optimize the vaporization angle if location of operation is in the Philippines. The shortest time of vaporization was at 293 sec and the optimum angle of position of the lens was 36.42^0. The amount of condensate collected was directly proportional to the amount of salt water in the solar still.The design set-up for desalinating the salt water using the Fresnel lens was effective in removing the pollutants. The design set-up for the solar powered desalination system produced a condensate that passed the standards set by WHO for drinking water.

5. References

[1] Hammer M 2007 *Water and Waste-Water Technology 3rd ed.* (New York : John Wiley & Sons)
[2] Lipa M 2010 Water crisis in Bacolod City *Interview TV Patrol Bacolod City*
[3] Information on http://www.baciwa.com.ph
[4] Information on http://fuzo.com/science/RadIntro.htm#List
[5] Green D W and Perry R H 2008 Perry's Chemical Engineer's Handbook, 8th Ed. (New York : McGraw-Hill International)
[6] List R J 1971 *Smithsonian Meteorological Table* (Washington : Smithsonian Institution Press)
[7] Information on http://www.sciencedirect.com/science

Design of protonation constant measurement apparatus for carbon dioxide capturing solvents

S Ma'mun, E Amelia, V Rahmat, D R Alwani and D Kurniawan
Department of Chemical Engineering, Islamic University of Indonesia, Jl. Kaliurang Km. 14.5, Yogyakarta 55501, Indonesia

E-mail: sholeh.mamun@uii.ac.id

Abstract. Global warming phenomenon has led to world climate change caused by high concentrations of greenhouse gases (GHG), e.g. carbon dioxide (CO_2), in the atmosphere. Carbon dioxide is produced in large amount from coal-fired power plants, iron and steel production, cement production, chemical and petrochemical manufacturing, natural gas purification, and transportation. Carbon dioxide emissions seem to rise from year to year; some efforts to reduce the emissions are, therefore, required. Amine-based absorption could be deployed for post-combustion capture. Some parameters, e.g. mass transfer coefficients and chemical equilibrium constants, are required for a vapor-liquid equilibrium modeling. Protonation constant (pK_a), as one of those parameters, could then be measured experimentally. Therefore, an experimental setup to measure pK_a of CO_2 capturing solvents was designed and validated by measuring the pK_a of acetic acid at 30 to 70 °C by a potentiometric titration method. The set up was also used to measure the pK_a of MEA at 27 °C. Based on the validation results and due to low vapor pressure of CO_2 capturing solvents in general, e.g. alkanolamines, the setup could therefore be used for measuring pK_a of the CO_2 capturing solvents at temperatures up to 70 °C.

1. Introduction

Removal of carbon dioxide (CO_2) is an important industrial operation. Carbon dioxide is produced in large amount from coal–fired power plants, steel production, chemical and petrochemical manufacturing, cement production, natural gas purification, and transportation sectors. The reasons for the CO_2 removal are traditionally due to technical and economic concerns [1]. Carbon dioxide present in natural gas will reduce the heating value of the gas. Besides, as an acid gas, CO_2 has a potential to cause corrosion in pipes, process equipment, and catalyst poisoning in ammonia synthesis [2]. Natural gas pipe lines usually permit from 1 to 2% CO_2 and sometimes as high as 5% [3]. In the past decades, CO_2 removal from flue gas streams was mainly deployed for Enhanced Oil Recovery (EOR) operations. Moreover, CO_2 was also produced for other industrial applications, such as urea production, carbonation of brine, welding, food and beverage carbonation, dry ice, and soda ash industry [4, 5]. However, environmental concerns, such as the global climate change, are now focused as one of the most important and challenging environmental issues facing the world community, and have motivated intensive research on CO_2 capture, utilization, and storage. Carbon dioxide as one of the greenhouse gases (GHG) is currently responsible for over 60% of the enhanced greenhouse effect, while methane (CH_4) contributes for 20%, and the remaining 20% is caused by nitrous oxide (N_2O), a number of industrial gases, and ozone. Scientific evidence now strongly suggests that increased levels of GHG may lead to higher temperature, and cause climate change on a global scale. Various climate models estimate that the global average temperature may rise by 1.4 to 5.8°C by the year 2100 [1, 6].

A wide range of technologies currently exist for separation and capture of CO_2 from gas streams as given by [4]. Such systems have been used in the chemical industry and in the production of technical gases for industrial and laboratory use [7]. In principle, various aforementioned methods could be used for the removal of CO_2. Several process–related factors, according to [8], affect the selection of the appropriate method, such as CO_2 concentration in the feed stream, nature of other contaminants present in the feed stream such as H_2S and water in natural gas, pressure and temperature at which the feed stream is available, product considerations, geographical location, etc. In addition, other criteria that need to consider can also be found in [9]. Therefore, research on solvent development has been conducted intensively, e.g. amino acid salts [10, 11], to meet those criteria.

Absorption can be used for gas separation. It consists of physical and chemical absorptions. The physical absorption is generally favorable at high pressures, i.e. above 5 atm [12], while the latter is vice versa. The chemical absorption is, therefore, more suitable for the CO_2 removal process from exhaust gases where the gas stream is typically at atmospheric pressure.

Amine-based absorption is the most common technology for CO_2 removal today. It is a process with considerable inherent problems, particularly when used on large gas flows, e.g. exhaust from fossil–fuel–fired power stations. The processes are bulky, leading to large investment costs. They have high–energy requirements, and the absorbents in use today are not stable and form degradation products that need to be handled [1]. Aqueous monoethanolamine ($H_2NCH_2CH_2OH$, MEA), a primary amine, has been used extensively for post combustion CO_2 capturing process. It has several advantages over other commercial alkanolamines, such as high reactivity, low solvent cost, low molecular weight and, thus, high absorbing capacity on a mass basis, and reasonable thermal stability and thermal degradation rate. However, MEA has some drawbacks such as high enthalpy of reaction with CO_2 leading to higher regeneration energy consumption, formation of a stable carbamate and also formation of degradation products with COS or oxygen bearing gases, inability to remove mercaptans, and more corrosive than many other alkanolamines, thus it needs corrosion inhibitors at higher concentrations [13-16].

Mathematical modeling of vapor-liquid equilibria (VLE) is required in designing both absorber and desorber columns. The CO_2-amine-H_2O system is a reactive system where mass transfer and chemical reaction occur simultaneously [17]. To solve the VLE model, some parameters such as mass transfer coefficients and chemical equilibrium constants should be provided. However, some parameters could be determined experimentally and the rest could be regressed from the model. Protonation constant (K_a or pK_a), as one of the parameters, could then be measured experimentally.

The objective of this study is to design an apparatus to measure protonation constants of CO_2 capturing solvents, e.g. alkanolamines. However, the setup should previously be validated by measuring protonation constants of acetic acid (CH_3COOH) at a range of temperatures from 30 to 70 °C by a potentiometric titration method.

1.1. Protonation constant

The protonation constant, also known as the acid dissociation constant, can be calculated from the reaction of a weak acid and a strong base where K_a value is equal to the concentration of H^+ ions in the solution. This occurs when the total volume of titrant added is equal to half of equivalence volume (V_e). For a weak acid HA reacting with a strong base BOH, the protonation constant K_a can then be expressed as follows [18]:

$$K_a = \frac{[H^+][A^-]_{total}}{HA_{left}} \approx [H^+] \tag{1}$$

Equation (1) indicates that the K_a value of the weak acid HA is approximately equal to the concentration of H^+ ions or the pK_a equals the pH of the solution.

2. Materials and Method

The materials used in this work consist of glacial acetic acid, sulfuric acid (H_2SO_4), sodium hydroxide (NaOH), MEA, buffer solutions, and distilled water. The acetic acid, NaOH, and H_2SO_4 were obtained from Merck, while MEA was obtained from Sigma-Aldrich. Those chemicals were used without further purification.

Figure 1 shows a designed setup to measure the pK_a values of CO_2 capturing solvents. It consists of a 300-mL jacketed reactor, a pH meter (Lutron PH-201 with pH electrode PE-03 and accuracy up to 0.01), a TFA® digital thermometer with accuracy up to 0.1°C, a memmert water bath, an OHAUS® digital balance with readability up to 0.0001 g, a magnetic stirrer, and a 5-mL syringe.

The setup is designed to determine protonation constants of CO_2 capturing solvents. However, it should previously be validated. The setup validation was conducted by measuring the acetic acid protonation constants at a range of temperatures from 30 to 70 °C by the potentiometric titration method. Moreover, the setup was also tested to measure the pK_a of MEA at a certain temperature. A similar procedure to determine the pK_a using the potentiometric titration method can be found elsewhere [18].

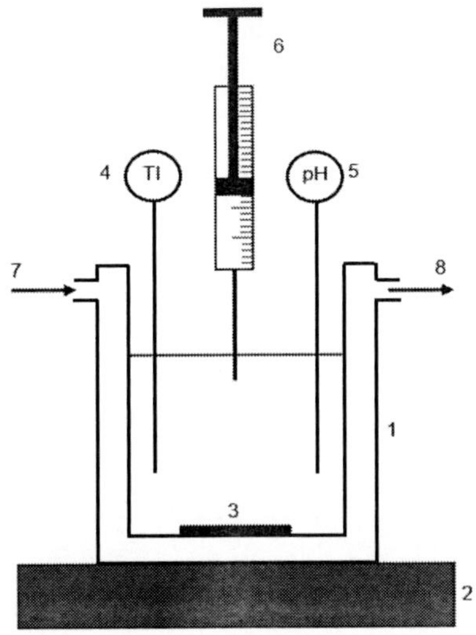

1: Jacketed reactor
2: Magnetic stirrer
3: Stirrer
4: Thermometer
5: pH meter
6: Syringe
7: Heating medium inlet
8: Heating medium outlet

Figure 1. Experimental setup for protonation constant measurements

3. Results and Discussion

Figure 1 shows the designed setup that shall be validated by measuring the acetic acid protonation constants at 30 to 70 °C by the potentiometric titration method with NaOH 0.1M solution as titrant.

The titration was conducted by discretely adding the titrant into the solution. The amount of the titrant added was calculated from the syringe weight difference. During the titration, there was no temperature rise observed significantly. This occurred due to very little heat that was released during neutralization the acetic acid by NaOH, i.e. 55.2 kJ mol^{-1} [19]. The titration was terminated after the solution reached pH about 12 in which the solution was in excess of NaOH. This is to ensure that all the acetic acid molecules have reacted with NaOH.

Figure 2 shows the titration results for Run 1 at 30 °C with 8.36 mmoles of acetic acid where the predicted equivalence point was reached at the titrant volume of about 83 mL. It is, however, relatively difficult to determine the precise equivalence point directly from the curve. Another way to determine the equivalence point precisely is, therefore, required. The equivalence point is then determined when the slope ($\Delta pH/\Delta V$) gives the highest value. By enlarging the slope around the equivalence point, it can be seen from Figure 3 that the equivalence volume is found to be 82.685 mL. Furthermore, the pK_a value occurs when the titrant volume equals half of the equivalence volume, i.e. 41.343 mL. At this point, pK_a is equal to pH and is found to be 4.77. In addition, the setup was also tested by measuring the pK_a of MEA at 27 °C which was found to be 9.65. The pK_a of MEA was measured by titrating MEA solution with H_2SO_4 0.1M as the titrant. A similar procedure was applied to other data and the results together with the literature data are then summarized in Table 1.

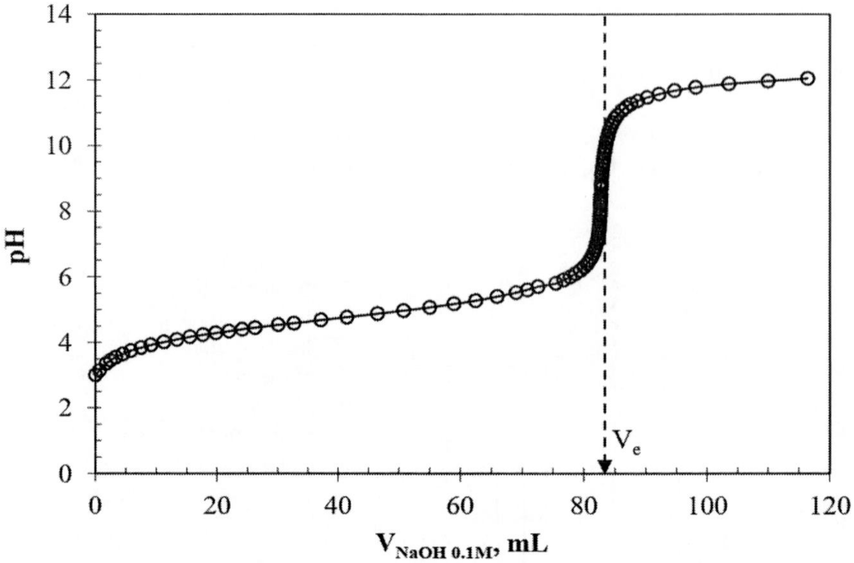

Figure 2. Titration curve of acetic acid at 30 °C (V_e = equivalence volume)

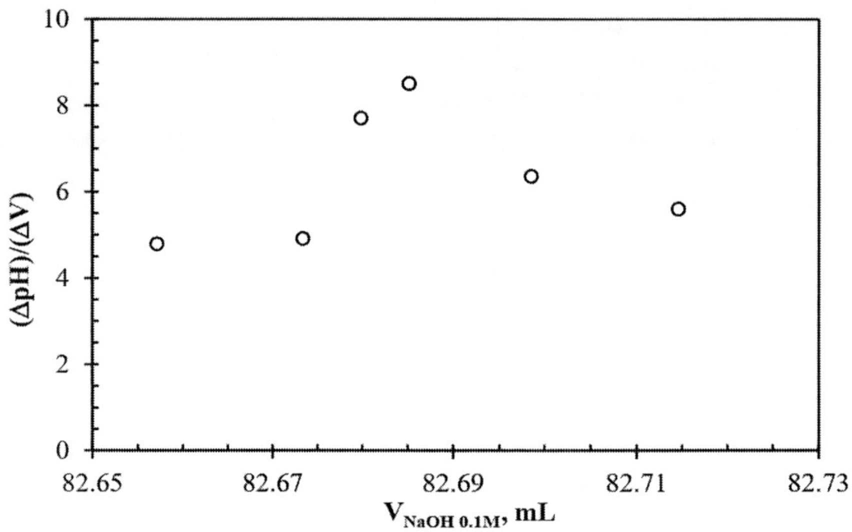

Figure 3. Titration slope (ΔpH/ΔV) of acetic acid at 30 °C

Table 1. pK_a values of acetic acid at various temperatures

Run	T (°C)	Number of data	V_e (mL)	pK_a	pK_a from [20]
1	30	93	82.685	4.77	4.76
2		68	81.302	4.75	
3	40	83	82.803	4.78	4.77
4		60	83.022	4.78	
5	50	93	80.980	4.83	4.79
6		44	83.735	4.86	
7	60	46	83.720	4.91	4.81
8		52	85.960	4.88	
9	70	74	78.027	5.00	
10		51	87.790	4.94	
11[a)]	27	50	62.380	9.65	9.44[b)]

[a)]MEA, [b)]interpolated from [21]

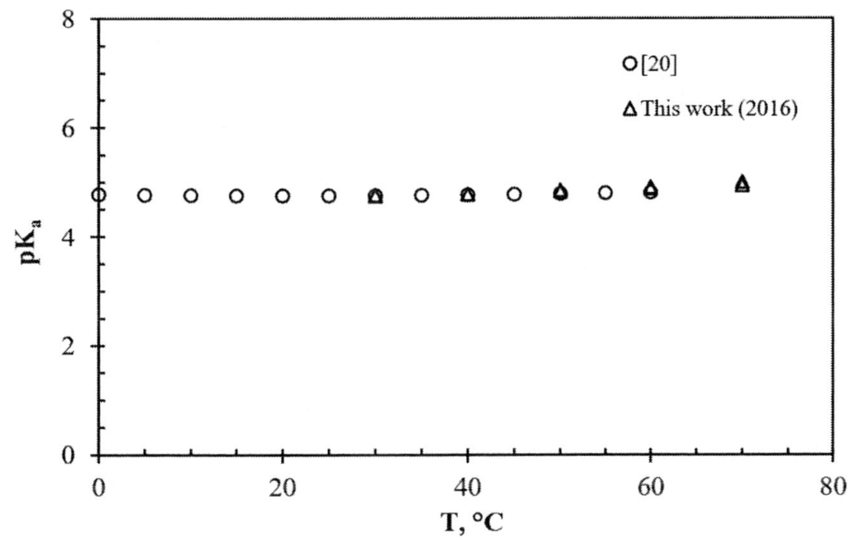

Figure 4. Comparison of pK_a values obtained from this work with those from literature

It can be seen from Figure 4 that the pK_a values of acetic acid obtained from this work, in general, agree with those of [20], except at 70 °C where the data obtained are slightly higher. This might occur due to evaporation loss during titration that took place for some time (approx. 2 hrs). Since the vapor pressure of the acetic acid at 70 °C is relatively high, i.e. 27.6 kPa [22], thus the evaporation loss during titration will significantly reduce the amount of acetic acid in the solution, thereby affecting the measurement result. In case of alkanolamines with somewhat lower vapor pressures, e.g. 1.25 kPa for MEA at 70 °C [22], the evaporation loss during titration might, therefore, be negligible. Based on the validation results, the setup could perform pK_a measurement of alkanolamines at temperatures up to 70 °C. At higher temperatures than 70 °C, the setup could give high measurement error due to evaporation loss problem.

4. Conclusion

An experimental setup for measuring the protonation constant of CO_2 capturing solvents was designed and validated by measuring the acetic acid protonation constants at 30 to 70 °C by the potentiometric titration method. The validation results, in general, agree well with the results published in the literature. In addition, the setup was also used to measure the pKa of MEA at 27 °C and the result showed in a good agreement with the published data. The setup could, therefore, be used to measure the protonation constants of CO_2 capturing solvents, e.g. alkanolamines, at temperatures up to 70 °C.

Acknowledgment
The authors acknowledge the Islamic University of Indonesia for the financial support through the Department of Chemical Engineering Research Grant 2016.

References

[1] Ma'mun S, Svendsen H F, Hoff K A, Juliussen O 2007 *Energy Convers. Manage.* **48** 251
[2] Astarita G D, Savage W, Bisio A 1983 *Gas treating with chemical solvents* (New York: John Wiley and Sons)
[3] Buckingham P A 1964 *Hydrocarbon Process.* **43** 113
[4] Rao A B, Rubin E S 2002 *Environ. Sci. Technol.* **36** 4467
[5] Suda T, Iijima M, Tanaka H, Mitsuoka S, Iwaki T 1997 *Environ. Prog.* **16** 200
[6] Williams M 2002 *Climate change: Information kit* (Geneva: the United Nations Environment Programme and the United Nations Framework Convention on Climate Change)
[7] Desideri U, Corbelli R 1998 *Energy Convers. Manage.* **39** 857
[8] Shaw T P, Hughes P W 2001 *Hydrocarbon Process.* **5** 53
[9] Hoff K A, da Silva E F, Kim I, Grimstvedt A, Ma'mun S 2013 *Energy Procedia* **37** 292
[10] Ma'mun S, Kim I 2013 *Energy Procedia* **37** 331
[11] Ma'mun S 2014 *Energy Procedia* **51** 191
[12] Kohl A, Nielsen R 1997 *Gas Purification* (Houston - Gulf Publishing Company)
[13] Bucklin R W 1982 *Oil Gas J.* **11** 204
[14] Li Y G, Mather A E 1994 *Ind. Eng. Chem. Res.* **33** 2006
[15] Jou F Y, Mather A E, Otto F D 1995 *Can. J. Chem. Eng.* **73** 140
[16] Liu Y, Zhang L, Watanasiri S 1999 *Ind. Eng. Chem. Res.* **38** 2080
[17] Ma'mun S, Jakobsen J P, Svendsen H F, Juliussen O 2006 *Ind. Eng. Chem. Res.* **45** 2505
[18] Ma'mun S, Kamariah, Sukirman, Kurniawan D, Amelia E, Rahmat V, Alwani D R 2016 *SICEST 2016*, 9-10 November 2016, Bangka, Indonesia
[19] Jain A K, Singh N B, Das S S 2013 *Physical Chemistry Volume 2* (New Delhi - New Age International Ltd.)
[20] Harned H S, Ehlers R W 1933 *J. Am. Chem. Soc.* **55** 652
[21] Bates R G, Pinching G D 1951 *J. Res. Natl. Bur. Stand.* **46**349
[22] Sinnot R K 2005 *Coulson & Richardson's Chemical Engineering Volume 6 - Chemical Engineering Design* (Oxford - Elsevier Butterworth-Heinemann)

Adsorption laboratory experiment for undergraduate chemical engineering : Introducing kinetic, equilibrium and thermodynamic concepts

S Muryanto and S Djatmiko Hadi
Department of Chemical Engineering, UNTAG University in Semarang, Bendhan Dhuwur Campus, Semarang, 50233 INDONESIA

E-mail : stmuryanto@gmail.com

Abstract. Adsorption laboratory experiment for undergraduate chemical engineering program is discussed. The experiment demonstrated adsorption of copper ions commonly found in wastewater using bio-sorbent, i.e. agricultural wastes. The adsorption was performed in a batch mode under various parameters: adsorption time (up to 120 min), initial pH (2 to 6), adsorbent dose (2.0 to 12.0 g L^{-1}), adsorbent size (50 to 170 mesh), initial Cu^{2+} concentration (25 to 100 ppm) and temperatures (room temp to 40°C). The equilibrium and kinetic data of the experiments were calculated using the two commonly used isotherms: Langmuir and Lagergren pseudo-first-order kinetics. The maximum adsorption capacity for Cu^{2+} was found as 94.34 mg g^{-1}. Thermodynamically, the adsorption process was spontaneous and endothermic. The calculated activation energy for the adsorption was observed as high as 127.94 kJ mol^{-1}. Pedagogically, the experiment was assumed to be important in increasing student understanding of kinetic, equilibrium and thermodynamic concepts.

1. Introduction

Adsorption is one of the methods frequently used in pollution prevention processes, especially in wastewater treatment, as it is considered to have several advantages compared to other methods such as chemical precipitation, coagulation-flocculation, ion exchange etc [1]. Wastewater may contain hazardous metal ions such as Cu^{2+}, Zn^{2+}, Pb^{2+}etc., and adsorption using agricultural wastes has been an extensive area of research to remove the metals [1-7]. In such a case, the adsorption method is seen as environmentally friendly.

Students may have a basic understanding of adsorption [8]. However, they may have not realized that adsorption involves basic concepts such as equilibrium, kinetic and thermodynamic. In this paper, a laboratory experiment on adsorption of Cu^{2+} using powder of dried teak (*Tectona grandis sp*) leaves is discussed. The experiment was carried out with the intention to enhance student understanding on the concept of equilibrium, kinetic and thermodynamic. Two undergraduate students conducted the experiment, analyzed the data obtained, plotted the graphs etc under the supervision of the authors. The experimental results were discussed in terms of two adsorption isotherms, i.e. Langmuir and Freundlich models, which are regarded as convenient in describing experimental results over a wide range of concentrations [9].

2. Materials and Methods

2.1. Dried powder leaf preparation

Fallen leaves of *Tectona grandis sp* were collected locally, i.e. from UNTAG University yard and gardens. These samples were washed extensively with tap water to remove dust and other particulates, followed by rinsing with distilled water. The cleaned leaves were then oven-dried (60°C, 24 h) until constant weight was obtained. The resulting samples were pulverized using a household blender and sieved to obtain powder of the following sieve-cuts: 50-80, 80-100, and 100-170 mesh. The sieved powder, named teak leaves powder (= TLP), was stored in air-tight plastic containers according to their size for subsequent used as adsorbents.

2.2. Preparation of stock solutions
Stock solution of 1,000 mg L^{-1} concentration of Cu^{2+} was prepared by dissolving 4.785 g of $CuCl_2 \cdot 6H_2O$ in doubly distilled water to obtain a volume of 1,000 ml. The experimental solutions of varying concentrations were obtained by successive dilution of the stock solution. The pH of the experimental solutions was adjusted to the desired values with 0.1 M HCl. All chemicals used were analytical grade and used without further purification.

2.3. Adsorption experiment
Batch adsorption experiments were carried out to investigate the effect of initial pH of solution, adsorbent dose, adsorbent particle size, initial metal ion concentration, contact time and temperature on the adsorption of Cu^{2+} on TLP.

In all experiments, fixed volume (100 mL) of Cu(II) were placed in conical flasks and thoroughly mixed with TLP of varying sizes and dose and stirred with varying speed (100 to 250 rpm). The experiments were done at different temperatures (ambient, 30, 40°C) for up to 120 min. Different conditions of initial pH of the solution (2.0; 4.0; 6.0) were evaluated. At the end of the adsorption run the solutions were filtered through ash-less filter paper.

The concentrations of Cu^{2+} remaining in the filtrates were analyzed using flame atomic absorption spectrophotometer (AAS-Varian, AA240). The percentage adsorption of Cu^{2+} adsorbed from the solution by TLP was calculated using the following (Eq. 1):

$$\% \text{ adsorption} = \frac{(C_0 - C_e)}{C_0} \times 100 \tag{1}$$

where C_0 and C_e are the initial and equilibrium Cu^{2+} concentration (mg L^{-1}), respectively.

The amount of Cu^{2+} adsorbed at time t, i.e. qt (mg g^{-1}) was calculated as follows (Eq.2):

$$q_t = \frac{(C_0 - C_t)V}{m} \tag{2}$$

where
C_t = concentration of Cu^{2+} at time t, (mg L^{-1}),
V = volume of solution, L,
m = mass of TLP, g.
The adsorption at equilibrium, q_e (mg g^{-1}) was measured using Eq (3) as follows:

$$q_e = (C_0 - C_e)\frac{V}{m} \tag{3}$$

Calibration curves of standard solutions of Cu^{2+} were used to calculate the Cu^{2+} concentrations. The experimental data were fitted to two adsorption models namely, Langmuir and Freundlich.

3. Results and discussion

3.1. Effect of initial pH, adsorbent size, agitation and adsorbent dose
Variation in pH leads to variation in the metal ion speciation and the dissociation of active functional sites on the adsorbent [3,4]. Further, different metal ions would show different optimum pH for adsorption, as would different biomass [5]. The effectiveness of Cu^{2+} adsorption by TLP as a function of pH is plotted (Figure 1-a).

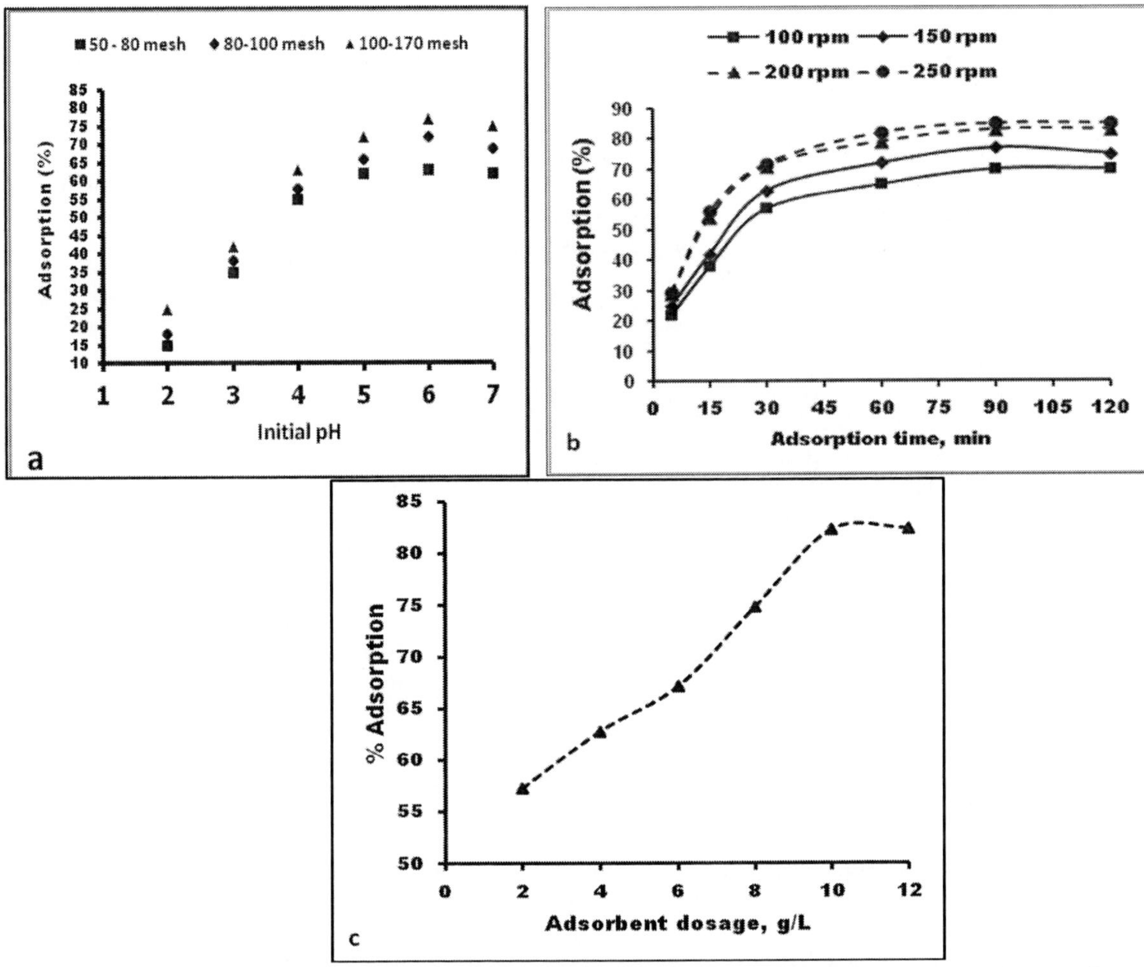

Figure 1 Effect of initial pH (a); effect of agitation (b) and effect of adsorbent dose (c) on adsorption Cu^{2+}

As can be seen (Figure.1-a), for all adsorbent sizes (50-80 mesh; 80-100 mesh; 100-170 mesh), a steady increase in adsorption with increasing pH is observed up to pH 6.0. Beyond 6.0, the adsorption capacity decreases, indicating that pH = 6.0 is the optimum. Further, as adsorption is a surface phenomenon, the smaller adsorbent sizes (= higher mesh) offer higher adsorption capacity due to the availability of relatively larger surface area. Depending on the adsorbent size, the increase in adsorption rises from 15% (pH 2.0) up to 77% (pH 6.0).

The effect of agitation speed on the adsorption of Cu^{2+} was carried out for pH =6.0, with TLP size cuts of 100 to 170 mesh, for a fixed initial concentration of 100 mg $L^{-1}$of Cu^{2+} at room temperature (Figureure 1-b). As shown, the adsorption capacity of TLP increased from 21% up to 85%. The effect of increased agitation speed can be attributed to the decrease in the boundary layer thickness surrounding TLP particles [4, 6], hence increased more adsorption.

For a fixed particle size, the amount of adsorbent correlates linearly with available surface area for adsorption. As can be seen the percentage of adsorption of Cu^{2+} increases dramatically with increasing TLP dose up to 10 g L^{-1}(Figure 1-c). This can be due to the availability of more adsorption sites with the increased dosage. However, beyond 10 g L^{-1}, the percentage removal remained relatively constant. This phenomenon can be attributed to the attainment of equilibrium in the system [3].

3.2. Adsorption models

Isotherm models would show the distribution of adsorbate (= Cu^{2+}) between the solution and the TLP particles. The isotherm models most widely used to describe the adsorption processes, i.e. the Langmuir and Freundlich isotherms relate the rate adsorption density, qe, (= Cu^{2+} uptake per unit weight of TLP) to equilibrium Cu^{2+} concentration in the bulk liquid phase, Ce.

The Langmuir adsorption isotherm model suggests a uniform surface and that the adsorbate uptake occurs by monolayer sorption without interaction among the adsorbate.

The linearized form of Langmuir isotherm equation is as follows [4]:

$$\frac{1}{q_e} = \frac{1}{Q^0} + \frac{1}{bQ^0C_e} \tag{4}$$

where,

q_e = adsorbate adsorbed per unit weight of adsorbent at equilibrium, mg g^{-1},
C_e = equilibrium concentration of the adsorbate in solution, mg L^{-1},
Q^0 = monolayer adsorption capacity, mg g^{-1},
b = a constant related to free energy or net enthalphy of adsorption.

When 1/qe was plotted against 1/Ce, a straight line with slope $1/bQ^0$ and intercept $1/Q^0$ is obtained (Figure 2-a), which shows that the adsorption of Cu^{2+} by TLP follows the Langmuir isotherm. The Langmuir parameters, Q^0 and b, are calculated from the slope and intercept of the graphs (Figure 2-a) and are given in Table 1.

Table 1. Langmuir and Freundlich constants for copper ion adsorption

Langmuir parameters	Adsorption maxima	Binding energy	R^2 coef.
	Q^0 [mg g^{-1}]	constant b [mg $L^{-1}]^{-1}$	
	94.34	79.97	0.993
Freundlich parameters	Adsorption capacity	Adsorption intensity 1/n	R^2 coef.
	K_F [mg g^{-1}]		
	2.36	1.61	0.987

The Freundlich isotherm describes the equilibrium of an adsorption system where the adsorption surfaces are heterogeneous. Hence the surface does not assume monolayer characteristics. As compared to the Langmuir model, the energy term *b* in the Langmuir equation (see Eq 4) varies as a function of surface coverage [4].

The linearized Freundlich equation is as follows:

$$\log q_e = \log K_F + {}^1\!/_n \log C_e \tag{5}$$

where q_e is the amount of adsorbate adsorbed per unit weight of adsorbent (mg g^{-1}), C_e is the equilibrium concentration of the adsorbate in solution (mg L^{-1}), K_F is the relative adsorption capacity constant of the adsorbent (mg g^{-1}) and 1/n is adsorption intensity.

Plots of Freundlich isotherm (log q_e against log C_e) would yield a straight line with slope 1/n and intercept log K_F. (Figure 2-b). The Freundlich parameters for the adsorption of Cu^{2+} (Table 1) indicate that the adsorption followed both the Freundlich (R^2 = 0.98) and the Langmuir isotherms (R^2 = 0.99), with the Langmuir isotherm best fits the current experiment.

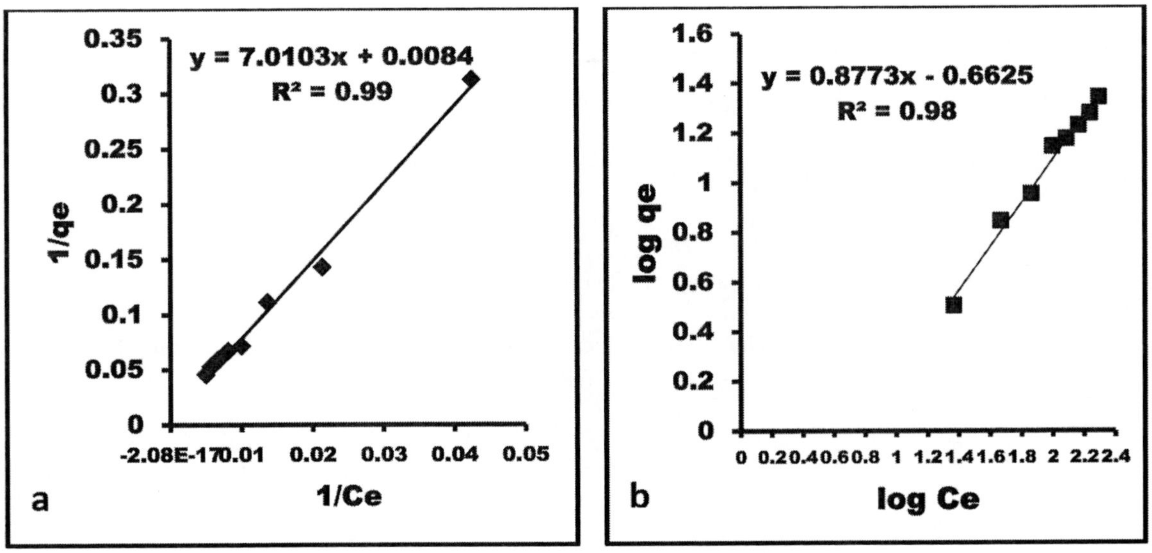

Figure 2. The Langmuir (a); and the Freundlich (b) adsorption isotherms obtained

3.3. Thermodynamic parameters

Figure 3-a depicts the effect of temperatures on Cu^{2+} adsorption carried out in the current study. As shown, higher temperatures enhanced the metal ions adsorption. It can be argued that temperature increases the mobility of ions by contributing additional energy, hence more adsorption. Further, higher temperature may generate a swelling effect so that Cu^{2+} may permeate further into the internal structure of the TLP.

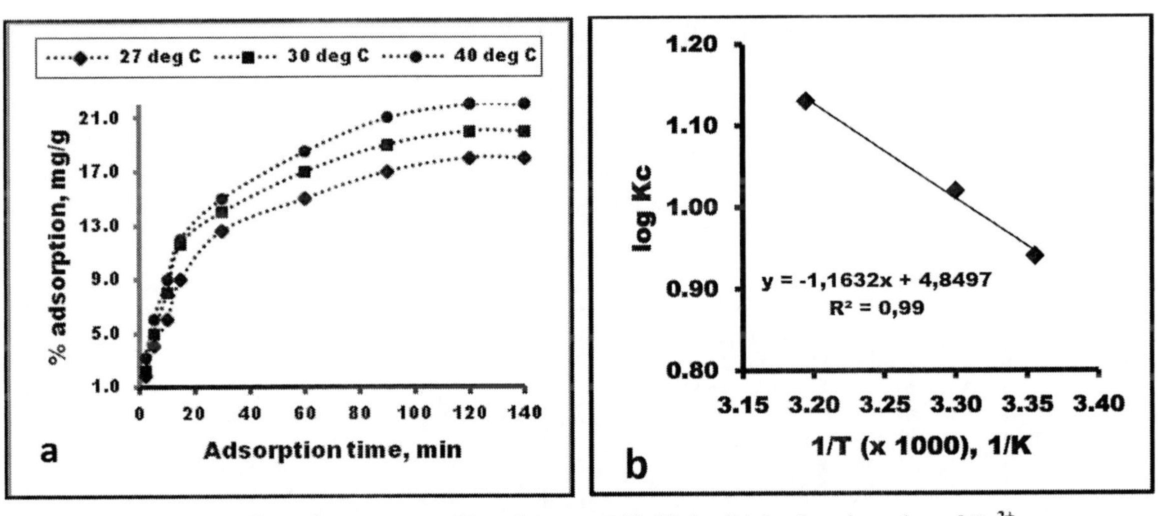

Figure 3 Effect of temperature (a) and the van't Hoff plot (b) for the adsorption of Cu^{2+}

Adsorption processes can be expressed in terms of changes in thermodynamics parameters: Gibbs free energy (ΔG^o), enthalpy (ΔH^o) and entropy (ΔS^o). These parameters are related as follows.

$$\Delta G^o = -RT \ln K_c \quad (6)$$

$$\Delta G^o = \Delta H^o - T\Delta S^o \quad (7)$$

where

G^o = change in free energy, kJ mol^{-1}; H^o = change in enthalpy, kJ mol^{-1}; S^o = change in entropy, J $mol^{-1}K^{-1}$;T = absolute temperature, K; R = gas constant = 8.314 x 10^{-3}; and Kc = equilibrium constant.

Next, K_c can be defined as:

$$K_c = C_{Ae} C_e \tag{8}$$

where C_{Ae} and C_e are the equilibrium concentration (mg L^{-1}) of the adsorbate, on TLP and in the solution respectively.

Further arrangemens of Eqs 5 and 6 would yield the following:

$$\log K_c = \frac{\Delta S^o}{2.303R} - \frac{\Delta H^o}{2.303RT} \tag{9}$$

The adsorption thermodynamic parameters were calculated by plotting log Kc versus 1/T (known as van't Hoff plot), as shown in Figure 3-b, where the slope and intercept of the plot equal ΔH°/2.303 R and ΔS°/2.303 R, respectively. Hence, the values of ΔH° and ΔS° were obtained (Table 2).

Table 2 Thermodynamic parameters of Cu^{2+} adsorption

Temperature [°C]	Kc	ΔG [kJ mol^{-1}]	ΔH [kJ mol^{-1}]	ΔS [kJ mol^{-1}]
27	0.94	-27.84	0.02	0.09
30	1.02	-28.11		
40	1.13	-29.04		

The positive values of ΔH° indicates that adsorption of Cu^{2+} on TLP is endothermic, and thus the adsorption performance would tend to increase with temperature and may involve a chemisorption mechanism [4]. On the other hand, the overall free energy changes, ΔG°,are negative indicating a spontaneous and thermodynamically feasible process for Cu^{2+} adsorption. The absolute values of ΔG° increases with temperature which implies that the spontaneous nature of adsorption of Cu^{2+} is inversely proportional to temperature [4]. The entropy of adsorption, ΔS°, is positive, suggesting an increase in randomness at the solid-solution interface during the adsorption process.

3.4. Adsorption kinetics

In the present experiment the simple pseudo-first order kinetic model, i.e the Lagergren was employed, and shown in Eq (10) [4]:

$$\log(q_e - q_t) = \log q_e - \left(\frac{k_{ad}}{2.303}\right)t \tag{10}$$

where k_{ad} is the Lagergren rate constant for adsorption (min^{-1}), q_t and q_e are the amounts of adsorbate adsorbed (mg g^{-1}) at time t and equilibrium time, respectively.

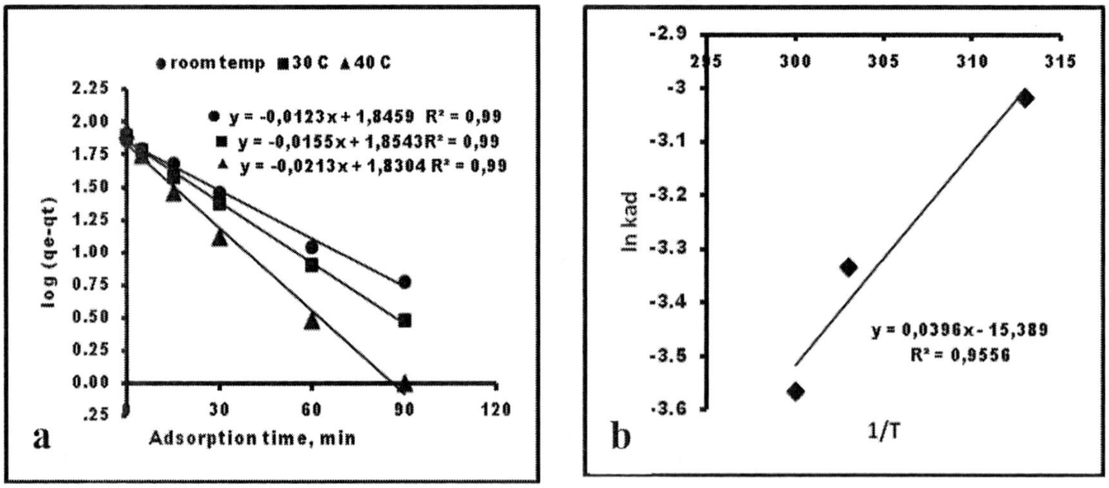

Figure 4 Pseudo-first order kinetic of Lagergren for different temperatures and Ea determination

The plots of Eq 10 were drawn for different temperatures (Fig 4-a) and indicate the applicability of the equation. The values of k_{ad} at room temperature, 30°C and 40°C were calculated from the slope of the linear plots and were found to be: 0.028; 0.036 and 0.049 per min, respectively. The increase in rate constants with temperatures denotes that the rate-limiting step is surface adsorption and the process is endothermic[4]. The adsorption process was also investigated in relation to its activation energy Ea, using Arrhenius equation:

$$\ln k_{ad} = \ln A - \frac{E_a}{RT} \tag{11}$$

where Ea the activation energy (kJ mol^{-1}), T the temperature in Kelvin, and R is the gas constant = 8.31 kJ mol^{-1} K^{-1}. The plot of Eq 11 yields a straight line with slope – Ea/R (Figure 4-b), and the activation energy was found as 127.94 kJ mol^{-1}.

4. Conclusion

Dried teak (*Tectona grandis sp*) leaves were pulverized and subsequently tested as adsorbent for Cu^{2+} adsorption from aqueous solutions. The equilibrium of the adsorption process was reached after approximately 200 min. The maximum adsorption capacity was observed at the following conditions: pH = 6.0; adsorbent sizes: 100-170 mesh, and adsorbent dose: 10 g L^{-1}. Langmuir isotherm and Lagergren pseudo first order kinetic were found to be best matched with the experimental data. Calculations of the thermodynamic parameters indicates that the adsorption was endothermic and spontaneous. The activation energy of the adsorption was high i.e.127.94 kJ mol^{-1}.

The adsorption experiment was easy and safe to conduct. The utilization of agricultural wastes as adsorbent would show to students that the adsorption is environmentally friendly. The students were able to analyze the experimental data and satisfactorily complete the lab report, indicating that their understanding of the underlying concepts of adsorption: equilibrium, kinetic and thermodynamic improved.

5. References

[1] Martin-Lara M A, Blazquez G, Trujillo M C, Perez A, Calero M 2014 New treatment of real electroplating wastewater containing heavy metal ions by adsorption onto olive stone *J.Cleaner Production* **81** 120-129
[2] Munagapati V S, et al. 2010 Biosorption of Cu(II), Cd(II) and Pb(II) by *Acacia leucocephala* bark powder: Kinetics, equilibrium and thermodynamics *Chem.Eng.J.* **157** 357-365
[3] Vilvanathan S and Shantakumar S 2012 Removal of Ni(II) and Co(II) ions from aqueous solution using teak (*Tectona grandis*) leaves powder: adsorption kinetics, equilibrium and thermodynamics study *Desalin.Water Treat.* DOI:10.1080/19443994.2014.989913
[4] Vankar P S, Sarswat R, Sahu R 2012 Biosorption of zinc ions from aqueous solutions onto natural dye waste of *Hibiscus rosa sinensis*: thermodynamic and kinetic studies *Environ. Prog.Sustain.* **31**(1) 89-99
[5] Ucun H, et al. 2002 Biosorption of chromium (VI) from aqueous solution by cone biomass of *Pinus sylvestris Bioresour. Technol.* **85** 155-158
[6] Azouoau N, et al. 2010 Adsorption of cadmium from aqueous solution onto untreated coffee-grounds: Equilibrium, kinetics and thermodynamics. *J.Hazard.Mater.* **184** 126-134
[7] Gashemi M, et al. 2015 Microwave-assisted functionalization of *Rosa canina L* fruits activated carbon with tetraethylenepentamine and its adsorption behavior towards Ni(II) in aqueous solution: Kinetic, equilibrium and thermodynamic studies *Powder Technol.* **274** 362-371
[8] Piergiovanni P R 2016 Introductory adsorption laboratory experiment, Paper presented at 2012 ASEE Annual Conference, San Antonio, TX, USA. https://peer.asee.org/21610 (accessed 12 May 2016).
[9] Perić J, Trgo M, Medvidović N V 2004 Removal of zinc, copper and lead by natural zeolite – a comparison of adsorption isotherms *Water Res.* **38** 1893-1899

IOP Conf. Series: Materials Science and Engineering **162** (2017) 012005 doi:10.1088/1757-899X/162/1/012005

Evaluation of phase separator number in hydrodesulfurization (HDS) unit

A D Jayanti, A Indarto
Department of Chemical Engineering, Institut Teknologi Bandung, Labtek X, Jalan Ganesha 10, Bandung, Indonesia 40132

E-mail : antonius.indarto@che.itb.ac.id

Abstract. The removal process of acid gases such as H_2S in natural gas processing industry is required in order to meet sales gas specification. Hydrodesulfurization (HDS)is one of the processes in the refinery that is dedicated to reduce sulphur.InHDS unit, phase separator plays important role to remove H_2S from hydrocarbons, operated at a certain pressure and temperature. Optimization of the number of separator performed on the system is then evaluated to understand the performance and economics. From the evaluation, it shows that all systems were able to meet the specifications of H_2S in the desired product. However, one separator system resulted the highest capital and operational costs. The process of H_2S removal with two separator systems showed the best performance in terms of both energy efficiency with the lowest capital and operating cost. The two separator system is then recommended as a reference in the HDS unit to process the removal of H_2S from natural gas.

1. Introduction

Hydrotreating process (HDT) is one of the most important processes in the refinery to reduce sulfur, nitrogen, oxygen, metals and other contaminants from the oil fraction at high operating conditions [1]. When the process is dedicated to remove sulfur only, the process known as hydrodesulfurization (HDS). One of the common schematic diagram of HDS in the petroleum refinery unit is shown in Figure 1. Currently, there were *ca.* 30 licensed HDS processes and most of them have the same basic process principles.

Various studies have been conducted to improve the HDS unit to produce the desired product while increasing profit, *eg.*[2] or other promising technologies, *eg.* [3-8]. However, none of studies was found in the literature that focuses on the separation configuration (see the configuration inside the box of Figure 1). In the HDS, phase separator serves to separate fluid in different phases, *ie.*oil, water, and gas, at a certain pressure and temperature. Some of the refining industries, the HDS units were equipped with one separator while others have two separators [9,10]. These two different configurations should produce different result in term of utility consumption and economic consideration. To understand the influence of the number of separator on the HDS unit, in terms of both performance and cost HDS unit, optimization of the number of separator in HDS unit were required to be done. In this study, we evaluated the separation process modelling (process configuration inside the box of Figure 1) by varying the number of phase separators in the HDS unit.

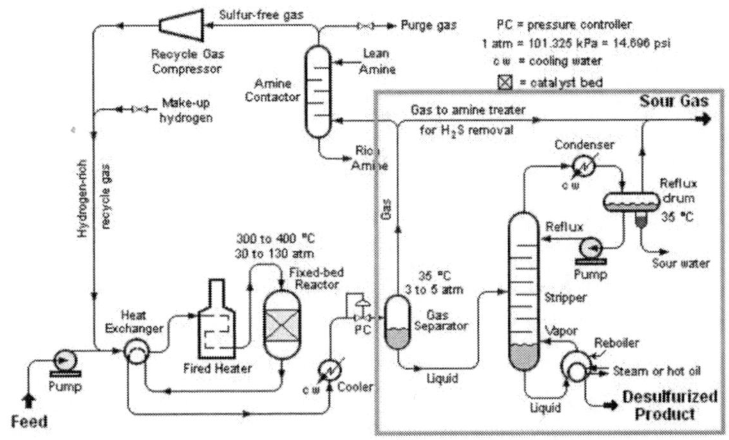

Figure 1. Schematic process flow diagram of HDS in petroleum refinery.

2. Methodology

The quantitative analysis was initiated by the construction of the process system model. The established process models that have been made were then validated by data literature or actual data from the field. This step is necessary to allow the model to be correct and could be used for the next step of the simulation.

2.1. Simulation Process Specification

In this study, two scenarios have been made to consider the H_2S concentration in the feed. Scenario 1 simulated the H_2S concentration of 0.59%-mol while Scenario 2 was conducted by using H_2S concentration of 10%-mol. Detail feed and product specification used in this model as well as the feed process condition are shown in Table 1 and 2.

Table 1. Feed composition and properties.

Process Properties	Scenario 1	Scenario 2
Pressure	689 psig (47.5 barg)	689 psig (47.5 barg)
Temperature	550°F (287.8°C)	550 F (287.8°C)
Flow rate	2355 kmol/h	2355 kmol/h
Component		
H_2	82.58%-mol	73.31%-mol
H_2S	0.59%-mol	10%-mol
Ammonia	0.09%-mol	0.09%-mol
C1	4.09%-mol	4.06%-mol
C2	0.43%-mol	0.43%-mol
C3	0.16%-mol	0.16%-mol
iC4	0.04%-mol	0.04%-mol
nC4	0.04%-mol	0.04%-mol
iC5	0.05%-mol	0.05%-mol
nC5	0.1%-mol	0.1%-mol
Diesel	2.31%-mol	2.30%-mol
Gas Oil	8.53%-mol	8.48%-mol
Naphtha	0.98%-mol	0.98%-mol

IOP Conf. Series: Materials Science and Engineering **162** (2017) 012005 doi:10.1088/1757-899X/162/1/012005

Table 2. Product composition and properties

Process Properties	Naptha	Diesel	Gas Oil
Molecular weight	152,3	210,6	282,7
Boiling point (°C)	197,5	277	335,8
Density (kg/m^3)	803,6	843,3	886,8
Critical temperature (°C)	388,2	465,6	526,5
Critical pressure (bar_g)	21,97	17,24	15,65
Component			
H_2S (ppm)	< 650	< 10	< 10
H_2O (ppm)	< 300	< 50	< 50

2.2. Model Development

Three different simulation models were built to evaluate to suitable number of separators, *ie.*one (HDS-1), two (HDS-2), and three (HDS-3) phase separator, used in HDS unit. Figure 2 shows the three simulation models of separator in HDS unit.

Feed is cooled to a temperature of 122°F prior to entry into the separator. The temperature of 122°F is the optimal temperature for the desired separation. Wash water is added to the flow to reduce the concentration of unwanted gas in the recycle gas (increasing purity) and to dissolve NH_4Cl that may form and cause deadlock. Vapor result from separator flow into the amine treating gas scrubber, while the liquid heated to 413°F before entering stripper that uses steam as a medium discharger. At the bottom of the cold separator is the output of sour water. Stripper aims to control the sour water until the content of H_2S and NH_3 in the water effluent are quite low and safe for the environment to be disposed. This unit also serves to sharpen separation of components; so that it can improve the quality of a product by separating unwanted light fraction in the product. Stripper top output is sour gas and naphtha while in the bottom is the desired product. In the bottom of the stripper is injected steam at a pressure of 50 psig and temperature of 446°F. Steam injection aims to lower the partial pressure above the liquid surface, so that the light fraction is entrained to the bottom of stripper column will be more volatile and returned to the fractionation column. Stripper top output is sour gas and naphtha while in the bottom is the product.

All simulation was conducted by process simulator Aspen Hysys v.8.8 as well as Aspen Economic Evaluation for plant cost estimation. In some cases, process simulator could calculate and predict to evaluate the plant process performance, *eg.* [11]. From the simulation results of each scenario, it will be observed product specifications, system performance, sizing equipment (separator, stripper, reboiler, and condenser), as well as capital and operating costs of the system. Once all steps have been done, the final step is to evaluate the final result and determine which model is most suitable to be applied to the process of H_2S degassing.

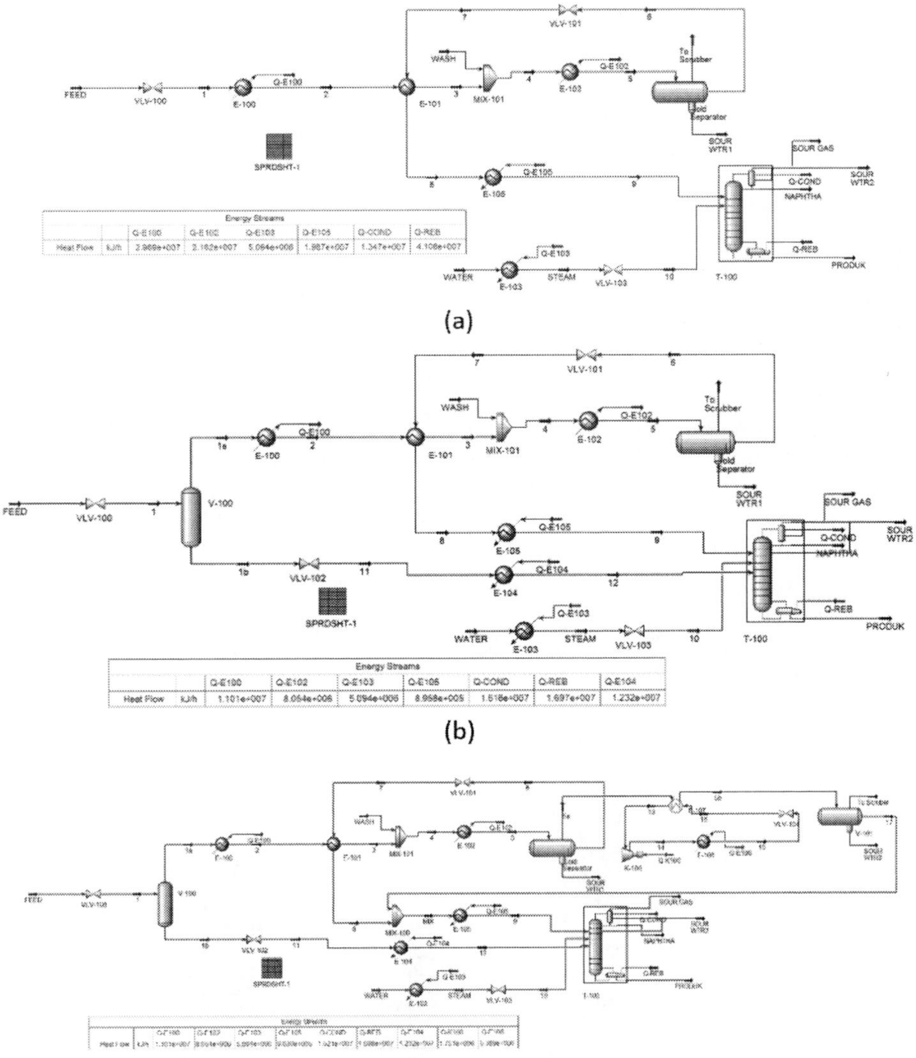

Figure 2. Process Flow Diagram of three HDS simulation models: (1) One separator system or HDS-1; (2) Two separator system or HDS-2; and (3) Three separator system or HDS-3.

3. Result and Discussion

The simulation result of three models are tabulated in Table 3. All separator systems were capable to remove H_2S and H_2O below the limit value in the product. The product has a high purity >99.5% and naphtha produced from all separator system contains H_2S below the limit value. The content of H_2O in all separator system exceeds a predetermined limit, each 610.87 ppm; 497.78 ppm; and 496.84 ppm. As the H_2O content is still very high (above 400 ppm), dehydration is required before naphtha is supplied to the consumers.

Table 3. Comparison simulation result of three HDS with different separator systems

Process Properties	HDS with 1 Separator	HDS with 2 separators	HDS with 3 separators
	Diesel Product		
Flow rate (kg/h)	68,323.25	68,322.23	68,323.25
Content of H_2S (ppm)	0	0	0
Content of H_2O (ppm)	27.85	27.93	27.85
Purity (%-wt)	99.88	99.88	99.88
	Naptha Product		
Flow rate (kg/h)	3,449.75	3,437.84	3,449.75
Content of H_2S (ppm)	333.74	332.82	333.74
Content of H_2O (ppm)	496.84	497.78	496.84
Purity (%-wt)	98.79	98.80	98.79

As shown in Figure 3, HDS unit with one separator system (HDS-1) has the highest workload of heating and cooling among other separator systems. One separator system used a heater that requires high energy consumption, as well as its cooler that has very large workload than any other separator systems. At one separator system, all feed cooled from 550°F into 122°F. The large of feed flow rate which is still in two phases increases the workload on the heat sink. This could be the reason why workload on the cooling unit is very high. In two (HDS-2) and three (HDS-3) separator system, the feed was separated first at hot separator. Vapor outlet from separator then cooled from 550°F to 122°F. Although the temperature difference among HDS-1, HDS-2, and HDS-3 are same, but the flow rate of one separator system (HDS-1) is the highest than others. The fluid flow rate of one separator system which shall be cooled was ca. 2,355 kmol/h while two and three separator system was only 2,095 kmol/h.

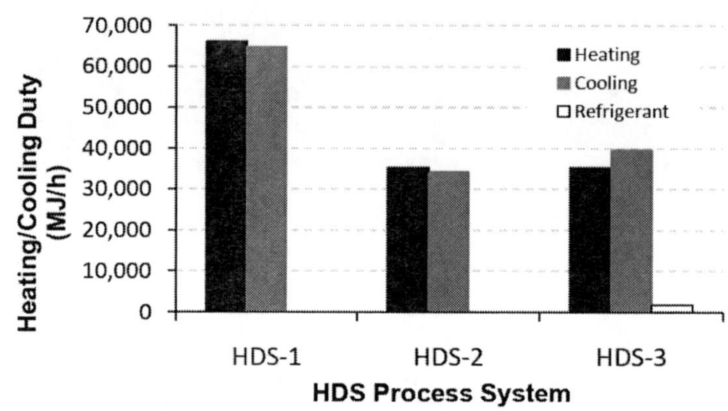

Figure 3. Utility workload comparison of the three HDS units.

At one separator system (HDS-1), the fluid has to be heated from the cold separator is very large. While the two other separator systems, the fluid to be heated is divided into two streams. The first stream coming from the hot separator and the second comes from the cold separator. This results the load heating of HDS-2 and HDS-3 were not as big as one separator system (HDS-1).

Although it seems one separator system is a simple process, it shows that the capital cost (CAPEX) to build the system is the highest among others. Figure 4 shows that the HDS unit with one separator (HDS-1) requires capital and operational costs the highest among the other separator systems. Total cost for the HDS one separator system reached 26.98 million USD while HDS two separators system (HDS-2) have the lowest investment costs 15.64 million USD.

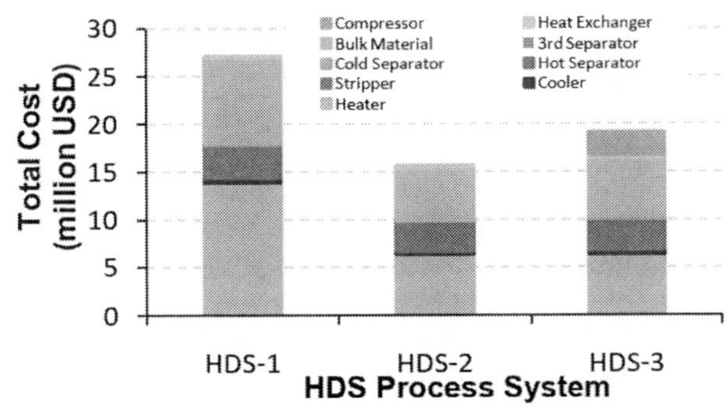

Figure 4. Comparison of Equipment cost estimation of the three HDS units.

The biggest capital cost of one separator system (HDS-1) is on the use of heater. These heaters have the largest portion of the capital costs. The high of workload on the heater, making the heater size becomes larger than the other separator systems. In three separator system (HDS-3), refrigeration system and one additional separator raise the capital cost for this system.

One separator system has very large steam utility thus affecting the total cost of utilities. At one separator systems, fluid to be heated all come from the cold separator enters the heater so the steam needs to be very large. While the two other separator systems, the fluid to be heated is divided into two streams. The first stream coming from the hot separator and the second comes from the cold separator. So the need of steam for both separator is not as big as one separator system.

HDS one separator system has large cooling utilities. The amount of fluid to be cooled in the separator affects the utility costs on cooling. The more fluid to be cooled, the more water cooling needed to cool the fluid. As a result, the cost is greater. The more amount of fluid to be cooled in one separator system due to the fluid from reactor flow into cooler directly while the others separator systems need to be flow into the hot separator and then the vapor which produced flow into cooler.

4. Conclusion

The simulation of three different separator systems in Hydrodesulfurization (HDS) unit has been conducted. It shows that HDS unit with two separator system (HDS-2) has the lowest workload (heating and cooling) while the highest work load at one separator system. From the results of the simulation and calculation of investment costs, HDS unit with one separator system (HDS-1) requires the greatest capital and operational costs while two separator system requires the lowest capital and operating costs. Taking into account the total investment costs both for CAPEX and OPEX as well as the utility workload the two separator system (HDS-2) selected as the best system in the hydrodesulfurization (HDS) unit.

References

[1] Gary J H, Handwerk G E, Kaiser M J and de Klerk A 2007 *Petroleum refining: technology and economics* (New York: CRC Press)
[2] Dahlberg A, Mukherjee U and Olsen C W 2007 *Hydrocarbon Process* Sep 111.
[3] Hyung H K, Choi J W, Lee H and Indarto A 2005 *Toxicol. Environ. Chem.* **87(4)** 509.
[4] Indarto A, Choi J-W, Lee H and Song H-K 2006 *J. Environ. Sci.* **18(1)** 83.
[5] Indarto A, Choi J W, Lee H, Song H K and Palgunadi J 2006*J. Rare Earth.* **24(5)**513.
[6] Indarto A, Yang D R, Choi J W, Lee H and Song H K 2007 *Chem. Eng. Comm.* **194(8)** 1111.
[7] Indarto A, Choi J W, Lee H and Song H K 2008 *Chin. Sci. Bull.* **53(18)** 2783.
[8] Indarto A 2012 *Environ. Technol.* **33(6)**663-666.
[9] Harwell L, Thakar S, Polcar S, Palmer R E and Desai P H 2003 *Oil Gas J.*
[10] Milosevic Z and Shire T 2011 *Hydrocarbon Process*. June ed.
[11] Qirom A, Indarto A and Putrawan I A 2014 *Appl. Mech. Mat.* **493** 27.

IOP Conf. Series: Materials Science and Engineering **162** (2017) 012006 doi:10.1088/1757-899X/162/1/012006

Continuous absorption of CO_2 in packed column using MDEA solution for biomethane preparation

A Mindaryani*, W Budhijanto, S S Ningrum
Chemical Engineering Department, Universitas Gadjah Mada, Yogyakarta, Indonesia

E-mail : amindaryani@ugm.ac.id

Abstract. Nowadays, the energy consumption in Indonesia is increasing. Raising of energy consumption force Indonesia to find other energy resources. Biogas is one of the renewable energy, which was developed in anticipation to the fossil energy reduction. Reducing the content of impurities in biogas may reduce the corrosion impact and increase the combustion efficiency. The biomethane can be utilised as fuel for generator in small and medium scale industries (IKM). Continuous CO_2 absorption in packed column using MDEA solution as absorbent is studied for biomethane preparation. CO_2 absorption experiments was performed continuously in the packed absorption column with a diameter of 6 cm and 75 cm length. Gas is sparged from the bottom of the column while the liquid is pumped through the top of the column. The concentration of CO_2 at exit gas is analysed by GC and recorded as a function of time. The flowrate of the inlet gas was varied at 1 LPM; 1.5 LPM; and 1.8 LPM. Variation of MDEA solution concentration used was 20% and 35.31%. Mathematical model for unsteady state CO_2 absorption in packed column was developed. The reaction rate constant (k) and mass transfer coefficient KGa were determined by fitting the outlet CO_2 concentration data as a function of time to the model solution with smallest Sum of Square of Errors (SSE). The experimental data shows that absorption of 1 LPM gas flow rate with 0,15 LPM MDEA solution flow rate may reduce 40 % CO_2 to be 17 % CO_2 in outlet gas. The steady state process reaches at 10 minutes. Increasing gas flow rates shows the higher overall mass transfer coefficient. The reaction rate constant is not affected by gas flow rate variation.

1. Introduction

Energy scarcity is a latent risk from our dependence in nonrenewable energy source. In order to anticipate energy scarcity, renewable energy source development is a must. Biogas promises a sustainable energy source, is produced from biomass anaerobic bacterial digestion. Biogas consists of methane, carbon dioxide, hydrogen sulfide and some trace element. Biogas quality is determined by methane content. Generally CO_2 is the biggest impurity of biogas, hence biogas purification is best achieved by CO_2 removal. The chosen absorbent must be economic, nontoxic, having high CO_2 absorption ability, and easy to regenerate. Abharchaei, 2010 studied the absorption of CO_2 by 2(Methyl)-Aminoethanol (MEA) solution, but the solution is corrosive[1]. Normal methyldiethanolamine (MDEA) is a widely used in industry due to high absorption capacity, especially for acid gases including CO_2. In addition to the large absorption capacity, MDEA also has degradation proof property due to exposure to heat and direct contact with chemicals [4]. Another reason that encourages the use of MDEA extensively on industrial scale is a low corrosivity and low vapor pressure of MDEA [2]. The advantage of chemical absorption is the bond formed between CO_2 with MDEA is stronger compared to physical absorption. But the chemical absorption has shortcomings such as high energy requirements for regeneration of solvent [5]. CO_2 absorption with MDEA is an exothermic reaction with heat of reaction -72 kJ / mol. During absorption process CO_2

IOP Conf. Series: Materials Science and Engineering **162** (2017) 012006 doi:10.1088/1757-899X/162/1/012006

rich gas is contacted directly with absorbent. During this process, CO_2 is transferred from gas bulk into liquid body [6]. This condition continues until liquid reaches a saturated point where absorption can no longer take place since absorbent has already lost its absorption capability. In this paper effect of gas flow rate dan MDEA concentration on absorption process of CO_2 in MDEA solution is studied.

The chemical absorption of CO_2 in MDEA solution follows this reaction (Gao et al, 2013).

$$MDEA + H_2O + CO_2 \rightarrow MDEAH^+ + HCO_3^- \quad (1)$$

Mathematical model of unsteady state continuous absorption in packed column is developed by applying mass balance.

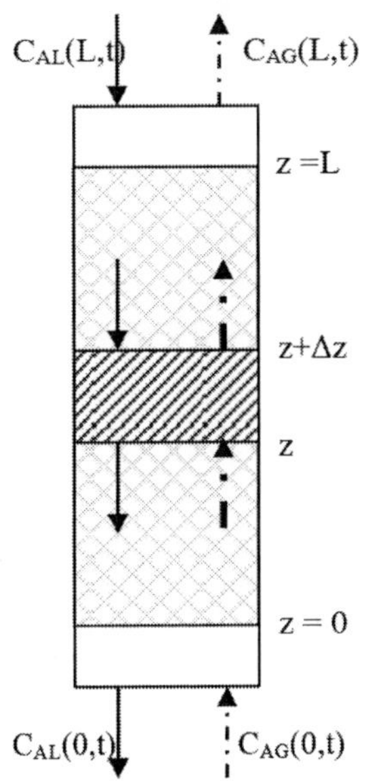

Mass balance of CO_2 in gas phase:

$$\frac{\partial C_{AG}}{\partial t} = -\frac{Q_G}{S.\varepsilon}.\frac{\partial C_{AG}}{\partial z} - K_G a.(C_{AG}.R.T - H.C_{AL}) \quad (2)$$

Mass balance of CO_2 in liquid phase :

$$\frac{\partial C_{AL}}{\partial t} = \frac{Q_L}{S.\varepsilon}.\frac{\partial C_{AL}}{\partial z} + K_G a.(C_{AG}.R.T - H.C_{AL}) - k.C_{AL}.C_{BL} \quad (3)$$

$$\frac{\partial C_{BL}}{\partial t} = \frac{Q_L}{S.\varepsilon}.\frac{\partial C_{BL}}{\partial z} - k.C_{AL}.C_{BL} \quad (4)$$

IC : $t = 0, z = z, C_{AL} = 0, C_{AG} = 0$
BC : $t = t, z = 0, C_{AL} = C_{AL}(0,t), C_{AG} = C_{AG}(0,t)$
$z = L, C_{AL} = C_{AL}(L,t), C_{AG} = C_{AG}(L,t)$

Differential Equations (2), (3) and (4) with boundary conditions was solved using MATLAB.

Figure 1. Continuous absorption of CO_2 by MDEA solution in packed bed column

The solution are C_{AG} and C_{AL} as function of time, for a certain value of k and $K_G a$. The values of k, $K_G a$ and H will be applied to absorber design.

2. Methodology

2.1. Material

1. MDEA is provided by PT.Pupuk Kujang
2. Gas CO_2 (99%) is supplied by CV Gas Sumber Agung Sukses, Yogyakarta
3. Mixed gas (CO_2 and N_2 40 %, 74 % v/v) are supplied by PT Samator Gas, Jln Ringroad utara, Yogyakarta.

2.2. Method

Continuous absorption experiments was conducted in the absorption column as shown in Figure 2. MDEA solution with a certain concentration was flowed from the top of the packed column (D = 6 cm, L = 75 cm), then the mixed gas was sparged from the bottom of the column. At a certain time the gas and liquid samples were taken. Gas samples were analyzed using GC while the liquid sample was analyzed by conductivitymeter. The same experiments were conducted for variation of MDEA concentration, gas flow rate.

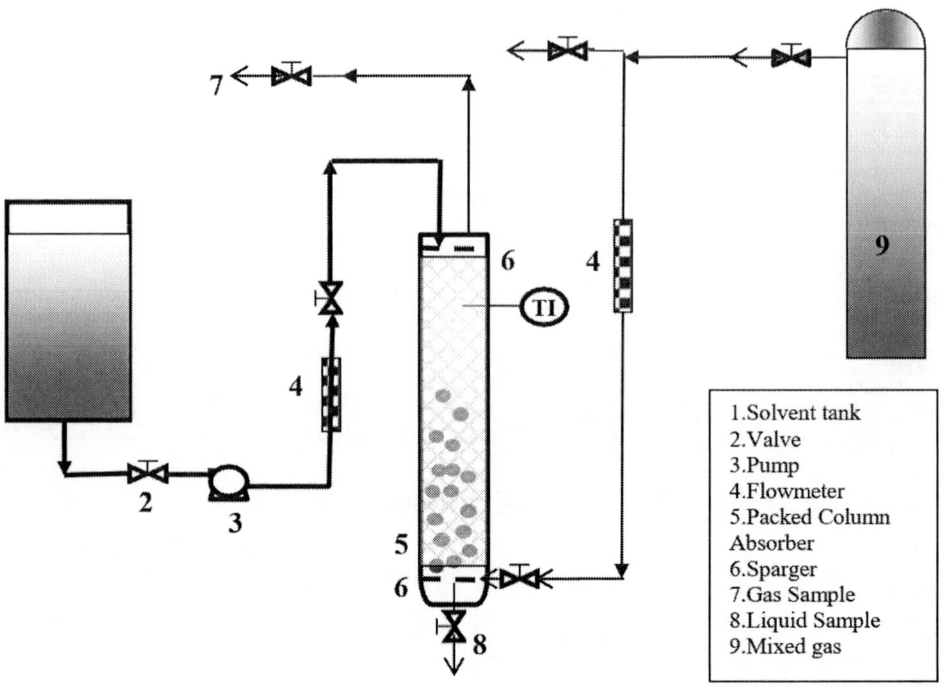

Figure 2. Experimental rig

The unsteady state absorption of CO_2 mathematical model was developed. Together with Initial and boundary condition, the differential equations were solved. The solution were outlet CO2 concentration as a function of time. Equilibrium parameter (H) was taken from previous experiment which are 37.64 atm.L/mol. Reaction rate constant (k) and mass transfer coefficient K_Ga were determined by fitting the outlet CO2 concentration data as a function of time to the model solution with smallest Sum of Square of Errors (SSE).

3. Result and Discussion

Figure 3 and Figure 4 show the experimental data for CO2 absorption for initial gas content 40% at various gas flow rate which are 1 LPM, 1.5 LPM and 1.8 LPM with 0.15 LPM MDEA solution flow rate at different MDEA solution concentrations which are 20 % and 35.3 %. The outlet CO2 concentrations for different MDEA concentration have only 1 % difference, it is not significant. The absorption experimental data for various gas flow rate is shown in Figure 5 for 40 % inlet CO2 concentration and Figure 6 for 74 % inlet CO2 concentration. At 0.15 LPM MDEA flowrate with inlet gas flow rate 1 LPM, the outlet CO2 concentration drop from 40 % to 17 %. It meets the requirement to be the fuel of electric generator which is less than 20 % CO2. By increasing the flow rate of MDEA solution and conduct the absorption at several columns, it is possible to get the composition of biomethane more than 90%, and fulfill the requirement as generator fuel.

IOP Conf. Series: Materials Science and Engineering **162** (2017) 012006 doi:10.1088/1757-899X/162/1/012006

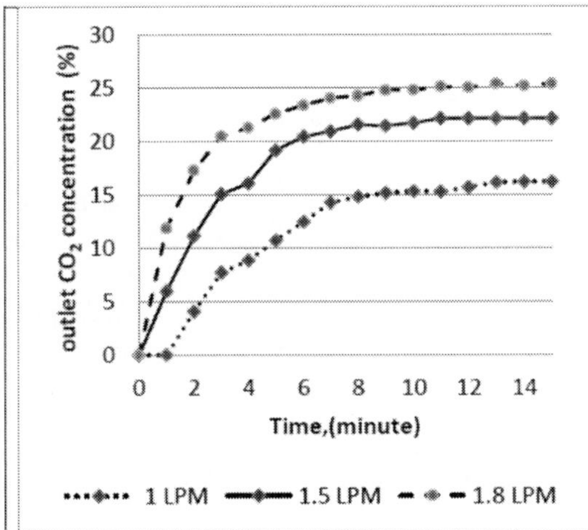

Figure 3. Outlet CO_2 concentration for MDEA 20 %

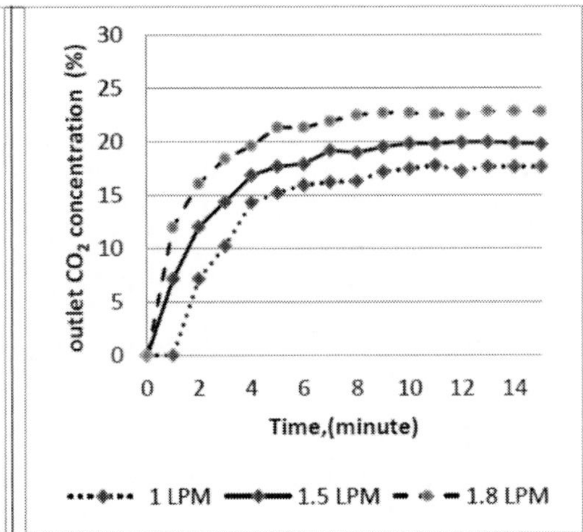

Figure 4. Outlet CO_2 concentration for MDEA 35,3 %

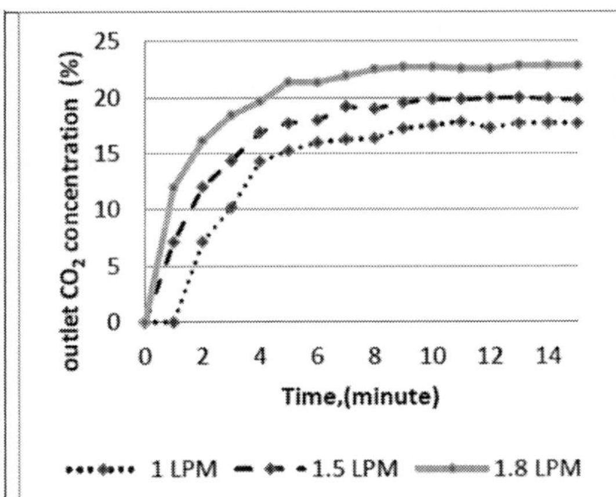

Figure 5. Outlet CO_2 concentration for MDEA 20 %, inlet CO_2=40%

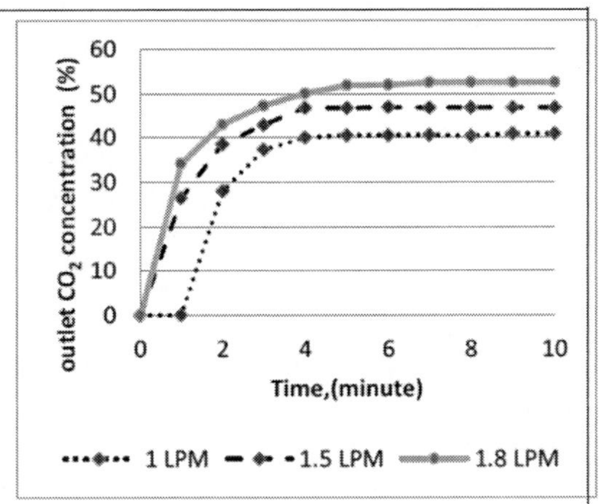

Figure 6. Outlet CO_2 concentration for MDEA 20 %, inlet CO_2=74%

Table 1 and table 2 show the parameters determined from the data obtained, the parameters are reaction rate constant and overall mass transfer coefficient at various gas flow rate. Henry constant (H) was taken from previous experiment, H = 37,64 (atm.L/mol).

Table 1. Gas mass transfer coefficient of CO_2 at various gas flowrates for MDEA concentration of 20% dan 35,31 %, inlet CO_2 concentration 40 %

MDEA 20 %			MDEA 35,31 %		
Q_G (L/menit)	k (L/mol.min)	K_Ga (mol/atm.min.L)	Q_G (L/menit)	k L/(mol.min)	K_Ga (mol/atm.min.L)
1	0,302	0,0367	1	0,301	0,0402
1,5	0,291	0,0412	1,5	0,307	0,0510
1,8	0,281	0,0415	1,8	0,292	0,0535

Table 2. Gas mass transfer coefficient of CO_2 at various gas flowrates for MDEA concentration of 20% dan 35,31 %, inlet CO_2 concentration 70 %

MDEA 20 %			MDEA 35,31 %		
Q_G (LPM)	k (L/mol.min)	K_Ga (mol/atm.min.L)	Q_G (LPM)	k L/(mol.min)	K_Ga (mol/atm.min.L)
1	0,199	0,0167	1	0,282	0,032
1,5	0,152	0,0201	1,5	0,285	0,047
1,8	0,197	0,022	1,8	0,294	0,0455

The experiments were conducted at isothermal condition, the values of reaction rate constant is almost constant. The overall mass transfer coefficient are affected by gas flow rate. The higher the flow rate, the higher the mass transfer coefficient.

4. Conclusion

Continuous absorption of CO_2 using MDEA solution at ambient temperature in packed column is possible to reduce the 40 % CO_2 concentration to 17 % CO_2. By applying some stages of this absorption method, it is possible to get biomethane from biogas, and to be utilised as fuel for electrical generator in the small and medium scale industries. 20 % MDEA solution is not significantly different from 35.3 % MDEA solution in absorbing CO_2, therefore based on economical point of view, 20 % MDEA is more feasible.

Acknowledgement

The authors would like to thank General Directorate of Higher Education of Indonesia through the Penelitian Unggulan Perguruan Tinggi research grant for the financial support, and to PT Pupuk Kujang for the MDEA supply.

Nomenclature

Q_G, Q_L	Gas flow rate, Liquid flow rate, L/min (LPM)
k	reaction rate constant (L/mol.min)
K_Ga	volumetric mass transfer coefficient (mol. atm/min/L)
C_{AG}	Concentration of CO2 in gas phase, (mol/L)
C_{AL}	Concentration of CO2 in solution, (mol/L)
C_{BL}	Concentration of MDEA in solution, (mol/L)
H	Henry constant, (atm.L/mol)

IOP Conf. Series: Materials Science and Engineering **162** (2017) 012006 doi:10.1088/1757-899X/162/1/012006

References

[1] Abharchaei P R 2010 Kinetic Study of Carbon Dioxide Absorption by Aqueous Solutions of 2(Methyl)-Aminoethanol in Stirred Tank Reactor: Experimental and Numerical Study of Absorption Process and Mass Transfer Phenomena, M.S. Dissertation, Chalmers University of Technology, Göteborg, Sweden.

[2] Bullin J A, Polasek J C, Donnelly S T 2006 The Use of MDEA and Mixtures of Amines for Bulk CO_2 Removal *Proc. Of 69th Gas Processors Association Convention* 1-2

[3] Fu D, Zhang P 2014 Investigation of the Absorption Performance and Viscosity for CO_2 Capture Process Using [Bmim][Gly] Promoted MDEA (N-Methyldiethanolamine) Aqueous Solution, *J. Energy Eng.*, ***87***, 167-168.

[4] Guo X P ,Tomoe Y 1994 Difference in Corrosion Behavior between Carbon Steel in CO_2-Saturated MDEA and DGA Solutions, *J. Soc. Corros. Eng.*, ***48***, 5-6

[5] Plasynski S I, Chen Z Y 2008 Review of CO_2 Capture Technologies and Some Improvement Opportunities, *Energ. Fuel*, ***45-4***, 644-646.

[6] Treybal R E 1981 Mass-Transfer Operations 3 ed., 3, Mc-Graw Hill, Inc., Singapore.

IOP Conf. Series: Materials Science and Engineering **162** (2017) 012007 doi:10.1088/1757-899X/162/1/012007

Influence of mid-point temperature of heavy hydrocarbons separator to the liquefaction process for small LNG plant

H A Hakim, A Indarto
Department of Chemical Engineering, Institut Teknologi Bandung, Labtek X, Jalan Ganesha 10, Bandung, Indonesia 40132

E-mail : antonius.indarto@che.itb.ac.id

Abstract. In liquefied natural gas (LNG) process production, one of the important units is heavy hydrocarbon removal unit to prevent freezing during liquefaction. For small scale of LNG plant, this unit is usually integrated with main heat exchanger. Feed is obtained from main heat exchanger then flows to separator to separate liquid from gas. The separator operating condition is called as Midpoint condition. Selecting Midpoint conditions have impact to light hydrocarbon losses, Specific Brake Horse Power (SBHP) process, and heating value of LNG. Hence understanding of selecting this condition and its effect to light hydrocarbon losses, SBHP process, and HHV of LNG will help to design more efficient LNG plant. According to study, the lower of Mid-Point temperature will result in lower SBHP, lower of light hydrocarbon losses, and increase LNG of HHV value. Meanwhile, the higher Mid-Point pressure will result in lower SBHP, higher light hydrocarbon losses, and lower LNG of HHV value. The change of Mid-Point pressures have more impact to light hydrocarbon losses than SBHP process.

1. Introduction

Liquefied Natural Gas (LNG) is one of efficient natural gas transportation and trading methods [1]. Recently, Small Scale LNG Plant (SSLNG) is becoming very popular as a concept for liquefying gas fuel in remote area which is not connected to the gas pipe [2]. This method is much easier compared to converting the gases into other forms, such as syngas or methanol [3-6]. Some SSLNG plants obtained the gas feed from gas pipelines that have not been meet the specifications yet for direct liquefaction process because it still has a high content of heavy hydrocarbons, such as butane, pentane, hexane, and another long-chain hydrocarbons. Thus, the heavy hydrocarbons are required to be separated prior to the liquefaction process.

In SSLNG, the separation process is preferably using a simpler and cheaper separator unit (see read box in Figure 1). However, this method results in loss of light hydrocarbons such as methane and ethane that affects the LNG plant efficiency. This single-stage separation is proposed to knock-down heavy hydrocarbon at suitable operating conditions, *ie.* temperature and pressure, so called Midpoint separation. The discussion in this paper is trying to evaluate the influence of midpoint condition to the Specific Brake Horse Power (SBHP) of refrigeration process for LNG production and the heating value (HHV) of LNG product. In many references of LNG design, process optimization is necessary to improve the efficiency and well as the economic point of view [7,8]. Moreover, further suggestion to determine the temperature and pressure of midpoint separator could be obtained.

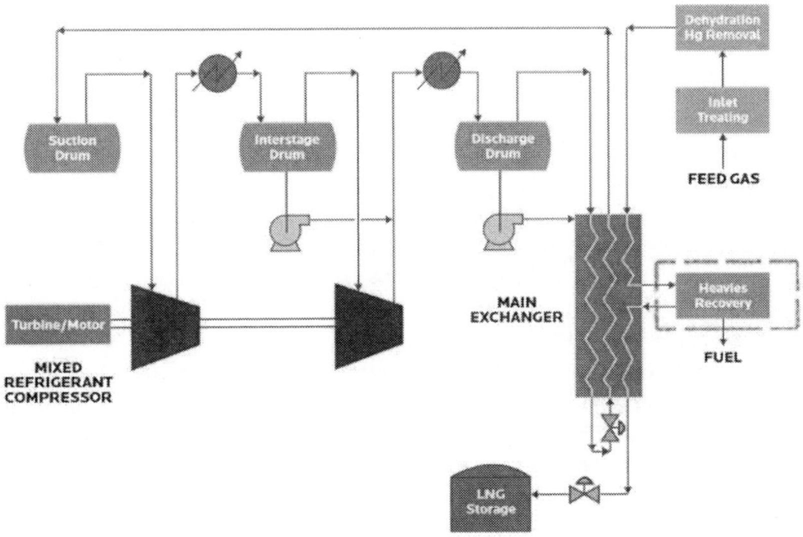

Figure 1. Schematic process flow diagram of PRICO and single stage separator [9].

2. Methodology

LNG process simulation was done using the PRICO liquefaction scheme (see Figure 1) as a tool to evaluate the parameters of the study. This process was selected because it has been widely applied to SSLNG plant and has been proven to operate more than 40 years [9]. Simulations include the removal process of CO_2 and water from gas.

In this study, the feed gas was 50 MMSCFD. Process simulation was conducted by using Aspen Hysys while freezing point was calculated using the software GPSA Luks Solubility Solid. In some cases, process simulator could calculate and predict very well to evaluate the plant process performance, *eg.* [10].

2.1. *Simulation Process Specification*

In this study, two feed compositions have been simulated. Detail feed gas composition used in this model is shown in Table 1.

Table 1. Feed composition.

Components	Composition (%)	
	Feed 1	Feed 2
Nitrogen	0.8821	0.2130
CO_2	0.3678	0.6090
Metane	86.0445	88.1134
Etane	62.5170	52.8430
Propane	3.6982	3.3952
i-C_4	0.8079	0.6710
n-C_4	1.0085	0.8090
i-C_5	0.3369	0.3070
n-C_5	0.2195	0.1680
C_{6+}	0.3831	0.4300
H_2O	7 lbs/MMSCF	

The composition of the refrigerant consists of nitrogen, methane, ethane, propane and isopentane[1]. The pressure loss at pre-treatment (CO_2 and water separation) area and main heat exchangerwas assumed to be 43 psi. After the pretreatment, feed gas went tophase separator to knock down the heavy hydrocarbons or called NGL.The gas outlet from the separator then routed back into the main heat transfer equipment to be liquefied into LNG. The feed gas composition and properties

IOP Conf. Series: Materials Science and Engineering **162** (2017) 012007 doi:10.1088/1757-899X/162/1/012007

become the main criterion for determining the condition of the midpoint, as also mentioned by [11]. This gas should have a lower freezing point than the desired final temperature. With a final temperature of LNG at -260°F (-162°C) and the value for safety taken by 10°F, the highest temperature of the freezing point of the gas should be -270°F. Based on the brief simulation, the midpoint temperature has to be above -20°F to allow the freezing point of the gas is higher than -270°F.

In this study, the simulated feed gas pressures were 500 psig (low), 700 psig (medium), and 800 psig (high) with the separator temperaturewere varied at -30°F, -50°F, -70°F, and -90°F. Simulated pressures have been selected based on the condition of most gas natural gas pipelines in Indonesia.

2.2. *Observed Parameters*

In this study, the influence of midpoint condition changes has been explored and evaluated based on three observed parameters, *ie.* SBHP, amount of heavy hydrocarbons, and heating value.

1. SBHP (Specific Brake Horse Power)
   SBHP is the energy required to produce a certain number of LNG. SBHP value indicates the efficiency of the LNG process. The lower SBHP value, the better energy efficiency of the LNG process.
2. The amount of separated heavy hydrocarbons fraction
   In this study, the term separated heavy hydrocarbons refers to hydrocarbons that change into a liquid phase and separated by separator. For this study, its value is a percentage of the amount of feed gas into the natural gas liquefaction process.
3. Higher Heating Value (HHV) of LNG product
   HHV is the energy content of fuel gas. HHV of the gas is obtained from HYSYS software in BTU/Scf.

3. Result and Discussion

In order to rigorously evaluate the impact to pressure and temperature changes on observed parameter, the evaluation and discussion were divided into three feed gas pressure variation, *ie.* 500 psig (low), 700 psig (medium), and 800 psig (high).

3.1. *Feed Gas Pressure of 500 psig*

At the feed gas pressure of 500 psig and the pressure loss in the pretreatment unit was 38 psi, the pressure of gas goes into the main heat exchanger were 462 psig. Furthermore with the pressure loss in the main heat transport equipment by 5 psi, then the pressure control valve in the upstream toward the midpoint was at 457 psig.

Following the result of Table 2, the mole fraction of methane in heavy hydrocarbons (NGL) at a midpoint temperature of -90°F reaches 48% greater than the midpoint temperature of -30°F. It is certainly not favorable condition since methane was expected to be converted into LNG. Meanwhile for the different feed composition (Feed 2 of Table 1), the correlation trend between midpoint temperature and percentage of separated heavy hydrocarbons was similar the same.

Table 2. Composition of NGL at two midpoint temperatures.

Composition	T = -30°F	T = -90°F
Nitrogen	0.0007	0.0012
CO_2	0.0000	0.0001
Methane	0.2689	0.4806
Ethane	0.1357	0.1910
Propane	0.2449	0.1787
i-butane	0.0865	0.0427
n-butane	0.1221	0.0542
i-pentane	0.0484	0.0184
n-pentane	0.0326	0.0120
n-hexane	0.0602	0.0211

A surprising result was obtain in the case of Specific Brake Horse Power (SBHP) correction. At lower midpoint temperatures, lower SBHP values were obtained that means the energy efficiency of the process is getting worse. When the midpoint temperature was decreasing, the amount of separated heavy fraction increased and lower the LNG production. The total energy supplied to the refrigerant compression is not only utilized for liquefaction process but also used to cool the heavy fraction of hydrocarbons. When most of the butane compound was required to be knocked-down from the feed gas by lowering the separation temperature, pentane and other heavier hydrocarbonswere also cooled although these compounds have already been in liquid condition. This could impact the increasing value of power consumption.The second reason was due to the gas composition after phase separation. At higher midpoint temperature, the gas composition tends to be leaner means the gas has a higher concentration of nitrogen and light hydrocarbonsthan initial condition. This condition will significantly change the composition of refrigerant allows to match the liquefaction process.In general, leaner condition will reduce the concentration of iso-pentane in the refrigerant and increase the concentration of nitrogen and methane. As higher heat capacity compounds was reducing, the circulation of refrigerant and power increased.

Table 3. The influence of Midpoint pressure variation to SBHP, NGL fraction, and HHV.

Midpoint Pressure (psig)	Feed 1			Feed 2		
	SBHP (kW/ton/d)	NGL (%-wt)	HHV (Btu/Scf)	SBHP (kW/ton/d)	NGL (%-wt)	HHV (Btu/Scf)
367	17.48	5.328	1107	17.00	4.235	17.00
387	17.43	5.563	1106	16.91	4.423	16.91
407	17.30	5.797	1104	16.82	4.609	16.82
427	17.18	6.028	1103	16.47	4.793	16.47
447	16.98	6.259	1102	16.37	4.976	16.37

Table 3 shows the trend of SBHP, NGL, and heating values with variation in midpoint temperaturesat feed pressure of 500 psig. The SBHP trend is nearly linear with a small deviation. This means SBHP value can be assumed directly proportional with the variation of midpoint temperature. In the other hand, by comparing the curve for all pressure values we can conclude that increased value of pressure will decrease SBHP. The main reason for this fact is that the higher the midpoint pressure value, the smaller the heat was required for gas condensing process.

In contrary to the SBHP trend, lower midpoint pressureswere favored for higher heavy hydrocarbon production.With increasing midpoint pressure of 5%, SBHP value has changed 0.53% lower while the heavy hydrocarbon fraction increased by 4.43%. From this discussion, increasing midpoint pressure in order todecrease SBHP value is not a good operation choice because the significant increased value of NGL which led to lower amount of LNG product.

Regarding the heating value (HHV), all variation could produce HHV values that in the normal range of LNG (1100 – 1200 Btu/Scf). Small differences were found among the values that means the different pressure did not give significant influence to the heating value of LNG.

3.2. *Feed Gas Pressure of 700 psig*

At the feed gas pressure of 700 psig, different situation was found to the linearly trend of SBHP and LNG products. Although the trend was similar to 600 psig condition, at current pressure, the trends have a bending point as shown in Figure 2.

Another important finding that shown by Figure 2b clearly mentioned that at critical temperature of above 70°F, the pressure changes give more influence to the SBHP rather than temperature effect. This could drive to critical operation decision of selecting the suitable pressure and temperature of the Midpoint separator. As this point could shift to other different temperatures, the decision should be made based on the feed composition and pressure. In the case of heating value (HHV), the current pressure (700 psig) also resulted almost linear trend as shown in Figure 3.

IOP Conf. Series: Materials Science and Engineering **162** (2017) 012007 doi:10.1088/1757-899X/162/1/012007

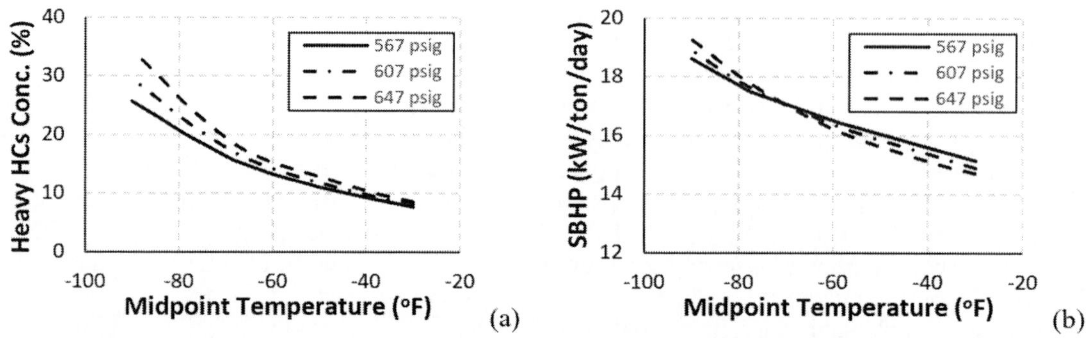

Figure 2. Influence of Midpoint temperature to (a) fraction of NGL and (b) SBHP value.

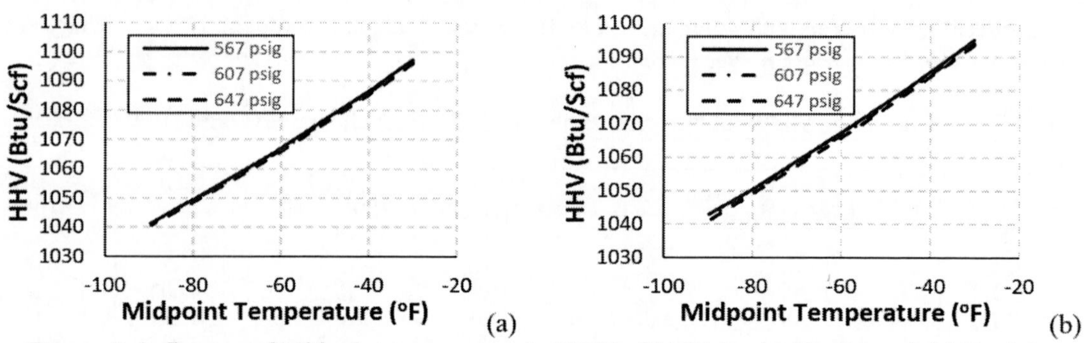

Figure 3. Influence of Midpoint temperature to HHV of LNG for (a) Feed 1 and (b) Feed 2.

3.3. *Feed Gas Pressure of 800 psig*

A similar situation as feed gas pressure of 700 psig was found in the case of feed gas pressure of 800 psig (Figure 4). At midpoint temperature lower than -60°F, the amount of heavy hydrocarbon (NGL) increased due to more carryover of lighter hydrocarbons. In this case, the influence of pressure gives more significant changes to the SBHP compared to feed pressure of 700 psig at temperature below -70°F. Again, this drive to the conclusion that optimization of Midpoint is necessary to be conducted based on feed composition and pressure to obtain the best process result and economics.

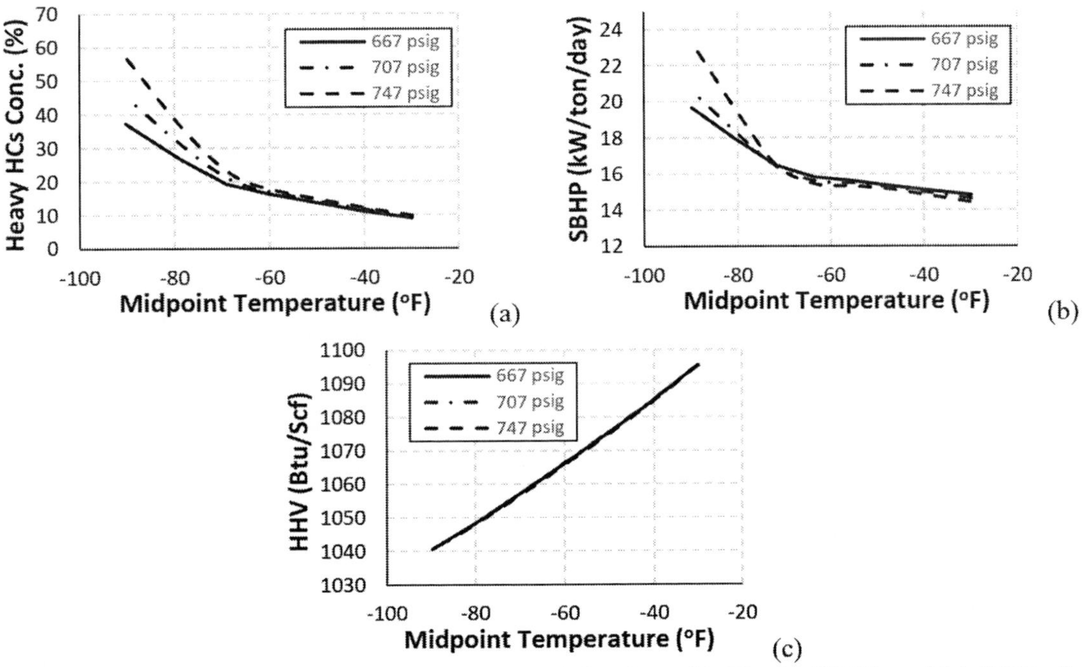

Figure 4. Influence of Midpoint temperature to (a) fraction of NGL; (b) SBHP; and (c) HHV of LNG.

4. Conclusion

The simulation and evaluation of Midpoint phase separator condition, *ie*. pressure and temperature has been conducted. From this research, regarding the effects of midpoint condition to liquefaction of natural gas using PRICO process, the conclusions can be concluded as:

1. Midpoint condition (temperature and pressure) has significant impact to SBHP, fraction of heavy hydrocarbon (NGL), and HHV of LNG. Therefore, optimization is necessary to be conducted during the process design based on feed composition and pressure.
2. Some interesting points that were obtained in this study are:
   a. Higher midpoint temperature will cause smaller SBHP/increased process efficiency, smaller percentage of heavy hydrocarbon, and higher HHV.
   b. Higher midpoint pressure will cause smaller SBHP, higher percentage of heavy hydrocarbon, and smaller HHV.
   c. Effect of midpoint pressure to SBHP is less significant than the increase of percentage of heavy hydrocarbon.
   d. Effect of midpoint pressure to SBHP is higher significant than the change of HHV.

References

[1] Mokhatab S, Mak J Y, Valappil J V and Wood D A 2014 *Handbook of Liquefied Natural Gas* (United States: Gulf Professional Publishing).
[2] Semple N 2012 *Pricing of LNG from Small Scale Facilities – Some Examples from Indonesia and Thailand*, LNG Global Congress Asia Pacific.
[3] Indarto A, Choi J-W, Lee H, Song H K and Palgunadi J 2006 *J. Rare Earth*. **24** 513.
[4] Indarto A, Lee H, Choi J-W and Song HK 2008 *Energy Source A* **30** 1628.
[5] Indarto A, Choi J-W, Lee H, Song HK 2008 *Chin. Sci. Bull.* **53** 2783.
[6] Indarto A, Choi J-W, Lee H, Song HK 2008 *IEEE Trans. Plasma Sci.* **36** 516.
[7] Austbo B, Lovseth S W and Gundersen T 2014 *Comp. Chem. Eng.* **71** 391.
[8] Xu X, Liu J, Jiang C and Cao L 2013 *Appl. Energy* **102** 1127.

[9] Black & Veatch 2016 *World Class LNG Capabilities* http://www.bv.com/docs/energybrochures/lng.
[10] Qirom A, Indarto A and Putrawan I A 2014 *Appl. Mech. Mat.* **493** 27.
[11] Uwitonze H, Han S, Jangryeok C and Hwang K S 2014 *Chem. Eng. Process.* **85** 187.

Comparison on pore development of activated carbon produced by chemical and physical activation from palm empty fruit bunch

A Hidayat and B Sutrisno
Chemical Engineering Department, University of Islam Indonesia
Jalan Kaliurang km 14,5 Yogyakarta 55584, Indonesia

E-mail : arif.hidayat@uii.ac.id

Abstract. It is well-known that activated carbon is considered to be the general adsorbent due to the large range of applications. Numerous works are being continuously published concerning its use as adsorbent for: treatment of potable water; purification of air; retention of toxins by respirators; removal of organic and inorganic pollutants from flue gases and industrial waste gases and water; recuperation of solvents and hydrocarbons volatilized from petroleum derivatives; catalysis; separation of gas mixtures (molecularsieve activated carbons); storage of natural gas and hydrogen; energy storage in supercapacitors; recovery of gold, silver and othernoble metals; etc. This work presents producing activated carbons from palm empty fruit bunch using both physical activation with CO_2 and chemical activation with KOH. The resultant activated carbons were characterized by measuring their porosities and pore size distributions. A comparison of the textural characteristics and surface chemistry of the activated carbon from palm empty fruit bunch by the CO_2 and the KOH activation leads to the following findings: An activated carbon by the CO_2 activation under the optimum conditions has a BET surface area of 717 m^2/g, while that by the KOH activation has a BET surface area of 613 m^2/g. The CO_2 activation generated a highly microporous carbon (92%) with a Type-I isotherm, while the KOH activation generated a mesoporous one (70%) with a type-IV isotherm, the pore volumes are 0.2135 and 0.7426 $cm^3.g^{-1}$ respectively. The average pore size of the activated carbons is 2.72 and 2.56 nm for KOH activation and CO_2 activation, respectively. The FT-IR spectra indicated significant variation in the surface functional groups are quite different for the KOH activated and CO_2 activated carbons.

1. Introduction

Activated carbons are the most versatile and commonly used adsorbents because of their extremely high surface areas and micropore volumes [1], large adsorption capacities, fast adsorption kinetics, and relative ease of regeneration [2]. They are produced from a variety of carbonaceous source materials. The choice of precursor is largely dependent on its availability, cost, and purity, but the manufacturing process and intended application of the product are also important considerations [4].

Precursors used for the production of activated carbons are organic materials that are rich in carbon, such as coal, lignite, and wood. Although coal is the most commonly used precursor, agricultural waste in certain condition is a better choice [1]. Activated charcoal produced from residues would reduce the pressure on forests since wood is also commonly used for this purpose [5]. Many agricultural by-products such as coconut shell [2,5,6], grain sorghum [7], coffee bean husks [8], rubber wood sawdust [9], chestnut wood [10], and fruit stones [11], have been discovered to be suitable precursors for activated carbon due to their high carbon and low ash contents. Agricultural wastes are considered to be a very important feedstock because of especially two facts: they are renewable sources and low cost materials [4].

Empty fruit bunches (EFBs) are major waste products generated from the palm oil industry. Every year, 25.9 and 19.8 million tons of oil palm EFB are produced in Indonesia and Malaysia, respectively [12]. EFB, which consists of high cellulose and hemicellulose contents and is available as a low-cost material, has gained attention as biomass feedstock for producing activated carbon [14,15]. Conversion of EFB to value-added product such as activated carbon would directly solve part of the environmental problems while utilizing abundant and low cost biomass.

The preparation of activated carbons consists of two main steps: carbonization of a carbonaceous raw material and char activation [16]. Activation can be done using gases such as carbon dioxide, steam, and others (physical activation) or dehydrating agents and oxidants, such as H_3PO_4, H_2SO_4, HNO_3, NaOH, KOH, and $ZnCl_2$ (chemical activation). Chemical activation is largely employed in the preparation of activated carbons. Among its advantages over physical activation processes are: (i) lower activation temperatures, between 600 and 800°C, as compared to the activation temperature between 700 and 1100°C used in physical activation [17], (ii) shorter activation times [19], (iii) single activation step (carbonization activation simultaneously), (iv) higher yields, and (v) better surface area. Basta et al. (2009) have reported that two-step KOH activation was more efficient in producing activated carbons than one-step KOH activation, as it gave better carbon yields [18]. In this process, KOH reacts with char, not with biomass, producing carbons with high surface areas. In contrast, some disadvantages are: (i) need for intense washing to remove process generated impurities (Na_2CO_3 and others), and (ii) general process corrosiveness [20].

The most important characteristic of an activated carbon is its adsorption performance whichis highly influenced by the preparation conditions, as the activated carbon preparation variables will influence the pore development and surface characteristics of the activated carbon produced. The aim of this study is to prepare an activated carbon from palm empty fruit bunch (PEFB) that are abundant Indonesia by-products by using the physical activation with carbon dioxide (CO_2). The effect of activation temperature and time were investigated. The resultant activated carbons were characterized by several techniques.

2. Experimental

2.1. Apparatus

The schematic diagram of the experimental setup is shown in Figure 1.

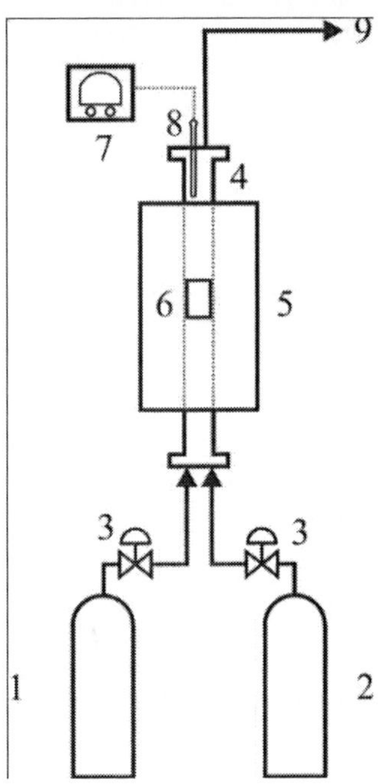

Legends:
1. CO_2 gas tank
2. N2 gas tank
3. Valve
4. Reactor
5. Tubular Furnace
6. Sample
7. Temperature controller
8. Thermocopel
9. Exhaust gas

Figure 1. Schematic diagram of the experimental setup.

2.2. Preparation of porous carbon by KOH activation

Palm empty fruit bunch (PEFB) was obtained from Oil Palm plantation on the Riau province in Indonesia. Two parts of PEFB of shaggy and core were prepared for the activated carbon presursors. PEFB was dried in an oven at 110°C for 6 h. It was then ground and sieved to a 20 or 30 mesh particle size prior to activation. PEFB with a particle size of 30 mesh was used for characterization and the production of activated carbons.

The dried PEFB was placed in a muffle furnace maintained at 400°C for 2 h. The carbon thus obtained was kept at 400°C in the muffle furnace and simultaneously purged with N_2 for 2 h. Then the temperature was reduced to room temperature to get the char. Finally the black residue was sieved to get PEFB having average particle size of 0.096 mm and was stored in air tight plastic container.

For chemical activation, the char was impregnated with a solution of the potassium hydroxide (KOH). Then the mixture was stirred at 80 °C for 2 h in a stainless steel beaker. The temperature of the solution was later increased to the boiling point until complete dryness, which forced the incorporation of all chemical into the particles; at this stage, the temperature reached about 150°C. After being dried, the carbons were placed in sealed ceramic oven and heated at a rateof 10 ◦C/min to 800 °C. In the meantime, N_2 gas in a tubewas let into the oven by a rate of 200cm^3/min. The time foractivation at 800 ◦C was 1 h. The activated products werecooled to room temperature and washed with deionized water. The carbonized sampleswere repeatedly washed with water. Subsequent to this, a 5 M solution of HCl was used to remove the activating agent, followed bydistilled water rinsings until free of chloride ions. The presence ofchloride ions was verified by silver nitrate solution. Consequently, the samples were dried at 110 ◦C for 12 h.

2.3. Preparation of porous carbon by CO_2 activation

The dry PEFB sample (10 g) was loaded into a tubular furnace under N_2 flow (120 mL/min) and heated up to a carbonization temperature of 800 °C at a heating rate of 20 °C/min. The sample was held at 800 °C for 1 h. Then N_2 flow was switched to CO_2 flow with a rate of 50 mL/min for 2 hours for activation and finally CO_2 flow was switched back to N_2flow and cooled down to room temperature under N_2 flow.

2.4. Characterization

The physico-chemical characteristics of activated carbon produced in this experiment were determined. Adsorption characteristics of the activated carbon samples were determined by nitrogen adsorption. The specific surface areas were determined from the isotherms using the Brunauer–Emmett–Teller equation. The surface functional groups were analyzed by using a FT-IR spectroscopy. The analysis of surface functional groups was based on the Boehm titration method. The basic groups were neutralized with 0.1 mol/L HCl solution, and the acidic groups were neutralized with 0.1 mol/L NaOH, Na_2CO_3 and $NaHCO_3$ solutions. The various free acidic groups were determined with the assumption that NaOH neutralizes carboxyl, lactone and phenolic groups, Na_2CO_3 neutralizes carboxyl and lactone groups and $NaHCO_3$ neutralizes carboxyl groups.

3. Results and Discussion

3.1. Textural characterization

Figure 2 shows the Nitrogen adsorption isotherms of activated carbons from PEFB produced at different activation method. The different shape of the adsorption isotherm is an indication of the different pore size distribution of the samples. It can be seen that the volume of nitrogen adsorbed increases by physical activation. By physical activation, the Nitrogen adsorption isotherm showsan intermediate between types I and II of the referred IUPAC classification. Type I isotherms are given by predominantly microporous adsorbents having a relatively small external surface area. The adsorption capacity at relatively low relative pressures (P/Po) is very high, which indicates the presence of a well-developed microporous structure. Hysteresis is detected between adsorption and desorption isotherms, indicating the presence of a little proportion of mesopores and macropores. The isotherms of chars prepared by chemical activation belong to type II IUPAC classification. As shown in Figure 2, the initial part of the type II isotherm for the chars represents micropore filling, and the slope of the plateau at high relative pressure is due to multilayer adsorption on the nonmicroporous surface, i.e., in mesopores, in macropores and on the external surface. The desorption isotherms presents a hysteresis loop at high relative pressure. The desorption isotherms is useful for determination of the mesoporous nature of activated carbons, and the presence of a hysteresis between the adsorption and desorption curves implies the existence of mesopore pores.

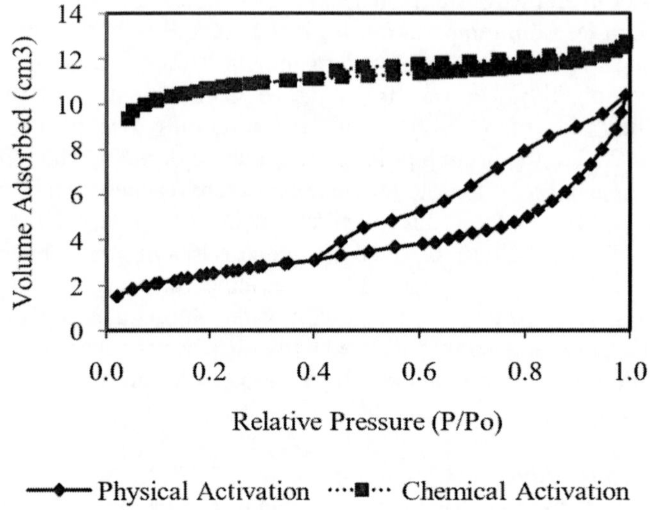

Figure 2. Adsorption/desorption isotherms of N_2 at 77 K for PEFB activated carbons prepared different activation method.

In accordance to the classification adopted by the International Union of Pure and Applied Chemistry (IUPAC), adsorbent pores are classified into three groups: micropores (size <2 nm), mesopores (2–50 nm), and macropores (>50 nm). Micropores can be divided into ultramicropores (width less than 0.7 nm) and supermicropores (width from 0.7 to 2 nm). The size of pores that are formed would have an effect on the porosity, the total surface area that is available for adsorption and most importantly, the size of molecules that can diffuse into the solid. Thus the development of an appropriate pore structure is necessary if the activated carbon produced is to be used for a particular application.

The effect of avtivation method on porosity development of activated carbon is shown in Table 1. The higher BET surface area is an indication of the existence of a significant amount of microporosity. By physical activation, the adsorption capacity of the activated carbon increased enhancing the amount of microporosity. A more ordered structure is likely to be developed in the char that leads to a slower rate of gasification in the interior of the particle. From Table 1, it was observed that the BET surface area, total volume and micropore volume of activated carbon prepared increased, while average diameter decreased. The highest the BET surface area, total volume and micropore volume were obtained by using physical activation.

Table 1. Characteristics of porosity in PEFB activated carbons prepared different activation method.

Parameter	Chemical Activation	Physical Activation
S_{BET} (m^2/g)	405.81	661.46
V_{tot} (cm^3/g)	0.1590	0.2455
V_{micro} (cm^3/g)	0.1177	0.1989
D_{avg} (nm)	3.63	2.48

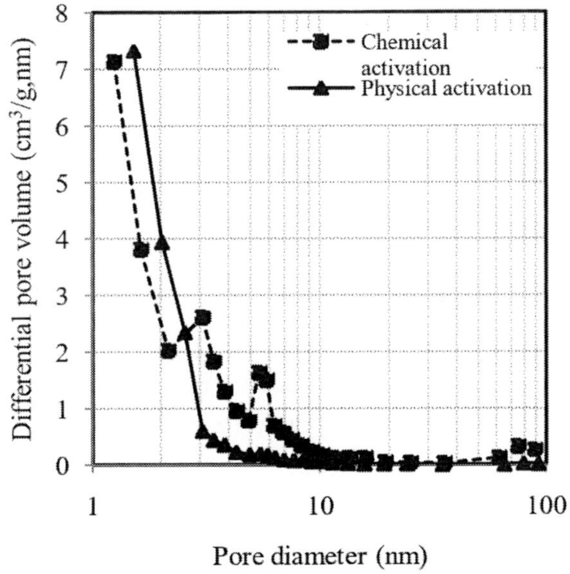

Figure 3. Pore size distributions of activated carbons prepared at different activation method.

Figure 3 illustrated the pore size distributions of activated carbons prepared at different By physical activation. Pore-size distribution is the principal factor determining the adsorption characteristics of activated carbons. The structural heterogeneity of porous materials is generally characterized in terms of the pore size distribution. The pore size distribution represents a model of solid internal structure and regularly shaped model pores which represent the complex void spaces within the real solid. This figure confirmed that the activated carbons prepared by physical activation had more microporous structures compared to activated carbons prepared by chemical activation.

3.2. Chemical surface characterization

Table 2 shows the elemental analyses of PEFB activated carbon produced at activation method. Results indicate the presence of many elements with predominance of carbon in both of the samples. The percentage of carbon increased after activation. These increases in carbon percentage are essentially due to the loss of volatile matter and the decomposition of cellulose and hemicellulose during pyrolysis, and, on the other hand, to the decomposition of lignin during activation. Table 2 also shows a decrease in the H/C and O/C ratios when using physical activation. Physical activationmay cause more heteroatoms to evaporate from the carbon rings and an increasing number of C-H bonds are ruptured as the activation of the char support progresses.

Table 2. Elemental analysis of the thePEFB activated carbons produced at different activation method.

Elemental (wt %)	Chemical Activation	Physical Activation
C	69.19	81.38
H	0.41	0.39
O	15.96	6.04
H/C	0.0059	0.0048
O/C	0.2307	0.0742

The oxygen functional groups are very important characteristics of the activated carbons because they determine the surface properties of the carbons and hence their quality as ion exchangers, adsorbents, catalysts, and catalyst supports. The contents of oxygen-containing functional groups with various acidity strength (total acidic groups: carboxyl, lactonic, hydroxyl and carbonyl) as well as, the total amount of the basic groups are presented in Table 3. It shown in Table 3, for activated carbons prepared by chemical activation, the total oxygen functional groups is 0.196 mmol.g^{-1} with a breakdown of 0.196 mmol.g^{-1}, 0.196 mmol.g^{-1}, and 0.196 mmol.g^{-1}, individually, for carboxylic, lactone and phenolic groups.When the activated carbons prepared by physical activation, the total oxygen containing functional group decreases to 0.240. Decrease in oxygen containing functional groups, takes mostly in the carboxylic, lactone and phenolic groups. The phenolic group decreases to 0.240. Meanwhile the lactone group decreases to 0.240. While the concentration of the carboxylic groups decrease to 0.240.

Table 3. Results of the Boehm and pH_{ZPC} methods for the PEFB activated carbons produced at different activation method.

Activation method	Acid (mmol.g^{-1})			Basic (mmol.g^{-1})	pH_{pzc}
	Carboxylic (mmol.g^{-1})	Lactonic (mmol.g^{-1})	Phenolic (mmol.g^{-1})		
Chemical Activation	0.664	0.012	1.143	1.635	5.36
Physical Activation	0.427	0.011	0.745	0.867	5.86

The FTIR spectroscopy provides information on the chemical structure and functional group of PEFB and its chars prepared at different activation method, shown in Figure 4. Typical infrared band assignments for PEFB indicate that, PEFB contains a number of atomic groups and structures. The functional groups O–H, aromatic C=C, aliphatic C–H, olefinic C=C, C–O, C–O–C, C=O, CH_3, C–O–H and Si–O–Si bands are contained in the PEFB. The band at 3415 cm^{-1} O–H is corresponding to (O–H) vibrations in hydroxyl groups. The position and shape of this band suggest that the hydroxyl groups are involved in hydrogen bonding. Water residual in PEFB could take part in the formation of hydrogen bonds. According to the four types of hydrogen-bonded structures, the pre dominant one in PEFB is self-associated OH groups. The fact that the band at 3415 cm^{-1} is somewhat broader towards lower wavenumbers suggests that PEFB also contains OH-ether hydrogen bonds. The band at 2919 cm^{-1} representing the presence of (C–H) vibrations. The appearance of (C–H) adsorption bands indicating the presence of methyl and methylene groups in PEFB. However, the peak at 1459 cm^{-1} was attributed to (CH_3) stretching vibration. The characteristic absorption of methyl group at 1375 cm^{-1} is valuable for methyl group estimation. Hence the band at 1459 cm^{-1} is slightly weaker than the band at 1375 cm^{-1}, proving the presence of acetoxy groups (–O–$COCH_3$) in PEFB. The band at 1732 cm^{-1} indicating the presence of (C=O) vibrations in carbonyl groups, shows the presence of acetyl derivative groups, aldehydes groups, and others. The olefinic (C=C) and aromatic (C=C) vibrations were detected at 1635 cm^{-1} and (1606–1459 cm^{-1}) region, respectively. While, alkyl groups were detected from the C–CH_3 blending vibration at 1326 cm^{-1}. Moreover, the bands between 1100 and 1300 cm^{-1}, the (C–O–H) vibrations in phenolic and (C–O–C) vibrations can be related to asymmetric vibrations in single graphitic sheet and between two such sheets; oxygen can act as a crosslinking agent between aromatic sheets. The absorbance band at 1000–1100 cm^{-1} are attributed to (C–O) vibrations in primary C–OH, secondary O–H and the presence of silica, Si–O–Si. Furthermore, some bands are observed in the region 470–900 cm^{-1}, which are band positions compatible with C–H vibrations in olefinic or aromatic structures.

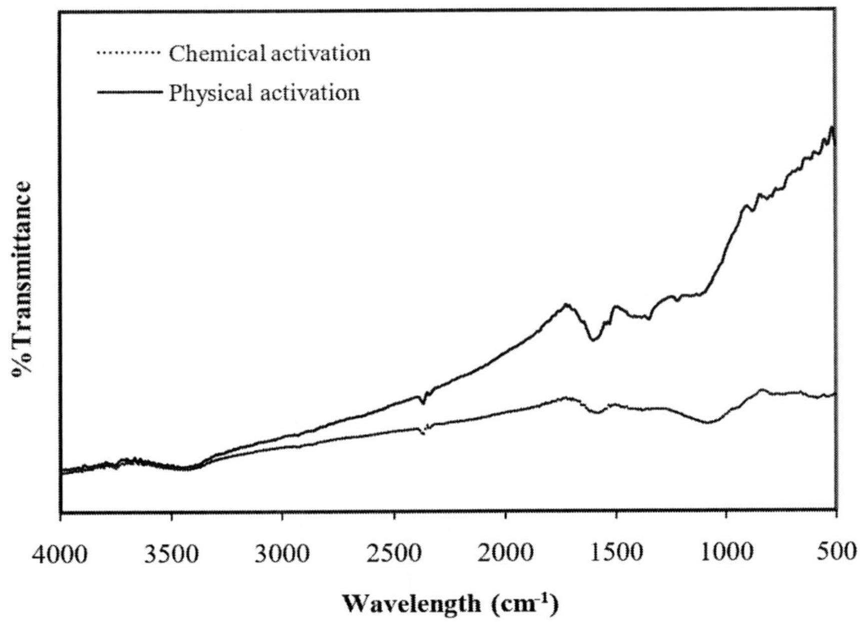

Figure 4. The FT-IR spectra of the PEFB activated carbons produced at different activation method.

Table 4. Activation conditions and characteristics of the activated carbon reported in the literature.

Material	Activation condition	S_{BET} (m^2/g)	V_{mic} (cm^3/g)	Vt (cm^3/g)	Pore size (nm)	References
Palm empty fruit bunches	KOH	488.51	0.2539	0.196	2.079	Arshadet al.
Peanut hull	Steam, 600 °C/2 h	253	0.09	0.22	–	Girgis et al.
Coconut shell	CO_2, 600 °C/2 h	1700	0.88	1.14	2.7	Guo et al.
Coconut shell	Steam, 1000 °C/120 min	1926	0.93	1.26	2.6	Li et al.
Sugarcane bagasse	CO_2, 850 °C/2 h	625.6	0.097	0.040	1.454	Goncalves
Peanut hull	Steam, 600 °C/2 h	253	0.09	0.22	–	Girgis et al.
Coconut shell	CO_2, 600 °C/2 h	1700	0.88	1.14	2.7	Guo et al.
Palm shell	$ZnCl_2$	1118	0.42	0.51	2,24	Arami-Niya
Palm shell	CO_2, 800 °C/4 h	544	0.26	0.27	1.95	Arami-Niya
Bagasse	$ZnCl_2$	923	–	0.53	0.8	Boonpoke et al.
Rice husk	$ZnCl_2$	927	–	0.56	0.8	Boonpoke et al.
Palm empty fruit bunches	CO_2, 800 °C/2 h	717	-	0.7426	2.56	This paper
Palm empty fruit bunches	KOH	613	-	0.2135	2.72	This paper

4. Conclusions

A comparison of the textural characteristics and surface chemistry of the activated carbon from palm empty fruit bunch by the CO_2 and the KOH activation leads to the following findings: An activated carbon by the CO_2 activation under the optimum conditions has a BET surface area of 717 m^2/g, while that by the KOH activation has a BET surface area of 613 m^2/g. The CO_2 activation generated a highly microporous carbon (92%) with a Type-I isotherm, while the KOH activation generated a mesoporous one (70%) with a type-IV isotherm, the pore volumes are 0.2135 and 0.7426 $cm^3.g^{-1}$ respectively. The average pore size of the activated carbons is 2.72 and 2.56 nm for KOH activation and CO_2 activation, respectively. The FT-IR spectra indicated significant variation in the surface functional groups are quite different for the KOH activated and CO_2 activated carbons.

Acknowledgement

The author would like thank to Direktorat Riset dan Pengabdian Masyarakat, Direktorat Jenderal Penguatan Riset dan Pengembangan, Kementerian Riset, Teknologi, dan Pendidikan Tinggi for financial support on this research.

References

[1] Do D D 1998 Adsorption Analysis: Equilibria and Kinetics, Imperial College Press, London
[2] Hu Z , Srinivasan M P 1999 *Microporous Mesoporous Mater* **27** 11-18
[3] Hidayat A, Rochmadi, Wijaya K, Hinode H, Budiman A 2013 *Asian J Chem* **25**(3) 1569-1579
[4] Stavropoulos G G, Zabaniotou A A 2005 *Microporous Mesoporous Mater* **83** 79–85
[5] Mozammel H M, Masahiro O, Bhattacharya S C 2002 *Biomass Bioenergy* **22** 397–400
[6] Sekar M, Sakthi V, Rengaraj S 2004 *J Colloid Interface Sci* **279** 307–313
[7] Diao Y, Walawender W P, Fan L T 2002 *Bioresour Technol* **81** 45–52
[8] Baquero M C, Giraldo L, Moreno J C, Suarez-Garcia F, Martinez-Alonso A, Tascon J M D 2003 *J. Anal. Appl. Pyrol.* **70** 779–784
[9] Srinivasakannan C, Bakar M Z A 2004 *Biomass Bioenerg.* **27** 89–96
[10] Gomez-Serrano V, Cuerda-Correa E M, Fernandez-Gonzales M C, Alexandre-Franco M F, Macias-Garcia A 2005 *Mater. Lett.* **59** 846-853
[11] Puziy A M, Poddubnaya O I, Martinez-Alonso A, Suarez-Garcia F, Tascon J M D 2005 *Carbon* **43** 2857–2868
[12] Bardant T B, Sudiyarmanto S, Abimanyu H, Hanum A K 2013 *Indones. J. Chem.* **13** (1) 53-58
[13] Daud W R W and Law K N 2011 *Bioresources* **6** (1) 901-917
[14] Misson M, Haron R, Kamaroddin M F A, Amin N A S 2009 *Bioresource. Technol.* **100** 2867-2873
[15] Parshetti G K, Hoekman S K and Balasubramanian R 2013 *Bioresource. Technol.* **135** 683-689
[16] Ioannidou O, Zabaniotou A 2006 *Renew. Sust. Energ. Rev.* **11** 1966-2005
[17] El-Hendawy A A, Alexander A J, Andrews R J, Forrest G 2008 *J. Anal. Appl. Pyrol* **82** 272-278
[18] Basta A H, Fierro V, El-Saied H, Celzard A 2009 *Bioresour. Technol.* **100** 3941-3947
[19] Nowicki P, Pietrzak R, Wachowska H 2008 *Fuel* **87** 2037-2040
[20] Lozano-Castello D, Calo J M, Cazorla-Amoros D, Linares-Solano A 2007 *Carbon* **45** 2529-2536
[21] Girgis B S, Yunis S S, Soliman A M 2002 *Mater. Lett.* **57** 164–172
[22] Guo S, Peng J, Li W, Yang K, Zhang L, Zhang S, Xia H 2009 *Appl. Surf. Sci.* **255** 8443–8449
[23] Boonpoke A, Chiarakorn S, Laosiripojana N, Towprayoon S, Chidthaisong A 2011 *J. Sustain. Energ. Environ.* **2** 77–81
[24] Li W, Yang K, Peng J, Zhang L, Guo S, Xia H 2008 *Ind. Crops. Prod.* **28** 90–198
[25] Niya A A, Daud W M A W, Mjalli F S 2011 *Chem. Eng. Res. Des.* **8**(9) 657–664
[26] Goncalves G C, Pereira N C, Veit M T 2016 *Biomass Bioenerg.* **85** 178-186
[27] Arshad S H M, Ngadi N, Aziz A A, Amin N S, Jusoh M, Wong S 2016 *J. Energy Storage* Available online 8 October 2016

IOP Conf. Series: Materials Science and Engineering **162** (2017) 012009 doi:10.1088/1757-899X/162/1/012009

Synthesis of renewable diesel through hydrodeoxygenation reaction from nyamplung oil (*Calophyllum Inophyllum* oil) using NiMo/Z and NiMo/C catalysts with rapid heating and cooling method

B H Susanto, M B Prakasa and M H Shahab
Chemical Engineering Department, Faculty of Engineering, Universitas Indonesia, Kampus Baru 16424, Depok, Indonesia

E-mail : bambanghs@che.ui.ac.id

Abstract. The synthesis of metal nanocrystal was conducted by modification preparation from simple heating method which heating and cooling process run rapidly. The result of NiMo/Z 575 °C characterizations are 33.73 m^2/gram surface area and 31.80 nm crystal size. By used NiMo/C 700 °C catalyst for 30 minutes which had surface area of 263.21 m^2/gram, had 31.77 nm crystal size, and good morphology, obtained catalyst with high activity, selectivity, and stability. After catalyst activated, synthesis of renewable diesel performed in hydrogenation reactor at 375 °C, 12 bar, and 800 rpm. The result of conversion was 81.99 %, yield was 68.08%, and selectivity was 84.54 %.

1. Introduction

Due to the depletion of crude reserves, shortage in petroleum supply, and increasing demand for clean hydrocarbon fuels, the production of renewable feedstocks-based fuels from non-edible vegetable oil has emerged to solve at least partially in this problems. As a result, more consumers are now turning to biodiesel fuel instead of petroleum-based diesel for transportation. Unfortunately, biodiesel gave low oxidative stability, higher viscosity, higher cloud point and pour point, higher nitrogen oxides (NO_x) emissions, lower energy density, and higher injector/engine wear [7]. To avoid the disadvantageous physical properties of oxygenated fuels, hyrodeoxygenation of non-edible vegetable oil has been investigated as an alternative pathway ofbiofuel production.

Hydrodeoxygenation is a series of reactions that convert the molecular structure of triglyceride in non-edible vegetable oil to linear hydrocarbon and it meansremoval of oxygen from an oxygen-containing compound under hydrogenpressure at high temperatures with the help of heterogenous catalyst, which product known as renewable diesel [9]. The hydrodeoxygenation process consisted of decarboxylation, decarbonylation, and hydrogenation reaction as can be seen below.

Decarboxylation Reaction : $$C_{17}H_{35}COOH \rightarrow C_{17}H_{36} + CO_2 \quad (1)$$

Decarbonylation Reaction : $$C_{17}H_{35}COOH + H_2 \rightarrow C_{17}H_{36} + H_2O + CO \quad (2)$$

Hydrogenation Reaction : $$C_{17}H_{35}COOH + 3H_2 \rightarrow C_{18}H_{38} + 2H_2O \quad (3)$$

The main drawbacks of this reaction seem to be the large consumption of hydrogen during the process, which increase the production cost [10]. The aim of this experiment is to investigate the effect of temperature and calcination time during the catalyst preparation step on the activity, stability, selectivity, conversion, and yield of triglyceride content in nyamplung oil to renewable diesel product. Mo is found to facilitate the reduction, and promoting C-O activation while inhibiting the C-C breaking [5]. This NiMo/C and NiMo/Z catalysts synthesized through modification of simple heating method which heating and cooling process run rapidly [1]. Several studies using simple heating

IOP Conf. Series: Materials Science and Engineering **162** (2017) 012009 doi:10.1088/1757-899X/162/1/012009

method are: Liherlinah et al.[4] synthesizing CuO/ZnO/Alumina catalyststofuel cell applications for 30 minutes; Vioktalamo et al. [8] synthesize Y_2O_3/Gd catalysts todecompose of pollutant for 60 minutes with holding time of 30 minutes, Faisal et al. [11] synthesizing NiMo/Z catalyst for renewable diesel productions, and Fransisca et al.[2] synthesizing NiMo/Ccatalyst for renewable diesel productions. Studies abovewere using the same method but with different treatments. The products was characterized by GC-MS (Agilent) to find functional groups and components of renewable diesel.

2. Methodology

2.1. Materials

This study was conducted in two steps namely the preparation and characterization of nanocrystal NiMo/C and NiMo/Z catalysts and synthesis of renewable diesel in a hydrogenation reactor. The materialswhich needed in this study are:nyamplung oil (UNS Solo), $(NH_4)_6Mo_7O_{24}.4H_2O$ and $Ni(NO_3)_2.6H_2O$ (Merck), activated carbon and zeolite (Merck), HCl (Merck), Aquadest (CV. Dwinika), polyethylene glycol or PEG (Merck) and demineralized water (Bratachem Indonesia).

2.2. Catalyst preparation

Nanocrystal of nickel molybdenum on carbon and zeolite supports was prepared by rapid heating and cooling method adopted from Liherlinah [4] and Vioktalamo [8]. These catalysts were assigned as 5 wt% [6] and required amounts of the precursor salts and polyethylene glycol, i.e., $(NH_4)_6Mo_7O_{24}.4H_2O$, and $Ni(NO_3)_2.6H_2O$ were dissolved in demineralized water. Then, activated carbon or zeolite was addded dropwise with constant stirring followed by heating at 100 °C for 40 minutes in that solutions to evaporize water. Polymer solution, as continous media, was used to avoid agglomeration of particle group of catalyst. Therefore the polymer should be remained until the end of the process. Finally the mixture was heated in furnace to evaporize the PEG polymer and to produce the final nanocrystal NiMo/C and NiMo/Z catalysts at 525 °C , 575 °C, 600 °C, 700 °C for 30 minutes and 60 minutes with holding time 30 minutes. After that, these catalysts were rapidly cooled (quenching process) with ice blocks. The catalysts were removed and stored in bottle of sample.

2.3. Catalyst characterization

The catalysts were characterized byXRD (Shimidzu) to determine crystal size and type of crystal catalyst, BET (Micromeritrics ASAP 2020) to determine surface area, pore size and pore volume of catalyst and SEM-EDX (JED-2300 Analysis Station JEOL)to determine the morphology of the catalyst.

2.4. Catalyst activity test

Catalyst activation was carried out at 400 °C with internal temperature being measured by thermocouple. Before catalyst activation, the reactor was purged with nitrogen during experiment (30 minutes) to remove any oxygen content that might have been dissolved and present in nyamplung oil so that activity of catalyst will increase. After that, added hydrogen gas of 10 bar and the temperature was slowly decreased into 100 °C to provide activityperformace during the reaction.

2.5. Hydrodeoxygenation of nyamplung oil

The reaction wasperformed in a 330 mL batch stirred reactor integrated with a condenser and furnace. Extensive stirring of the oil and catalyst in the reactor was achieved using a magnetically driven stirrer with a DC geared motor. Nyamplung oil was used as feed and NiMo/C or NiMo/Z as a catalyst with a feed ratio of 100:1 wt. Nitrogen gas was purged for 30 minutes. Stirrer speed and pressure during the reaction was kept constant at 800 rpm and 12 bar respectively. The reactions were performed at temperature 375°C for 2.5 hours. When the inside temperature of the reactor reached the target value, the hydrogen pressure was adjusted to the desired value. From this time onwards, the reaction time was occured. In this step, the double bonds that were present in the tryglycerides were saturated with hydrogen gas. Every 15 minutes, the resulting gas was analyzed online by GC-TCD to identify the hydrogen, carbon dioxide, carbon monoxide and methane gasses. The gas products were condensed and collected as condensate. Figure1 shows a schematic diagram of the apparatus.

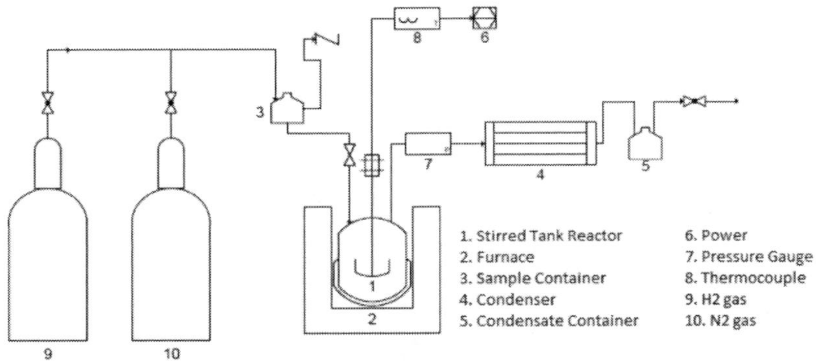

Figure 1. Scheme of Hydrodeoxygenation Reaction

The bottom product was distillated according to the boiling range of petroleum products in three categories i.e. IBP-200 °C, 200-300 °C, and 300-EP. Distillation products were characterized by GCMS to identify fraction component of sample, and also investigate the density, viscosity, and cetane index of the products.

3. Results and Discussion

3.1. Results of catalysts characterization

From BET characterization result, the best catalysts that gives highest surface area are NiMo/C with temperature 700 °C and NiMo/Z with temperature 575 °C for 30 minutes. These catalysts has surface area 263.21 m^2/gram and 33.73 m^2/gram respectively. Large surface area of the catalyst was needed to increase the collision between particles so it would increase the rate of reaction as well as the kinetics reaction [6]. This NiMo/C and NiMo/Z catalysts were continued confirmed through several tests by SEM-EDX and XRD to see morphology and crystal size of catalyst. Increasing temperature led to more rapid PEG vaporization and sintering of catalyst which occured during reaction in zeolite but due to activated carbon stable with high temperature, so that increasing of temperature can not effect of catalyst surface area. If temperature increased, surface area of catalyst will reduce due to the collapse of walls inside catalyst pore [9].

NiMo/C 700 °C and NiMo/Z °C catalysts also give largest pore volumewhich indicating that the support structure has a large enough pore spaces. In catalysis reaction, large volume is needed to achieve the complete reaction, as the site of the reaction medium and can be regarded as a mini reactor [3]. The reason for low surface area and pore volume for this catalyst could be attributed to the pore blocking by NiMo metal. For sample of NiMo/C and NiMo/Z that was not kept constant, had higher surface area and the sintering that occured on the catalysts to be weaker than other profil temperature.

The result of XRD characterization can be seen on Figure2. Carbon and zeolite which originally had a weak peak, after impregnated the catalysts on the surface has a sharp peak. The intensity of the peak represents the crystallinity. The sharper peak obtained, means particles tend to be better crystallinity. High crystallinity also will affect to the catalytic activity and stability at high temperatures.

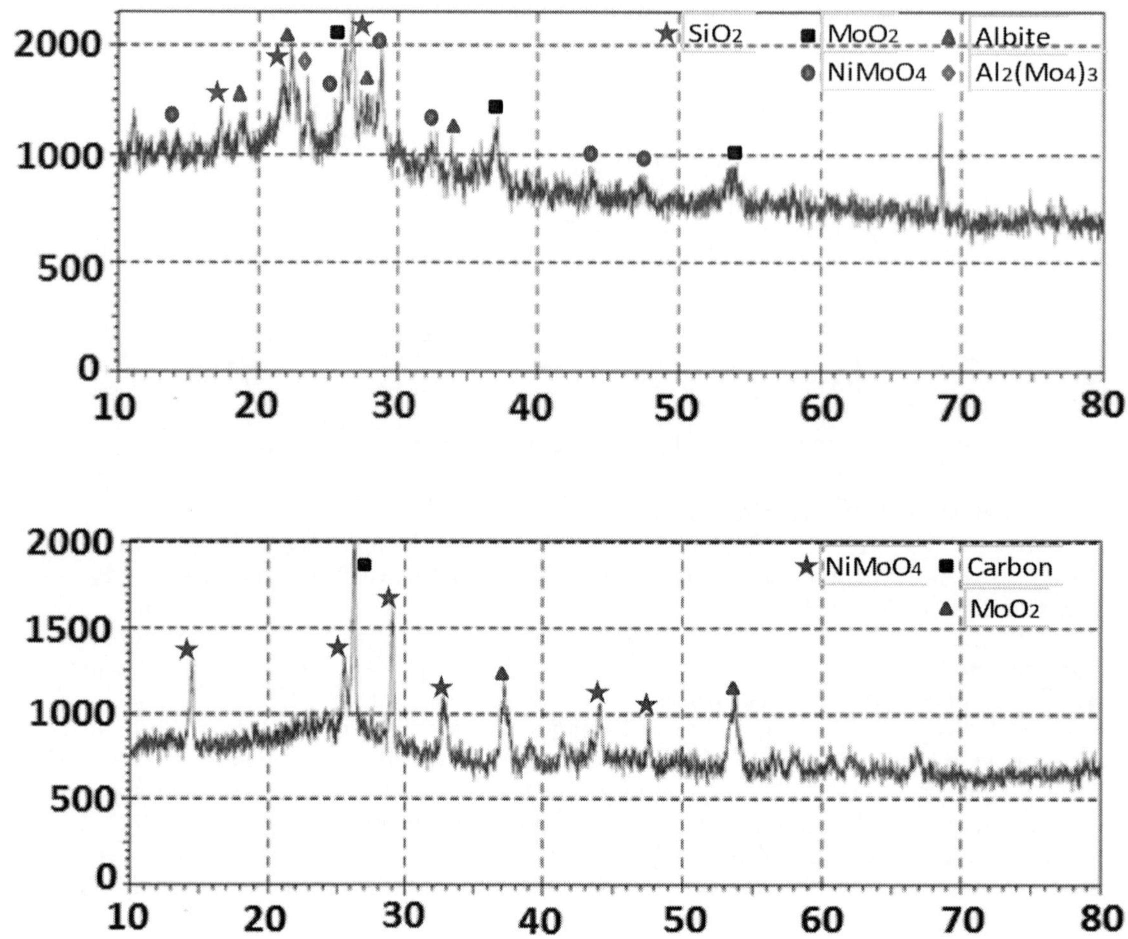

Figure 2. Diffractogram of (a) NiMo/Z and (b) NiMo/C Catalysts

Diffractogram also shows that crystal size of NiMo/C 700 °C which estimated by Scherrer formula is31.77 nm and for NiMo/Z 575 °C is 31.80 nm. From this result, both of catalyst crystals size which formed were nanocrystalline.

Morphology plays an important role in activity and selectivity of the catalyst. From Figure3 in SEM result, it can be seen that the magnification of 1000x and 10000x, these catalysts have the morphology which tend to be agglomerate and consist of small particles and large particles.

Based on EDX characterization result, the average yield active core NiMo content in carbon support was 6.76 wt% and 6.72 %wt in zeolite support. This number is almost close to the number of active nuclei was added during preparation which is 5 wt%. Increasing metal loading favors sintering due to weak interaction with supported phase, agglomeration and micropore blocking could also take place [5]. The SEM results can be seen in Figure3.

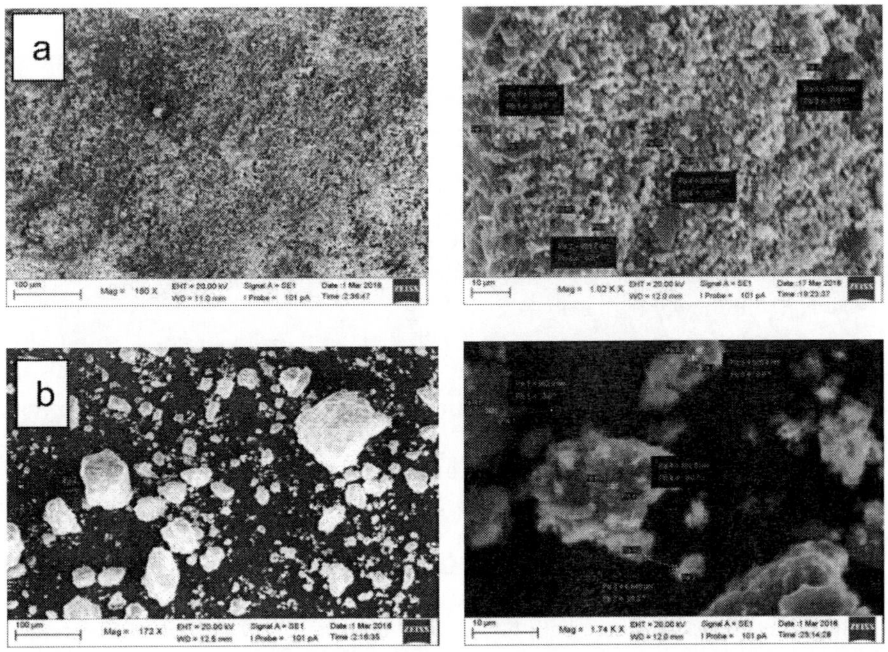

Figure 3. Catalyst Morphology of (a) NiMo/C 700 °C and (b) NiMo/Z 575 °C Catalysts at 1000x and 10000x Magnification

High *loading* which produce on both of catalyst happened due to percent weight of oxide element still high. Oxide element can be removed while catalyst activation next before hydrodeoxygenation reaction with hydrogen gas.

3.2. Results of catalysts characterization

Reactions of hydrodeoxygenation were carried out after the catalyst activation completed and the reactor temperature reaches 100°C. Hydrodeoxygenation of nyamplung oil was performed with four sample catalysts which had lowest crystal size, highest surface area, and good morphology namely NiMo/C, NiMo/C activated, NiMo/Z, and NiMo/Z activated. After HDO reaction, the products still in mixture and had to be purified through distillation process. Distillation products at temperature 300-EP were carried out to determine the physical and chemical properties of renewable diesel product.

*3.2.1. Results of Activation Catalyst Test on Hydrodeoxygenation Reaction.*GCMS shows that the distillation bottom products consist of hydrocarbon fractions (C_5-C_{23}) and oxygenatesuch as acetone, alcohol, aldehyde, carboxylic acid, etc. The percentage of oxygenate contained in the products less than 15% as can be seen on Table 2.

Table 1. Comparison Physical Properties of Renewable diesel with Other Products

Sample	Density (gram/mL)	Viscosity (cst)	Cetane Index
NiMo/C	0.86	1.37	54
NiMo/C activated	0.78	1.16	81.83
NiMo/Z	0.87	1.39	53.96
NiMo/Z activated	0.86	1.35	55.30
Nyamplung Oil	0.94	90.88	-
Biodiesel	0.85	2.30	51
Diesel	0.82	1.90	40
SNI	0.85 - 0.89	2 s/d 6	Min. 51

Table 2. Selectivity and Yield of Distillation Product (300-EP)

Sample	Product	% Selectivity	% Yield	%Conversion	%Oxygenate
NiMo/C activated	Biogasoline	12.32	9.92	81.99	10.29
	Biokerosene	2.35	1.89		
	Renewable diesel	84.54	68.08		
	Lubricant	0.79	0.64		
NiMo/Z activated	Biogasoline	10.17	8.11	80.95	10.71
	Biokerosene	4.43	3.53		
	Renewable diesel	74.54	59.46		
	Lubricant	10.86	8.66		

NiMo/C activated was a sample that produced the smallest percentage of oxygenate (10.29%). Based on this result, it can be seen that NiMo/C activated catalyst was the suitable catalyst to produce renewable diesel with the largest hydrocarbon content.

From Table 2 it can also be seen that the largest conversion of reactants occurred at NiMo/C activated catalyst with 81.99%. Similar with conversion, NiMo/C activated also has the largest selectivity and yield of C_{12}-C_{18} fraction (diesel fraction) which are 84.54 % and 68.08 %. To evaluate the effectiveness and selectivities of NiMo/C activated catalyst in desrided hydrodeoxygenation pathway, so analysis of gas products with GC-TCD were important as seen at Figure4.

From Figure 4, NiMo/C activated catalyst was selective to decarboxylation reaction as equation in (1). This reaaction can break oxygen bond so that alkane bond can be produced which indicates CO_2 gas was high approximately 80%. This NiMo/C activated catalyst also can minimize rhe production of methane and CO gasses, so that alkene bond which undesirable product was not formed. Therefore, can be conclued that NiMo/C activated catalyst which has highest loading, highest surface area, and lowest crystal size, can convert nyamplung oil to renewable diesel with yield was 68.08%, conversion was 81.99%, and selectivity was 84.54%.

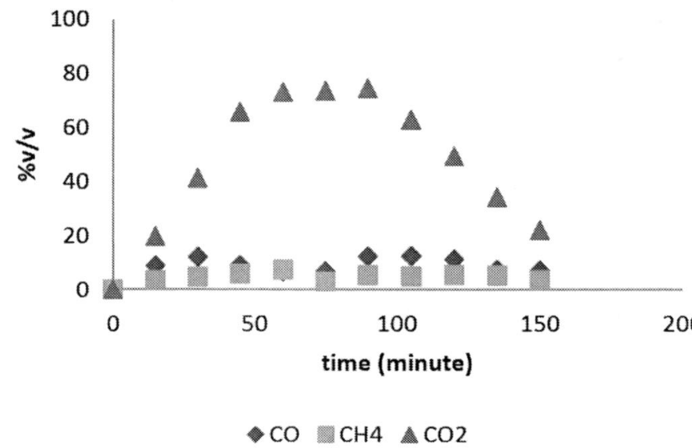

Figure 4. Gas Products Result of NiMo/C Activated Catalyst

4. Conclusion

In this study, NiMo/C and NiMo/Z catalysts were prepared by rapid heating and cooling method with polymer solution as a growth inhibitor. Characterization of catalysts have shown that NiMo/C catalyst has lowest crystal size of 31.77 nm and surface area of 263.21m^2/g which prepared at 700 °C when the heating temperature was not kept constant. The experimental conditions were 375 °C for average reaction temperature and 150 minutes for residence time. These conditions with NiMo/C activated led to dominant yield of 68.08% of diesel fraction (C_{13}-C_{18}), conversion was 81.99%, and selectivity was

84.54%. It means theoxygenation removal pathway of nyamplung oil through NiMo/C activated catalyst wassuccessfully converted into the renewable diesel which the spesification met the petroleum diesel standard.

Acknowledgments
Directorate General of Higher Education- Indonesia Ministry of Education and Culture through PKM-PE UI 2016 is acknowledged for the financial support to carry out the present study.

References
[1] Abdullah M, Khairunnijal, Maruly A R, Liherlinah and Sunny M 2008 *J. N. Sains N. Technol* **1** 1-25
[2] Fransisca 2012 *Sintesis Renewable diesel dari Minyak Jarak melalui reaksi Deoksigenisasi menggunaan Katalis Nano NiO/Alumina* (Depok: Thesis) pp 26-60
[3] Jankowska, Swiatkowski A and Choma J 1991 *Active Carbon* (New York: Ellis Horwood) p 79
[4] Liherlinah, Abdullah M A and Khairurrijal 2009 *J. N. Sains N. Technol* **2** 2-7
[5] Nasikin M and Susanto B H 2010 *Katalis Heterogen* (Depok: Universitas Indonesia Press) pp 42-64
[6] Sinaga A C 2011 *Preparasi, Karakterisasi, dan Uji Reaksi Katalis Nano NiO/*Al_2O_3 *untuk Sintesis Bahan Bakar Bio Non Ester dari Minyak Jarak Melalui Pirolisis Berkatalis* (Depok: Thesis) pp 28-54
[7] Lestari S, Palvi M A, Kari E, Beltramini J, Max L and Dmitry Y M 2010 *J. Cat. Let.* **134** 250-257
[8] Vioktalamo A S 2007 *Synthesis* Y_2O_3*: Gd Nanoparticles Using a Polymer Heating Methods* (Thesis) p 21
[9] Boyás R S, Yanyong L and Tomoaki M 2008 *Inst. Politéc. Nat.*2-7
[10] Veriansyah B and Han J Y 2012 *J. of Fuel* **94** 578-585
[11] Faisal A 2014 *Preparasi dan Karakterisasi Katalis NiMo/Zeolit dengan Metode Microwave Polyol Process untuk Sintesis Renewable Diesel Menggunakan Minyak Jarak Pagar* (Depok : Thesis) pp 41-74

IOP Conf. Series: Materials Science and Engineering **162** (2017) 012010 doi:10.1088/1757-899X/162/1/012010

Hydrogen production by plasma electrolysis reactor of KOH-ethanol solution

N Saksono [a*], T Batubara[b], S Bismo[c]
Chemical Engineering Department, Universitas Indonesia, Kampus Baru UI Depok, Depok 16424, Indonesia.

E-mail : [a]nelson@che.ui.ac.id, [b]taher.batubara@yahoo.com, [c]bismo@che.ui.ac.id

Abstract. Plasma electrolysis has great potential in industrial hydrogen production, chlor-alkali production, and waste water treatment. Plasma electrolysis produces more hydrogen with less energy consumption than hydrocarbon or Faraday electrolysis. This paper investigated the hydrogen production by plasma electrolysis of KOH-ethanol solution at 80 °C and 1 atm. The effects of voltage, KOH solution, ethanol addition, and cathode deep on plasma electrolysis performance were studied. The hydrogen production was analyzed using bubble flow meter and hydrogen analyzer. The electrical energy consumption was measured by a digital multimeter. The effectiveness of plasma electrolysis in terms of hydrogen production was evaluated by comparing it with Faraday Electrolysis. The results showed that hydrogen produced by plasma electrolysis is 149 times higher than the hydrogen produced by Faraday electrolysis. The optimum hydrogen production was 50.71 mmol/min, obtained at 700 V with 0.03 M KOH, 10% vol ethanol and 6.6 cm cathode deep, with energy consumption 1.49 kJ/mmol. The result demonstrates a promising path for hydrogen production by utilizing plasma electrolysis reactor.

1. Introduction

Hydrogen has been considered as a clean energy resource because its high energy density, environmental friendliness and safety. Hydrogen production by steam reforming process can produce pollutant gas. Other than fossil fuel, water electrolysis can also produced hydrogen. However, because of its high electrical energy requirements, producing hydrogen by water electrolysis is a much more expensive process than by using hydrocarbons as feedstock [1].

Plasma electrolysis is another alternative process that can be implemented in hydrogen production. Plasma electrolysis technology consists in an electrolysis conducted in high voltage that produces electrical spark, causing plasma to form in the electrolyte solution. Plasma electrolysis with KOH and Glycerol additive produces hydrogen 8.1 times higher than Faraday electrolysis [2]. Some researches have shown that hydrogen production in plasma electrolysis is affected by some variables such as voltage discharge, electrolyte solution, conductivity, and pH [3-4]. Adding several additives such as acetic acid to the electrolyte solution increased process productivity significantly [5].

This paper describes a reactor design – a scheme of a semi batch reactor for plasma electrolysis that includes a plasma generator that is able to generate voltage up to 700 V (DC). Using KOH as the electrolyte solution, ethanol as an additive, certain cathode configuration, and at high voltage, the reactor was expected to increase the process performance.

2. Experimental

There are three main components in this experiment; electrolysis reactor, coolant system, and purification system which explained as follow.

2.1. Electrolysis Reactor

The scheme of the reactor plasma electrolysis system is shown on Figure 1. The reactor was made of one liter capacity from transparent filter housing with 8 cm of diameter and 25 cm of height. On the top part, an output hole for the hydrogen and the oxygen was set as well as a thermometer and the anode. On the bottom was the cathode. The cathode was made of wolfram, while anode of SS 316G.

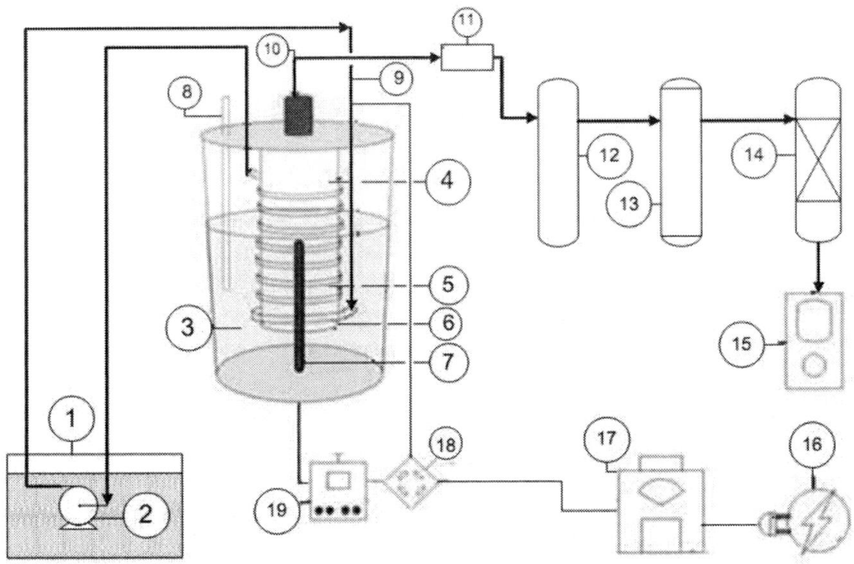

Figure 1. Experimetal set up: 1. Cooling water basin; 2. Peristaltic Pump; 3. Electrolyte Solution; 4. Hydrogen Outlet; 5. Cathode Room; 6.Acrylic Screen; 7.Cathode; 8. Thermometer; 9. Anode; 10. Hydrogen Outlet; 11. Flow meter; 12. Condenser 1; 13. Condenser 2; 14. Silica Absorber; 15. Hydrogen Analyzer; 16. Power Supply; 17. Slide Regulator; 18. Diode Bridge; 19. Digital Multimeter

2.2. Coolant System

The open circulation cooling system was implemented to reach a suitable reactor temperature, 80 – 85 °C, which is 5 °C lower than previous research [6]. A peristaltic pump inside the coolant water basin pumped to coolant through the anode coil all the way to the reactor. The anode coil inside the reactor was set to form a spiral, in order to extend surface area between electrolyte solution and anode.

2.3. Purification System

The reactor produced water vapour. Two condensers were set to separate the hydrogen and oxygen of the water vapor. Each condenser was made of a one liter capacity housing filter with 8 cm in diameter and 25 cm of height. A silica gel absorbent, 5 cm in diameter and 20 cm in height, was set as the final water vapor absorbent. GNL-400F Hydrogen Analyzer (ChangAi Electronic Science & Technology Co. Ltd China) was used to measure the composition of the hydrogen produced and its volume flow rate was measured by bubble flow meter. The volume of solution fed into the reactor for plasma electrolysis was 1L.

Hydrogen flow rate (F_{H_2}) is defined as :

$$F_{H_2} = \frac{V_{H_2} X_{H_2}}{24{,}45} \ mmol \cdot min^{-1} \quad (1)$$

where V_{H_2} and X_{H_2} denote hydrogen volume and hydrogen composition at 25°C and 1 atm.

Voltage (DC) and current (DC) were measured by a A830L digital multimeter. $G(H_2)$ is the mole number of hydrogen divided by the Faraday stipulated yield by passing the given electricity. In our experiments the $G(H_2)$calculation has been done by using the equation below [4] :

$$G(H_2) = \frac{V_{H_2}/22,4}{Q/2F} = \frac{V_{H_2}}{11.2Q}\ mol{\cdot}mol^{-1} \quad (2)$$

where Q, F, V_{H_2} denote the electric charge, Faraday constant, and hydrogen volume at 25°C and 1 atm respectively.

Wr is defined as the power depleted divided by the hydrogen mol. Wr is calculated by the equation:

$$W_r = \frac{VQ}{F_{H_2}}\ kJ{\cdot}mmol^{-1} \quad (3)$$

where V, Q, F_{H_2} denote the voltage across the circuit, the electric charge and hydrogen flow rate at 25°C and 1 atm respectively.

3. Results and discussion

This research describes the formation of plasma in the electrolyte solution during plasma electrolysis. It also observes the correlations of KOH concentration, ethanol additive percentage, voltage, cathode depth, and electrical energy consumption in the hydrogen production. It compares the effectiveness of plasma electrolysis with the Faraday Electrolysis. The result of this research can be seen in table 1 below.

Table 1. Result of plasma electrolysis on different variable at 500 V

	KOH Concentration (Molar)*			Ethanol Concentration (% vol) **				Cathode Deep (cm)***		
	0.03	0.05	0.1	0	5	10	15	0	3	6,6
H_2 Production (mmol/min)	15.97	37.80	35.98	2.89	29.18	35.46	33.03	18.06	35.23	44.36
G (mol/mol)	49.81	100.53	102.86	45.34	71.46	100.53	96.71	55.11	103.21	132.31
Energy (kJ/mmol)	3.68	1.25	2.83	12.42	3.23	2.31	2.55	1.24	1.42	2.66

* 10% ethanol, cathode deep 6.6 cm
** 0.05 M KOH, cathode deep 6.6 cm
*** 0.03 M KOH, 10% ethanol.

3.1. The Effect of KOH Concentration in Hydrogen Production

The effect of KOH concentration in hydrogen production can be seen in Table.1. This result were obtained using 10% ethanol as additive at 500 V and with cathode deep 6.6 cm. This research shows hydrogen generation at 0.05 M is higher than 0.03 M and 0.1 M KOH; it was 37.80mmol/min, 15.97 mmol/min, and 35.98 mmol/min, respectively. Energy consumption showed a similar pattern. The energy consumption at 0.05 M is lower than the energy consumption at 0.03 M and 0.1 M KOH; it was 1.25 kJ/mmol, 3.68 kJ/mmol, and 2.83 kJ/mmol, respectively. At 0.03M and 0.1M KOH, the normal electrolysis appeared to be more dominant, but at 0.05M KOH, plasma electrolysis is more dominant. The normal electrolysis is characterized by high electrical energy consumption and producing less hydrogen, while plasma electrolysis is characterized by low electrical energy consumption and producing more hydrogen.

Higher electrolyte concentration leads to higher conductivity and high conductivity increases hydrogen radicals forming [7-9]. In the course of the gas discharge or plasma electrolysis, the energy is primarily dissipated in the cathode fall section. When the cathode fall occurred on the solution side, the energy dissipated in the cathode fall section could be consumed primarily in solution vaporization. And then it is easy to get a thicker and bigger volume continuous gas envelope around the discharged electrode. When the thickness of the gas envelope is high, to develop a full glow discharge during plasma electrolysis is difficult and needs higher voltage [4].Higher voltage in high conductivity solution will produce a large amount of current. This large current destabilizes the plasma formation in the solution. Consequently, normal electrolysis becomes more dominant than plasma electrolysis.

3.2. The Effect of Ethanol Concentration in Hydrogen Production

Table 1 also shows the effects of ethanol concentration additive in hydrogen gas production. These results were obtained using 0.05 M KOH at 500 V and with cathode deep 6.6 cm. Table 1 shows that hydrogen production increased from 2.89 mmol/min to 29.18 mmol/min then to 35.46 mmol/min at 0%, 5%, and 10% ethanol, respectively. But at 15% ethanol, the hydrogen production decreased to 33.03 mmol/min. The energy consumption followed the same pattern. It decreased from 12.42 kJ/mmol to 3.23 kJ/mmol then to 2.31 kJ/mmol at 0%, 5%, and 10% ethanol, respectively. But at 15%, the energy consumption value increased to 2.55 kJ/mmol. It showed that at 0%, 5%, and 15% ethanol, normal electrolysis was more dominant, while at 10% ethanol plasma electrolysis was more dominant.

The plasma discharge is generated in the gas envelope between the electrode and the surface of the solution. Because ethanol is easier to vaporize than water, the gas enveloping around the discharged electrode is easier to form. In effect, it tends to be bigger in volume in ethanol solution than in aqueous solutions [4]. The bigger plasma volume, the higher the amount of hydrogen ion in the solution. As mentioned before, full glow discharge during plasma electrolysis in high concentration produces large current and this large current can destabilize plasma formation in solution.

3.3. The Effect of Cathode Deep in Hydrogen Production

Cathode Depth is the distance between the tip of cathode with surface of electrolyte solution, as shown on Figure 2. This research was conducted using 0.05 M KOH, 0.05% ethanol, and at 700V discharged voltage. The depths of cathode observed were 0 cm, 3 cm, and 6.6 cm.

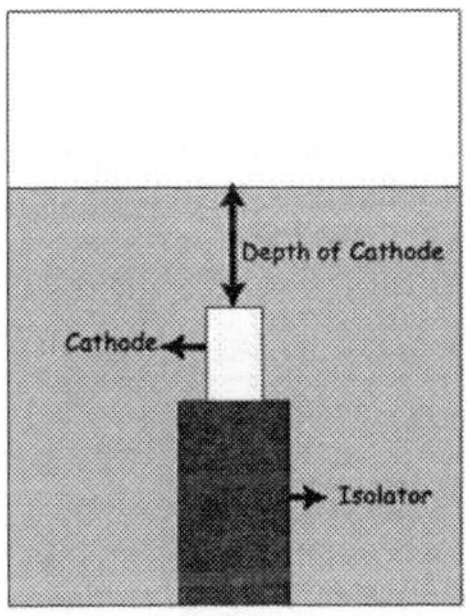

Figure 2. Cathode position at electrolyte solution

Figure 3 below shows that plasma emitted purple glow and its brightness was varied for every cathode depth. At a depth of 6.6 cm, the plasma glow that induced is brightest, while at a depth of 0 cm the glow was dimmest.

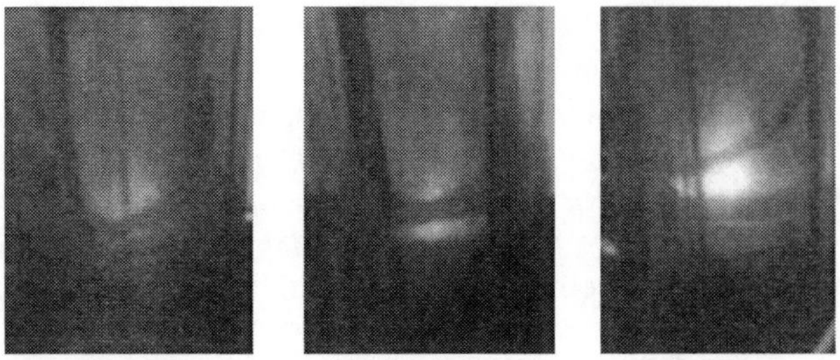

a. 0 cm b. 3 cm c. 6.6 cm

Figure 3. The colour of plasma at different cathode deep in 0.05 M of KOH solution, 10% ethanol, and 500V

The effects of cathode deep in hydrogen production are shown in Table. 1. Table 1 shown that the deeper the cathode was, the more hydrogen produced. The discharge plasma was generated in the gas envelope between the electrode and the surface of the solution. At the depth of 6.6 cm, more electrolytes were in contact with cathode, allowing more electrons to move freely to the cathode. Electron excitation intensified the plasma generation, thus increased the hydrogen production.

The increasing of cathode deep causes the increasing of energy consumption. Because the movement of electrons in the liquid zone is easier than in the gas zone, the electrical current increases. The formation of the gas sheath is the beginning of plasma formation due to Joule heating levels [10]. With the deeper cathode position, most of the energy was used for creating a continuous gas sheath. However, the presence of hydrostatic pressure hindered the stability of the gas, allowing electric current to flow easily to the cathode. Smaller amount of energy was used for gas formation, due to position of cathode that was close to the solution surface. Consequently, the plasma was formed and stabilized quickly due to the lower influence of hydrostatic pressure. The plasma stability can resist electrons movement effectively [11].

3.4. The Effect of Voltage in Hydrogen Production

This study conducted in 0.03 M KOH with 10% ethanol, and cathode deep 6.6 cm. Various voltage were given into the reactor such as 500 V, 600 V, 700 V. The results can be seen in table 2 below.

Table 2. Result of plasma electrolysis on different voltage in 0.03 M KOH with 10% ethanol and 6.6 cm cathode deep.

Voltage (v)	H_2 Production (mmol / min)	G (mol/mol)	Wr (kJ/mmol)
500	15.71	48.75	2.76
600	31.14	95.73	1.90
700	50.71	149.68	1.49

Table 2 shows that the higher the voltage, the more the hydrogen produced. It is due to increase of electron species. The electrons are responsible for creating the active species in plasma, that breaks down the water to produce hydrogen. Plasma in the electrolysis solution will increase the production of hydrogen by forming compounds of hydrogen radicals and hydroxide radicals [4]. High voltage excites the electron to attack OH- ion and H+, forming OH* and H*. The more electrons that attacked the vapour solution and got excited, the greater the plasma volume and the hydrogen produced. As the voltage increased, the amount of energy consumption decreased, as shown in table 2. The decrease in energy consumption is related to the stability of the plasma volume, as already described. At 700 Volt, hydrogen production is equal to 50.71 mmol/min and energy consumption is 1.49 kJ/mmol. In this condition, hydrogen production is 149 times higher than normal electrolysis.

4. Conclusion

Experiments have been conducted in order to produce hydrogen by plasma electrolysis and to study the influence of the variables during the hydrogen production by plasma electrolysis such as KOH concentration, ethanol additive, depth of the cathode, and the voltage. The result shows that KOH concentration and percentage of ethanol additive influence the level of of hydrogen production and electricity consumption during the plasma electrolysis. An increase in the voltage causes a decrease the electricity consumption and an increase in the hydrogen production. The more deeper the cathode submerged in electrolyte, the more hydrogen produced, which is characterized by plasma glow, and the greater the amount of electrical energy used. In this study, the optimum level of hydrogen production is 50.71 mmol/min and the optimum energy consumption is 1.49 kJ/mmol. Obtained by plasma electrolysis at 700 Volt, with 0.03 M KOH, 10% ethanol, and cathode 6.6 cm deep, this level of hydrogen production is 149 times higher than the level of hydrogen produced by normal electrolysis.

Referencees

[1] Holladay J D, Hu J, King D L, Wang Y 2009 An Overview of Hydrogen Production Technologies *Catalysis Today* **139** 244–260
[2] Saksono N, Ariawan B and Bismo S 2012 Hydrogen Productions System Using Non-Thermal Plasma Electrolysis in Glycerol-KOH Solution *International Journal of Technology* **1** 8-15
[3] Gao J Wang A, Fu Y and Wu J 2008 Analysis of Energetic Species Caused by Contact Glow Discharge Electrolysis in Aqueous Solution *Plasma Science and Technology* **10**
[4] Yan Z C, Li C, Wang H L W H 2008 Hydrogen generation by glow discharge plasma electrolysis of ethanol solutions *Journal of Physic* **41** 155205
[5] Saksono N, Faishal M, Merisa B, Setidjo B 2014 Hydrogen Production System with Plasma Electrolysis Method in Natrium Carbonate-Acetate Acid Solution *Proceeding the Regional on Chemical Engineering, Yogyakarta, Indonesia*
[6] Saksono N, Jainal A, Setidjo B 2010 Hydrogen Production Systems Design through Plasma Non-thermal Electrolysis Process *The 1st International Seminar on Fundamental & Application of Chemical Engineering, Bali, Indonesia*
[7] Mizuno T, Akimoto T, Azumi K, Ohmori T, Aoki Y and Takahashi A 2005 Hydrogen Evolution by Plasma Electrolysis in Aqueous Solution *Japanese Journal of Applied Physics* **44** (1A) 396- 401
[8] Chaffin J H, Bobio S M and Inyang H 2006 Hydrogen production by Plasma Electrolysis *Journal of Energy Engineering, ASCE* **132(**3) 104-108
[9] Slade A, Campbell S A, Ralph T R and Walsh F C 2002 Ionic conductivity of an extruded Nafion 1100 EW series of membranes *Journal of the Electrochemical Society* **149(**12) A1556-A1564
[10] Gai K 2006 Aqueous Diphenyl Degradation Induced by Plasma with Glow Discharge Electrolysis *Journal of the Chinese Chemical Society* **53** 627–632
[11] Saksono N, TienSeratri R, Muthia R and Bismo S 2015 Phenol Degradation in Wastewater with a Contact Glow Discharge Electrolysis Reactor using a Sodium Sulfate *International Journal Of Technology* **6**(7) 1153-1163

Improvement of bio-oil yield and quality in co-pyrolysis of corncobs and high density polyethylene in a fixed bed reactor at low heating rate

D Supramono, S Lusiani
Deparment of Chemical Engineering, Faculty of Engineering, Universitas Indonesia, Depok 16424, Indonesia

E-mail : dsupramo@che.ui.ac.id

Abstract. Over the past few decades, interest in developing biomass-derived fuel has been increasing rapidly due to the decrease in fossil fuel reserves. Bio-oil produced by biomass pyrolysis however contains high oxygen compounds resulting in low calorific-value fuel and therefore requiring upgrading. In co-pyrolysis of the feed blend of plastics of High Density Polyethylene (HDPE) and biomass of corn cob particles, at some compositions free radicals from plastic decomposition containing more hydrogen radicals are able to bond oxygen radicals originating from biomass to reduce oxygenate compounds in the bio-oil thus increasing bio-oil quality. This phenomenon is usually called synergetic effect. In addition to that, the pattern of heating of the feed blend in the pyrolysis reactor is predicted to affect bio-oil quality and yield. In a batch reactor, co-pyrolysis of corncobs and HDPE requires low heating rate to reach a peak temperature at temperature rise period followed by heating for some time at peak temperature called holding time at constant temperature period. No research has been carried out to investigate how long holding time is set in co-pyrolysis of plastic and biomass to obtain high yield of bio-oil. Holding time may affect either crosslinking of free radicals in gas phase, which increases char product, or secondary pyrolysis in the gas phase, which increases non-condensable gas in the gas phase of pyrolysis reactor, both of which reduce bio-oil yield. Therefore, holding time of co-pyrolysis affects the mass rate of bio-oil formation as the pyrolysis proceeds and quality of the bio-oil. In the present work, effects of holding time on the yield and quality of bio-oil have been investigated using horizontal fixed bed of the feed blends at heating rate of 5°C, peak temperature of 500°C and N_2 flow rate of 700 ml/minute. Holding time was varied from 0 to 70 minutes with 10 minutes interval. To investigate the effects of holding time, the composition of HDPE in the feed blend was varied 0, 50 and 100%, while the synergetic effect was investigated by varying the composition of HDPE in the feed blend 0, 25, 50, 75, and 100%. The results show that synergetic effect for non-oxygenate compound production started to work at 63% HDPE in the feed blend and beyond. It was observed that extension of holding time exceeding 0 minutes allowed increase of bio-oil production rate followed reduction of the rate. Pyrolysis of both the corncob feed and the feed blend containing 50% HDPE equally reached maximum bio-oil production rate at holding time of 50 minutes, while that of HDPE feed at 30 minutes. The result pertaining to holding time indicates that biomass in the feed blend governs crosslinking – secondary pyrolysis in the co-pyrolysis.

1. Introduction

Decreasing availability of fossil fuel sources such as coal, petroleum, and natural gas has encouraged research to develop new approaches to find new energy sources. Global primary energy consumption increased by just 0.9% in 2014 and global CO_2 emissions from energy use grew by just 0.5% in 2014 [1]. Beside the effect of decreasing of fossil fuels, the increase in environmental problems which are caused by the emissions of primary pollutants (sulfur dioxide, oxides of nitrogen, hydrocarbons, and carbon monoxide) from fossil fuel combustion also encourages the development of renewable energy.

The use of biomass as an energy source has been proven to have environmental benefits since it plays a role as a carbon-neutral energy source[2]. Besides, the availability of biomass is very abundant in nature and bio-oil can be derived from agriculture residues, wood residues, and municipal solid waste [3]. Biomass has the potential to serve as a new energy source because it contains organic compounds (cellulose, hemicellulose, and lignin) that can be converted into a source of energy in the form of gas, liquid or solid. Pyrolysis is generally chosen as a recommended process to achieve this goal. This method can produce liquid product highest up to 75 wt% of the pyrolysis feed at moderate temperature (500°C) and short hot vapor residence time (1s)[4,5]. Liquid from pyrolysis biomass is known as pyrolysis oil or bio-oil. Bio-oil may be routed to biofuels for many applications including boilers, furnaces, diesel engines, and turbines for the generation of electricity [6].

Corn is a significant crop with a total amount of annual production is about $520x10^9$kg worldwide. According to FAO, Indonesia ranks 8 out of 10 largest corn producer in the world [7]. Within five year span (2009-2013), Indonesia has harvested corn in areas of about 4 million hectares per year with production ranging between 17.6 and 19.4 million tons [8]. Of the amount of corn production in 2013(18.5 million tons), Indonesia could potentially produce corn cobs as much ± 3.7 million tons [9]. Bio-oil produced from biomass pyrolysis has high oxygenates level that causes a low calorific value, corrosion problems and instability. Several researchers have reported that bio-oil generally contains oxygen content of around 35–60 wt% [5,10,11,12]. Reducing oxygen content is becoming a priority to improve the quality of bio-oil and to provide a solution for several challenges in its applications.

Co-pyrolysis of biomass can be an optional technique for upgrading bio-oil. Co-pyrolysis is a process which involves two or more different materials as a feedstock. In co-pyrolysis of the feed blend of plastics and biomass, at some plastic compositions, hydrogen free radicals from plastic decomposition are able to reduce oxygenate compounds of the bio-oil thus increasing bio-oil quality [13,14,15]. This phenomenon is usually called synergetic effect. High Density Polyethylene (HDPE), which has higher hydrogen composition than biomass, is expected to transfer hydrogen atoms to bond free radical from biomass decomposition so that the yield of bio-oil and its non-oxygenate compound will increase [16].Polyethylene is one of thermoplastic types abundantly used in daily life. Polyethylene is white with melting point 110-137°C and molecular weight 50,000-30,000 [17]. HDPE has straight chain structures which makes it rather hard to decompose thermally.

In a fixed bed reactor, co-pyrolysis of corncobs and HDPE requires low heating rate to reach a peak temperature at temperature rise period followed by heating for some time at peak temperature called holding time at constant temperature period to decompose the remaining feedstock. Research on the effect of holding time on the bio-oil yield and quality has been focused on biomass pyrolysis carried out at low heating rate. This is attributed to the condition that at such low heating rate, the presence of oxygen radicals in the vapour phase may result in the crosslinking between radicals of the pyrolysis leading to the formation of more char [18]. Venderbosch *et al* [19] found that by upgrading biomass derived pyrolysis oil usingRu/C based-catalytic hydrotreatment reactions, repolymerisation of the pyrolysis oil leading to char formation was reduced in the presence of hydrogen gas and the hydrodeoxygenation reactions occur when the temperature of reactions more than 225°C up to 400°C. Tchapda and Pisupati [20] in their review on co-conversion of biomass and coal have identified that the hydrogen transfer from tetraline to coal in coal liquefaction allows production of more tar and gases and reduces the formation of char. These late two papers indicate that the presence of hydrogen has roles of both reducing the char formation as well as sequestering oxygen from the biomass oil to produce less oxygenate compounds in the bio-oil. Abnisa and wan Daud [21] in their review on biomass co-pyrolysis have identified that hydrogen transfer from plastic material to biomass during co-pyrolysis process may increase the bio-oil yield and reduce the oxygenate components in the bio-oil. Therefore, the presence of hydrogen radicals is of importance in both synergetic effect and reduction of char formation as a result of crosslinking. The present paper addresses research work with aim of investigating how long holding time is set in co-pyrolysis of plastic and biomass to obtain high yield and high quality of bio-oil. In biomass pyrolysis, holding time may affect either crosslinking of free radicals in gas phase, which increases char product, or secondary pyrolysis in the gas phase, which increases non-condensable gas in the gas phase of pyrolysis reactor, both of which reduce bio-oil yield [22]. However, in author understanding, no research has investigated holding time in co-pyrolysis. In the present work, where hydrogen is supplied by HDPE plastic decomposition, the function of plastic embedded in the feedstock in reducing char production and in improving non-oxygenate product will be investigated.

2. Experimental

Corncobs taken from Bogor, Indonesia, were used as a biomass sample, while HDPE waste was considered as a representative synthetic polymer. HDPE samples used in the experiments were supplied from plastic waste shelter in Depok, Indonesia. Feed preparation consist of reducing particle size of corncob and HDPE to size 500μm-1mm and drying corncobs in an oven at 105°C to reach moisture content of 10%. The results of ultimate and proximate analysis of corncobs and HDPE are shown in Table 1. The table shows that moisture contents in biomass and HDPE plastic materials were low, i.e. 11 and 15% weight, respectively. The presence of high moisture can increase the load of the heating and it may impure the bio-oil. Therefore, its content is restricted in favor of producing bio-oil and processing bio-oil towards bio-fuel [23].

Table 1. Proximate and Ultimate Analysis Result

Analysis (%wt)	HDPE	Corncob
Proximate Analysis		
Moisture Content	15.00	11.02
Ash	0.08	5.07
Volatile matter	84.90	69.31
Fixed Carbon	0.02	14.6
Ultimate Analysis		
Carbon	84.74	41.07
Hydrogen	11.65	6.54
Nitrogen	0.02	0.76
Sulphur	0.66	0.09
Oxygen	2.93	46.47

Co-pyrolysis process was carried out using a cylindrical fixed bed reactor placed horizontally as shown in Figure 1. The reactor used is made of stainless steel with a diameter of 60 mm and length of 250 mm. Feed composing corn cob and plastic particles was located in a boat made of half cylinder sizing 110 mm length and 50 mm diameter. The boat was put horizontally inside the reactor, which was also positioned horizontally. Front end of the boat was connected to a half cone-shaped end to allow smooth flow of gas N_2 over the outer wall of the boat thus expectedly improving convective heat transfer from hot nitrogen carrier gas to the side of the boat. On the top of the boat where there is direct heat transfer from nitrogen into the feed blend, it was expected that there was radiative heat transfer. The front end of the reactor is made of a cone with apex angle of 18 degrees to allow smooth flow enlargement from small nitrogen pipe to large cylindrical reactor. This construction has proved to make temperature of the cylinder uniform from the front end to rear end of the reactor. Nitrogen brought heat from a furnace in outer side of the reactor to the whole part of the reactor.

IOP Conf. Series: Materials Science and Engineering **162** (2017) 012011 doi:10.1088/1757-899X/162/1/012011

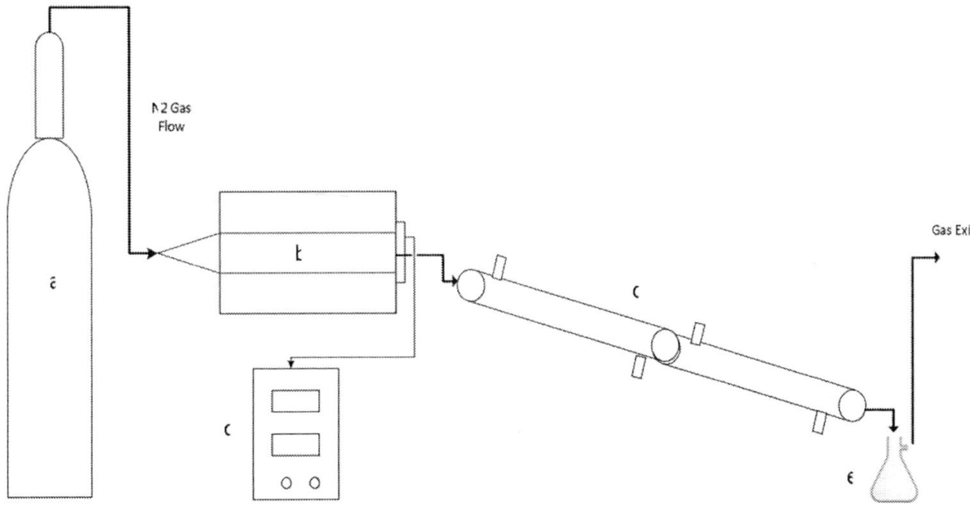

Figure 1. Schematic representation of experimental set-up for co-pyrolysis of Corncob with High Density Polyethylene (HDPE) at 500°C: (a) Cylindrical Tube of Nitrogen, (b) Fixed Bed Reactor, (c) Condenser, (d) Temperature Controller, (e) Erlenmeyer Flask as Bio-oil Container

Nitrogen flow was selected to provide maximum residence time of 8 seconds for vapour phase of the pyrolysis calculated from the front end of the boat to the exit point of the reactor in order to avoid the occurrence of secondary pyrolysis of plastic decomposition. This vapor phase residence time was short for the pyrolysis product of biomass, but long enough for plastics pyrolysis. Gas N_2 flowed through the reactor at 700 ml/minute. The heating rate of the pyrolysis was set at 5°C/minute to heat inside the reactor from initially ambient temperature to 500°C. Vapour produced by the pyrolysis reactor was delivered to a series of condensers for condensation to produce bio-oil with assistance of cooling water of about 10°C. This bio-oil dripped into an Erlenmeyer flask with uncondensed gas disposed to the ambient. The mass composition of plastic waste HDPE in the feed blend was varied at 0, 25, 50, 75 and 100% weight of the feed; while the variation of holding time was 0,10, 30, 50 and 70 minutes. Bio-oil collected in the flask was then characterized by GC-MS to identify components in the oil. Other measurements were viscosity and pH of the oil. The yields of bio-oil were weighed in order to correlate the composition of plastics in the feed blend and the yield of bio-oil at different holding times.

3. Results and Discussion

3.1. Effect of feedstock composition on physical properties

The oil produced from plastic was yellow and essentially transparent, while bio-oils derived from biomass pyrolysis and co-pyrolysis of the feed blend tend to have dark brown color. This is due to the high content of oxygenates in bio-oil at higher biomass content in the feed blends, causing bio-oil to be easily oxidized thus resulting in the brownish color[24].

The viscosity and pH number of the bio-oils with different plastic compositions in the feed blend of 0%wt, 50%wt, and 100%wt have been determined as shown in Table 2. It can be seen that bio-oil of 100% corn cob pyrolysis has higher viscosity of 7.1 cSt than 100% HDPE pyrolysis oil does with viscosity of 4 cSt. Greater amount of low molecular weight paraffins and olefins content in the plastic pyrolysis oil may affect its viscosity. The water content influences the thermal stability of the bio-oil during its storage and the oxygenated compound content may cause polymerization within bio-oil components resulting in the increase in viscosity [25].

Bio-oil obtained from pure corn cobs pyrolysis is very acidic (pH = 2.2). This was due to the carboxyl compounds found in the oil. Formic acid and acetic acid contribute most of the carboxyl group in bio-oil [23]. While the oil derived from pure plastic pyrolysis has a higher pH of 4. The cycloalkanes, alkanes and olefins dominate the oil content and the low presence of acid causes the plastic pyrolysis oil to have higher pH value [26]. The result shows that the addition of plastic wastes in the co-pyrolysis feed blend yields lower pH of the bio-oil (pH = 3).

Table 2. Viscosity and pH of Bio-oil Compared to the Commercial Fuel Oil

Characteristic	Bio-oil (corncob 100%)	Bio-oil (corncob 50% and HDPE 50%)	Bio-oil (HDPE 100%)	Diesel C_8-C_{25} (William et.al, 1990)	Gasoline C_4-C_{12} (William et.al, 1990)
Viscosity [cSt]	7.1	4.2	4	2.8-5	0.5-0.6
pH	2.2	3	4	5	5

3.2. Effect of feedstock composition on bio-oil yield

Figure 2 shows that with increasing HDPE composition in the feed blend, the yields of bio-oil and char produced decreased, while the yields of non-condensable gases and the waxy solids increased. Related to the competition of pyrolysis kinetics between biomass and plastic, Supramono [27] shows that conductive heat transfer from hot nitrogen gas to the plastic melt was lacking due to low thermal conductivity of the plastic PP melt, i.e. 0.25 to 0.32 W/(m.sec.°C) depending molecular weight of the plastic [28]. This figure is only 0.4 – 0.5 times the thermal conductivity of water. Friction of nitrogen gas on the plastic surface melt was weak because the melt surface in the boat was slightly below the height of the boat. Therefore, nitrogen gas was unable to transfer the heat convectively. On the other hands, during pyrolysis, biomass particles turned to form char, where emissivity reached 0.8 [29]. When the biomass particles formed char, radiative heat transfer became strong. Therefore, with the configuration of reactor used, the heat from nitrogen gas to the plastic melt and to biomass particles was transferred by heat radiation. With this nature, the heat transfer from nitrogen gas to biomass particles was expectedly higher than that from nitrogen gas to the plastic melt. In turn, the pyrolysis of biomass was stronger than that of plastic melt. As a result, the heat received by the plastic melt was not high enough to crack the plastic molecules. Moreover, low temperature of plastic melt maintained high intermolecular forces between long plastic molecules thus inhibiting the formation of a gas phase from the melting phase [32,33]. For the case of plastic pyrolysis, this lack of heat transfer caused higher composition of wax and in contrary lower composition of bio-oil at higher content of HDPE in the feed blend. Char was mostly contributed by the decomposition of lignin in biomass material [30], but it was much less produced by the decomposition of plastic as shown in the data associated with pyrolysis of pure plastic HDPE. This is confirmed by the finding of Kumar and Singh [31]. For this reason, co-pyrolysis process of corn cobs and plastic would have been better if it had been equipped with a stirring to improve convective heat transfer from the boat wall to the plastic melt.

Synergetic effects would occur if there was hydrogen transfer to free radicals from biomass pyrolysis initiated by free radical generation from plastic pyrolysis followed by the mid-chain beta-scissions of plastic molecules [34]. The reaction between the free radicals formed from biomass and plastics would increase the yield of the bio-oil [35]. Low convective heat transfer from nitrogen to the plastic melt hindered the production of hydrogen radicals sufficiently by plastic pyrolysis to allow synergetic effects. These can be evaluated by comparing the compositions of non-oxygenates in the bio-oil resulting from experiment and from theoretical calculation. If bio-oil of w_1 weight from pure biomass pyrolysis containing x_1 fraction of non-oxygenates and oil of w_2 weight from pure plastic pyrolysis containing x_2 fraction of non-oxygenates are mixed, theoretically the non-oxygenate composition y of the product mixture can be expressed by Eq. 1 as follows:

$$y = x_1.w_1 + x_2.w_2 \quad (1)$$

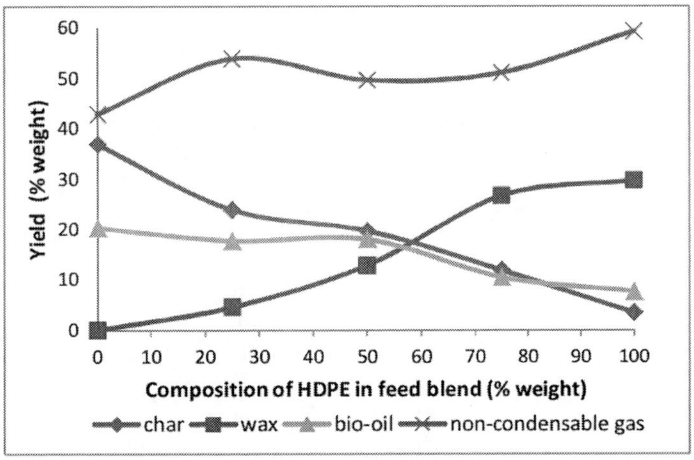

Figure 2. The Effect for Adding High Density Polyethylene (HDPE) to Co-pyrolysis Product Yield.

If the composition of non-oxygenates obtained experimentally from co-pyrolysis of a feedstock containing w_1 fraction of biomass and w_2 fraction of plastic is higher than the theoretical value y, then synergetic effect occurs during co-pyrolysis [37]. Based on Eq. 1 and GC-MS analysis of the bio-oil, Figure 3 has been constructed. The figure shows that HDPE synergetically increased the yield of bio-oil as the composition in the feed blend was more than 63%. It means that using the current reactor configuration, it requires more plastic weight fraction in the feed blend to achieve synergetic effect. The improvement of the reactor configuration is advocated to reduce this figure.

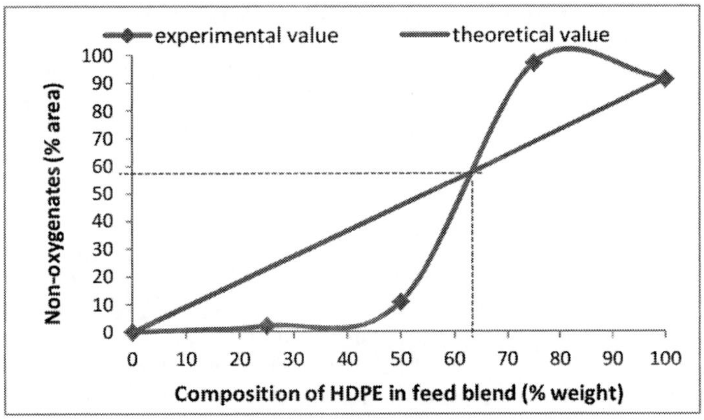

Figure 3. Experimental and Theoretical Non-oxygenate Compounds in Bio-oil from Co-pyrolysis Corn Cob and HDPE process

3.3. Effect of feed composition on compound composition in bio-oil

Figure 4a and 4b present the results of GC-MS analysis of bio-oil obtained from the co-pyrolysis of HDPE and corncobs with composition of HDPE in the feed blend varied at 0, 25, 50, 75, 100% weight of the feed. Figure 4b shows that by varying composition of HDPE in the feed blend, composition of oxygenates in the bio-oil predominated in the co-pyrolysis with HDPE contents in the feed blend less than 75%, while Figure 4a shows that the domination of non-oxygenates occurred in the co-pyrolysis with HDPE contents 75% and more. These results confirm the result described by Figure 3 where the synergetic effects occurred when HDPE composition was more than 63%.

The dominant composition of carboxylates in the bio-oil resulting from pyrolysis with HDPE contents in the feed blend less than 75% in Figure 4b is associated with the role of ash in the corn cob particles containing mostly potassium [38][39]. According to Table 1, dry corn cobs contain 5.07% ash. Yao *et al* have investigated ash composition in corn cobs and found that 29.74% of ash contains potassium oxide [39]. Kelkar found that switchgrass with high potassium level in its ash induces high composition of carboxyl and carbonyl group compounds [40]. This nature is similar to that in the corn cobs, which explains that co-pyrolysis with high HDPE composition feed blend produced high content of carboxyl groups in the bio-oil.

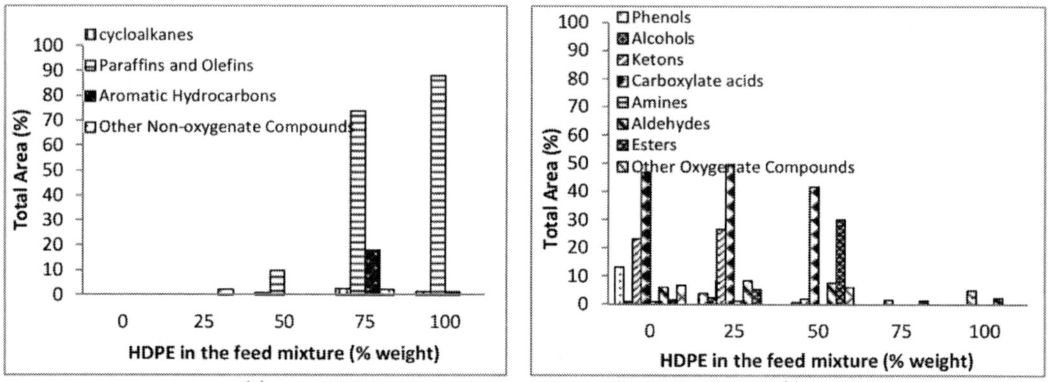

(a) (b)

Figure 4. Compounds Composition in Bio-oil Produced by Co-pyrolysis Corn Cob and HDPE; (a) non-oxygenate compounds; (b) oxygenate compounds

In general, Figure 4a shows that composition of paraffins-olefins tends to increase with increasing amount of HDPE in the feed composition. This trend can be caused by increased hydrogen donor of plastic free radicals which are able to bind free radicals derived from biomass pyrolysis and form non-oxygenate compounds [36].

3.4. *Effect of holding time on yield and composition of bio-oil*

Holding time is the additional time spent for continuing pyrolysis after the peak temperature has been reached. This time span is important to allow the pyrolysis of the remaining feedstock. Study on the effect of holding time aims at avoiding secondary reaction on pyrolysis of biomass which will increase the production of non-condensable gas or avoiding crosslink reactions leading to the formation of char, both of which reduce the yield of oil. Effects of holding time of pyrolysis of pure corn cobs, pure HDPE and the mixture of both on the mass rates at the pyrolysis are given in Figure 5. Mass rate of liquid generation is defined by Eq. 2 below :

$$\dot{m} = \lim_{\Delta t \to 0} \frac{\Delta m}{\Delta t} = \frac{dm}{dt} \quad (2)$$

Figure 5 shows the rates of mass change of bio-oil as a function of holding time, in which the rates increased initially until at a certain holding time it decreased. The negative values of mass rates show that there was mass reduction of bio-oil yields with holding time. The increasing mass rates at negative values show that the mass reduction of the yields became less and less with more holding time until they reached mass rates equal to zero indicating no reduction nor increase of bio-oil yields. Beyond this zero mass rates, they worked in positive values indicating there was increase of bio-oil yield with holding time, in which initially they underwent increasing mass rates followed by decreasing mass rates.

Figure 5 shows comparison of the shapes of these three mass rate curves in which the curve arising from corncob pyrolysis (Figure 5a) is similar to that of feed blend pyrolysis (Figure 5c). Related to Figure 4, where pyrolysis at 50% plastic in feed blend exhibits no synergetic effect, it shows that at this feed blend composition, the pyrolysis was dominated by the evolution of free radicals originating from biomass decomposition. Therefore, the shape of mass rate at Figure 5a resembles to that at Figure 5c and analysis of the curves may consider mass rates at Figure 5a and Figure 5b.

The shape of mass rate curve at Figure 5a demonstrates that at initial time (t = 0 minute), i.e. the time when temperature rise period reached the peak temperature, the mass rate was negative. In another word, there was mass rate reduction at t = 0 minute, but this reduction became less and less beyond t = 0 minute. At t = 22 minutes, positive mass rate started, which means that the mass rate became more and more beyond t = 22 minutes until t = 50 minutes, when the mass rate increase started to lessen approaching t = 75 minutes since the peak temperature has been reached. The shape of the curve indicates that for biomass pyrolysis beyond t = 0 the rate of sum of non-condensable gas and char generation became more and more with less and less rate. However, beyond t = 50 minutes when the feedstock remain was minimal, the generation of bio-oil became more and more restricted.

Figure 5b exhibits similar pattern but different in times when mass rate changed from negative to positive and when the peak of mass rate occurred. Comparison between Figures 5a and 5b indicates that plastic diminishes its decomposition much earlier than the decomposition of biomass. The low convective heat transfer from hot nitrogen gas to plastic melt seems unable to crack more plastic material into smaller carbon chains leaving high composition of wax in the boat indicated by high yield of wax in Figure 2.The problem in the plastic pyrolysis seems to be due to heat transfer caused by intrinsically low thermal conductivity of plastic melt. In plastic pyrolysis, holding time is not associated with char formation [31] indicated by low yield of char in plastic pyrolysis (see Figure 2), but may be related to non-condensable gas formation, which has not be investigated yet.

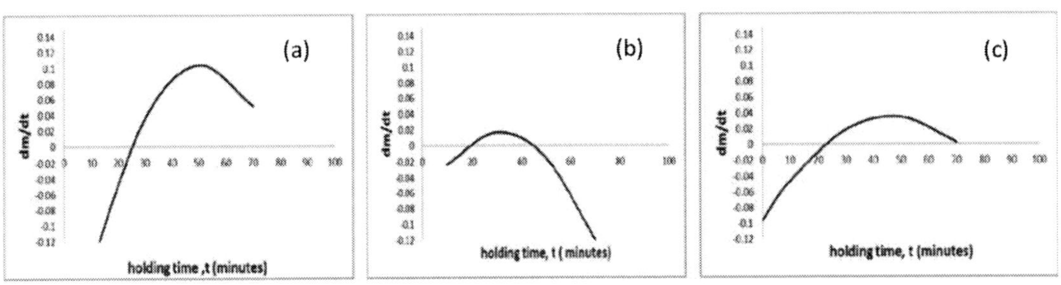

Figure 5. The Rate of Bio-oil Mass Generationas a Function of Time (Holding Time)
(a) Corncob; (b) HDPE; (c) Corncob and HDPE Mixture

Figure6 shows the compositions of non-oxygenates in co-pyrolysis oil with various holding times (10, 30, and 50 minutes). The changes of compositions of non-oxygenates from 10 minutes to 50 minutes show how the compositions accumulatively change as the pyrolysis was carried out at longer holding time. Fig 6a shows that when the biomass pyrolysis was extended its holding time, accumulatively the composition of dominant non-oxygenates changes from paraffins and olefins to aromatics and then to cycloalkanes. The smaller composition of non-oxygenates at longer holding time indicates that the oxygenate content became more dominant. It means that at longer holding time, which can be meant as longer interaction of free radicals in vapour phase, there could be changes in the tendency of non-oxygenate generation. However, its amount in the bio-oil was very low. By incorporating plastic in the feedstock, the amount of non-oxygenates slightly increased at holding times of 30 and 50 minutes compared to that in biomass pyrolysis (see Figure 6a), apparently contributed by plastic molecule cracking (see Figure 6c). In Figure 6b, for the pyrolysis of HDPE the longer holding time, the total amount of non-oxygenate compounds detected by GC-MS decreased with the decrease of cycloalkanes, while the composition of paraffins-olefins increased and then decreased. This indicates that more cycloalkanes were converted to olefins at longer holding time. The decrease of total non-oxygenated compound yield when the holding time was extended to 50 minutes correspond to the time when mass rate has deeply dropped according to Figure 5b.

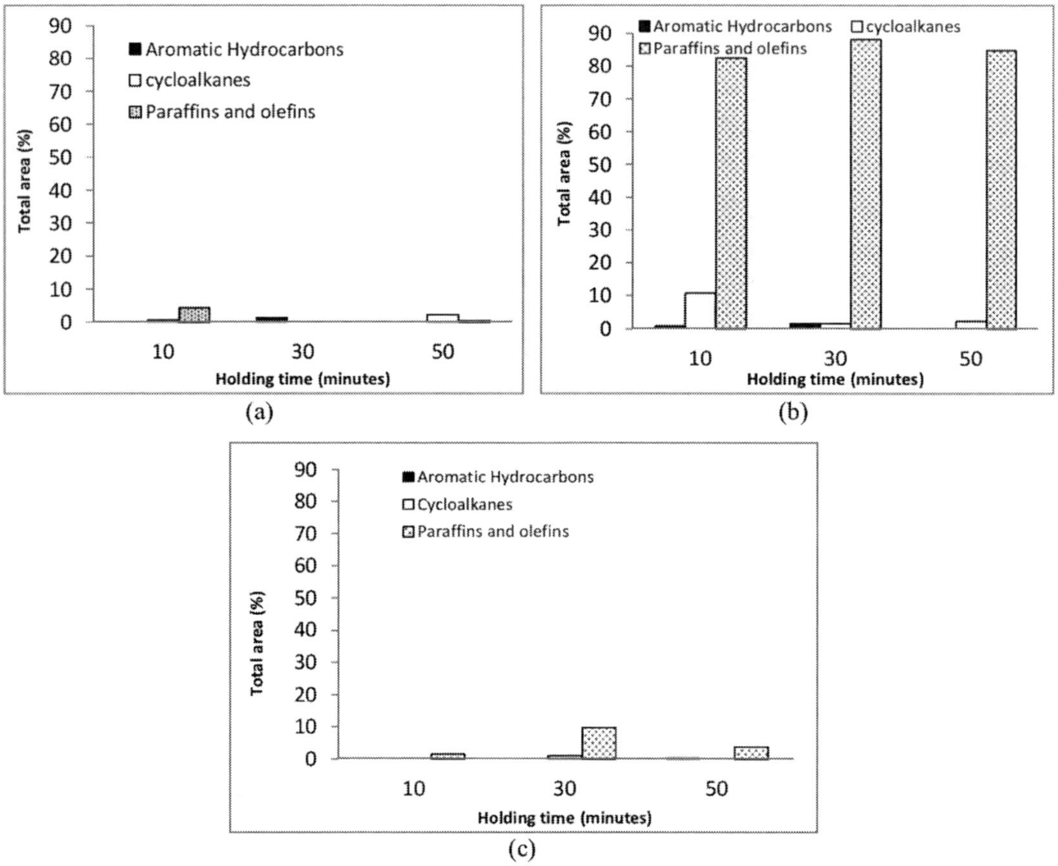

Figure 6. Non-oxygenate Compound of Bio-oil at Different Holding Times
(a) Corncob; (b) HDPE; (c) Mixture Corncob and HDPE

4. Conclusion

In the configuration of reactor used in the present work, due to lack of heat transfer from carrier gas nitrogen to the feed blend, the synergetic effect related to non-oxygenated compound content in the bio-oil is achieved when HDPE composition in the feed blend is high, i.e. more than 63%.

In the feedstock regime where the composition of HDPE is more than 63%, product of bio-oil contains mostly paraffins-olefins. In contrary, in the feedstock regime where the composition of HDPE is less than 63%, the product of bio-oil predominantly contains carboxylate functional group compounds.

With increasing HDPE composition in the feed blend, the yield of bio-oil decreases with viscosity and pH of bio-oil change to approach those of diesel oil, char yield decreases and those of non-condensable gas and wax increase.

During the holding time, pyrolysis with pure corn cob feed reaches the peak of mass rate of bio-oil formation later than that with pure HDPE feed. For the case of co-pyrolysis with 50% weight of corn cobs and 50% weight of HDPE in the feed blend, the pattern of mass trate of bio-oil formation is similar to that obtained by pure corn cob pyrolysis, because this feed blend composition is in the feedstock regime where synergetic effect does not occur.

Acknowledgement
The authors would like to express their gratitude to Ministry of Research, Technology and Higher Education Indonesia for funding their research on co-pyrolysis of biomass and plastic material through PUPT scheme. The authors are convinced that this research work is very worthwhile for Indonesia progress.

References
[1] Dudley B 2015 BP Statistical Review of World Energy, *BPstats.*
[2] Ahtikoski A, Heikkilä J, Alenius V, Siren M 2008 Economic viability of utilizing biomass energy from young stands – the case of Finland, *Biomass Bioenergy* **32** 988–96
[3] Easterly J L, Burnham M 1996 Overview of biomass and waste fuel resources for power production *Biomass Bioenergy* **10** 79–92.
[4] Bridgwater T 2006 Biomass for energy *J Sci Food Agric* **86** 1755–68
[5] Guillain M, Fairouz K, Mar S R, Monique F, Jacques L D 2009 Attrition-free pyrolysis to produce bio-oil and char *Bioresour Technol* **100** 6069–75
[6] Kim S, Dale B E 2004 Global potential bioethanol production from wasted crops and crop residues *Biomass Bioenergy* **26** 361–75
[7] Information on http://www.faostat.fao.org
[8] Information on http://www.bps.go.id
[9] Retnani Y, Herawati L, Widiarti W, Indahwati E 2009 Uji Sifat Fisik dan Palatabilitas Biskuit Limbah Tanaman Jagung Sebagai Substitusi Sumber Serat Untuk Domba *Buletin Peternakan* **33** 162-169
[10] Bridgwater A V, Meier D, Radlein D 1999 An overview of fast pyrolysis of biomass *Org Geochem* **30** 1479–93
[11] Oasmaa A, Czernik S 1999 Fuel oil quality of biomass pyrolysis oils-state of the art for the end users *Energy Fuel* **13** 914–21
[12] Parihar M F, Kamil M, Goyal H B, Gupta A K, Bhatnagar A K 2007 An experimental study on pyrolysis of biomass *Process Saf Environ Prot* **85** 458–65
[13] Abnisa F, Daud W M A W, Ramalingam S, Azemi M N B M and Sahu J N 2013 Co-pyrolysis of palm shell and polystyrene waste mixtures to synthesis liquid fuel *Fuel* **108**(0) 311-318
[14] Brebu M, Ucar S, Vasile C and Yanik J 2010 Co-pyrolysis of pine cone with synthetic polymers *Fuel* **89**(8) 1911–1918
[15] Panda A K, Singh R K, Mishra D K 2010 Thermolysis of waste plastics to liquid fuel: asuitable method for plastic waste management and manufacture of valueadded products - A world prospective *Renew Sust Energy Rev* **14** 233–48
[16] Sharpov V I, Berego N G and Kuznetso B N 2003 Co-pyrolysis of wood biomass and synthetic polymers mixtures Part III: Characterisation of heavy products *Journal of Analytical and Applied Pyrolysis* **67**(2) 325-340
[17] Curlee T and Das S 1991 Identifying and assessing targets of opportunity for plastics recycling *Resour. Conserv.Recycl.* **5**(4) 343–363
[18] Teng H, Serio M A, Wojtowicz M A, Bassilakis R, Solomon P R 1992 Reprocessing of used tires into activated carbon and other products *Am. Chem. Soc.* **37** 533-541

[19] Venderbosch R H, Ardiyanti A R, Wildschut J, Oasmaa A, Heeres H J 2010 Stabilisation of Biomass derived Pyrolysis Oils by Catalytic Hydrotreatment *Journal of Chemical Technology and Biotechnology* **85** 674-685

[20] Tchapda A H, Pisupati S V 2014 A Review of Thermal Co-Conversion of Coal and Biomass/Waste *Energie* **7** 1098-1148

[21] Abnisa F, Daud W M A W 2014 A Review on Co-Pyrolysis of Biomass: An Optional Technique to Obtain A High-Grade Pyrolysis Oil *Energy Conversion and Management*, **87** 71-85

[22] Wannapeera J, Fungtammasan B, Worasuwannarak N 2011 Effects of Temperature and Holding Time during Torrefaction on the Pyrolysis Behaviors of Woody Biomass *J. of Anal. and Appl. Pyrolysis*. **92**(1) 99-105

[23] Bridgwater A V 2012 Review of fast pyrolysis of biomass and product upgrading *Biomass and Bioenergy* **38** 68-94

[24] Lu Q, Yang X I and Zhu X F 2008 Analysis on chemical and physical properties of bio-oil pyrolyzed from rice husk *Journal of Analytical and Applied Pyrolysis* **82**(2) 191-198

[25] Vispute T 2011 Pyrolysis Oils : Characterization, Stability Analysis, an Catalytic Upgrading to Fuels and Chemicals *Dissertation Doctor of Philosophy, Chemical Engineering. University of Massachusetts Amherst* 22-24

[26] Sun J P, Sui S J, Zhang Z J, Tan S and Wang Q W 2013 Study on the pyrolyticbehavior of wood-plastic composites using Py-GC/MS *BioResources* **8**(4) 6196–6210

[27] Supramono D, Kusrini E, Yuana H 2016 Yield and Composition of Bio-oil from Co-Pyrolysis of Corn Cobs and Plastic Waste of HDPE in a Fixed Bed Reactor, *3rd Asian Conference on Biomass Science*, Niigata, Japan, 19 January, 2016.

[28] Ramsey J C III, Fricke A L, Caskey J A 1973 Thermal Conductivity of Polymer Melts *Journal of Applied Polymer Science* **17** 1597-1605

[29] Merrick D 1983 Mathematical models of the thermal decomposition of coal: 2, Specific heats and heats of reaction *Fuel* **62**(5) 540-546

[30] Dahlquist E 2013 Technology for converting biomass to useful energy: Combustion, Gasification, Pyrolysis, Torrefaction and Fermentation, *CRC Press*, Taylor and Francis Group

[31] Kumar S, Singh R K 2011 Recovery of Hydrocarbon Liquid from Waste High Density Polyethylene by Thermal Pyrolysis *Brazilian Journal of Chemical Engineering* **28**(04) 659-667

[32] Goosey M 1999 Plastics for electronics, 2nd edition, *Springer Science and Business Media, B.V.*

[33] Ziskind G, Fichman M and Gutfinger C 1995 Resuspension of particulates from surfaces to turbulent flows – review and analysis *Journal of Aerosol Science.* **26**(4) 613–644

[34] Levine S E, Broadbelt L J 2009 Detailed mechanistic modeling of high-density polyethylene pyrolysis : Low molecular weight product evolution *Polymer Degradation and Stability* **94**(5) 810–822

[35] Rutkowski P, Kubacki A 2006 Influence of polystyrene addition to cellulose on chemical structure and properties of bio-oil obtained during pyrolysis *Energy Conversion and Management* **47** 716–731

[36] Sharpov V I, Berego N G and Kuznetso B N 2003 Co-pyrolysis of wood biomass and synthetic polymers mixtures. Part III : Characterisation of heavy products *Journal of Analytical and Applied Pyrolysis* **67**(2) 325-340

[37] Brebu M, Spiridon I 2011 Thermal degradation of keratin waste *Journal of Analytical and Applied Pyrolysi,* **91** 288–295

[38] Patwardhan P R 2010 Understanding the product distribution from biomass fast pyrolysis *PhD thesis, University of Iowa*

[39] Yao X-w, Xu K-l, Wang W-j, Zhang X-m 2016 Physico Chemical Characteristics of Corn Cob Biomass Ash *Journal of Northeastern University : Natural Science* **37** Issue (1) : 100-104

[40] Kelkar S, Bovee J M, Thelen K D, Christopher M. S 2012 Pyrolysis of North-American Grass Species: Effect of Feedstock Composition and Location *National Conference Science for Biomass Feedstock Production and Utilization*, October 2-5, 2012, New Orleans, LA

Temperature effect towards methane gas production and performances of anaerobic fixed bed reactors

H Budiastuti[a*], D Widyabudiningsih[b] and D R D Kurnia[c]
Chemical Engineering Department, Politeknik Negeri Bandung, Indonesia

E-mail : [a]herabudi@polban.ac.id*, [b]Thewidyabudiningsih@yahoo.com, [c]diantyrosirda@gmail.com

Abstract. This study was conducted to observe the effect of temperature towards methane gas production and performances of Anaerobic Fixed Bed Reactors (AFBR) in degrading artificial wastewater. During seeding and acclimatization as well as operation at normal loading rates the reactor was controlled at 35 ± 1 ^{0}C. The reactor temperature was then switched to room temperature to compare its performance with performance of the temperature controlled reactor. Seeding was performed by wastewater feeding at concentration of 5,300 mgCOD/L. Seeding was continued to acclimatization after COD efficiency reached about 30% and mixed liquor volatile suspended solids (MLVSS) in the reactor were in the range of 28 to 29 g/L. Acclimatization was conducted by step increased feeding while obtaining organic loading rates (OLR) of 530 mgCOD/L/day and 40 days HRT, and it was stopped when COD efficiencies almost constant at about 80% were achieved. Normal loads at OLR of 530 mgCOD/L/day and HRT of 40 days resulted in COD efficiencies in the range of 80 to 92%, produced methane gas ranged from 80 to 170 ml/day, at pH around 7 during controlled temperature. At room temperature, COD efficiencies decreased to the fluctuated range of 78 to 84%, methane gas dropped to the highest of 144 ml/day but pH range was still at around 7. At OLR 1.5 times normal OLR (795 mgCOD/L/day) and HRT of 40 days, the controlled temperature reactor shows superiority by producing efficiencies in the range of 84 to 94% and 260 mL/day of CH_4 as the highest gas produced during observation. Efficiencies of COD obtained during uncontrolled temperature dropped to the range of 60 to 80%. The uncontrolled temperature reactor only produced the highest CH_4 production of 48 mL/day. pH from both reactors still ranged in the normal pH range about 7. It shows that AFBR has to be controlled at around 35 ^{0}C to produce higher COD efficiencies and methane gas production.

1. Introduction

Anaerobic digestion is a common technology used to anaerobically treat high organic contents of wastewater. Wastewater produced from sugar, poultry, pulp and paper, and pharmaceutical industries are examples of industrial wastewater to be degraded anaerobically. This type of wastewater is generally difficult to be degraded by using aerobic digestion system since this wastewater type contains high organic compounds. The application of anaerobic digestion to degrade high organic contents of such wastewater has widely been cited in literature [1].

There are three stages occurring during anaerobic digestion, i.e. hydrolysis, acetogenesis, and methanogenesis. During hydrolysis, organic compounds such as lipids, polysaccharides, proteins, and nitrates will be degraded into simpler organic compounds such as fatty acids, monosaccharides, and amino acids by cellulose, amylase, protease, and lipase enzymes. The simpler organic compounds will be converted into formic acid, hydrogen, and carbon dioxide during acetogenesis. During this stage, volatile fatty acids (VFAs) will also be produced. At the stage of methanogenesis, these VFAs will be degraded to methane by methane producing bacteria. Acetic acid, hydrogen, and carbon dioxide will be decomposed by methanogenic bacteria during methanogenesis stage to become methane and carbon dioxide. Methanogenic bacteria consume acid produced by acetogenic bacteria and acid producing bacteria provide ideal atmospheric conditions for methane producing bacteria [2, 3, 4]. However, if the content of VFA in the system is too high, the growth of methane producing bacteria will be disturbed so that the degradation will be affected. Therefore, to provide ideal conditions for consortium microorganisms to grow and properly function, all stages have to occur proportionally [5, 6].

Methane produced from the aforementioned three anaerobic stages of degradation is the main benefit of anaerobic degradation system since methane gas can be applied as an alternative energy in related industries. The other benefits include lower energy consumption as aeration is not required, production of low quantity of well stabilized sludge, and odor-free treatment as anaerobic digestion has to be carried out in sealed vessels.

Jerger and Tsao [7] obtained that *Alcaligenes sp, Aerobikacter sp, Clostridium, Escherichia sp, Methanobacteriaceae,* and *Propionibacterium* are among anaerobic microorganisms which have to be maintained in the reactors to produce methane gas. Attention has to be paid to provide ideal conditions for these anaerobic microorganisms, i.e. several factors effecting the growth of consortium microorganisms anaerobically. These include temperatures, pH, nutrients, toxic materials, etc. This study was conducted to observe the effect of temperature towards the growth of anaerobic microorganisms in terms of methane gas produced and performances of the reactor. The reactor type of Anaerobic Fixed Bed Reactors (AFBR) is used since this reactor type provides the bed in which microorganisms could attach on the surface of the carriers. Attachment of microorganisms on the carriers is to enhance biomass retention within the reactor so that de-linking of solid retention time (SRT) from hydraulic retention time (HRT) occurs. By de-linking SRT from HRT, the reactor could perform much better since more attached microorganisms separated from suspension microorganisms withdrawn in the effluent stream [8].

2. Materials and Methods

2.1. *Synthetic wastewater, inoculums, and reactor*

The synthetic wastewater consists of chemicals containing nutrients needed by growth of microorganisms and its composition follows wastewater composition applied Budiastuti et al. [9]. The wastewater represents high organic contents of wastewater to be degraded anaerobically. The inoculums were in the form of abattoir seed sludge, which was obtained from an abattoir house in Padalarang, West Java, Indonesia. The synthetic wastewater composition fulfills the requirement of COD:N:P for anaerobic conditions [10]. The reactor used was the Anaerobic Fixed Bed Reactors (AFBR) produced by Armfield [11]. The reactor contains bioball carriers where attachment of microorganisms is expected to occur on the surface of these carriers [9]. This reactor is completed with gas collection tanks in which measurements of resulting biogas take place. The biogas is measured by using a water displacement system. Water displaced by the gas overflows from the tank base.

About 2 L seed sludge was poured into the reactor (AFBR) maintained at 35 ± 1 ^{0}C and feeding with increases of influent flow rates at the feed concentration of 5,300 mg COD/L was then performed. Effluent from the reactor was not withdrawn during this feeding period until the active volume of reactor of about 4 L was obtained. Seeding was stopped and continued by acclimatization while obtaining organic loading rate (OLR) of 530 mg COD/L/day and at hydraulic retention time (HRT) of 40 days.

2.2. *Operations, HRT, and analysis parameters*

During normal operations, the OLR was maintained at 530 mg COD/L/d and HRT of 40 days. This OLR was defined as the normal load. Normal loads were initially conducted at 35 ± 1 ^{0}C by applying automatic control equipment equipped in the AFBR and then the reactor was switched to room temperature, with no control temperature equipment. The swiched temperature step was repeated

IOP Conf. Series: Materials Science and Engineering **162** (2017) 012012 doi:10.1088/1757-899X/162/1/012012

during loading rates of 1.5 times normal loading rates or at OLR of 795 mg COD/L/d and HRT of 40 days.

Parameters taken during room temperature and controlled temperature operations were chemical oxygen demand (COD), total gas, and reactor pH. Parameters taken during seeding and acclimatization were COD and mixed liquor volatile suspended solids (MLVSS). Analysis of total gas, conducted to observe the composition of methane gas and other gases produced, was done in the Chemical Engineering Department of ITB in Bandung, Indonesia by using gas chromatography Varian Star 3400. Measurements of influent COD and effluent COD were then converted to reactor efficiencies by substraction of both COD measurements devided by influent COD times 100%. From CH_4 produced and efficiencies obtained, the reactor performances were determined.

3. Results and Discussions

3.1. Operations during seeding and acclimatization

Performance of the AFBR during seeding is shown in Figure 1 whereas its performance during acclimatization is not shown but the acclimatization was performed by step increased feeding until COD efficiencies of about 80% were obtained. Figure 1 shows that the concentrations of microorganisms represented by MLVSS measurements were in the range of 28 to 29 g/L. This shows sufficient numbers of microorganisms to ensure that acclimatization performed by these mixed culture microorganisms would be successfully guaranteed [5, 9, 12]. The acclimatization, performed straight away after seeding, would probably be also successful by achieving constant increases of efficiency after day 1 during seeding (Figure 1).

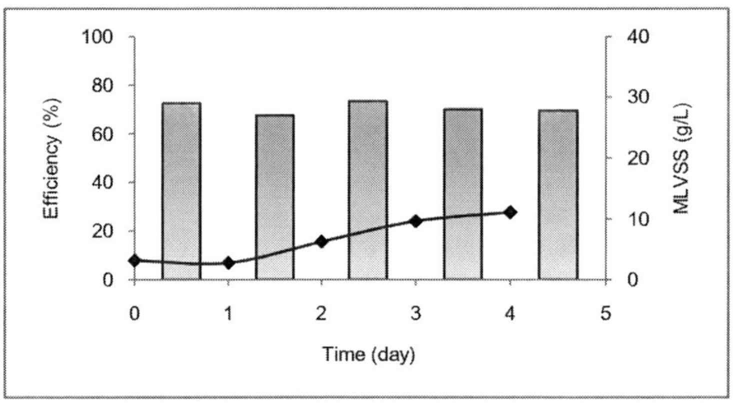

Figure 1. Reactor efficiency (♦) and MLSS (▒) during Seeding

3.2. Operations during normal loads of 530 mg COD/L/day

Operations during OLR of 530 mg COD/L/day were decided to be classified as normal loads to which higher organic loads were compared. Operations performed during this study were conducted at normal organic loads and at 1.5 times normal loads. Both operations were conducted at HRT of 40 days.

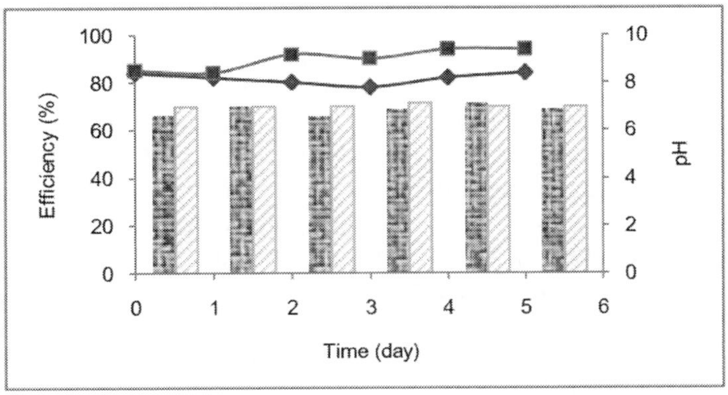

Figure 2. Reactor efficiency (♦ room temperature), (■ 35^{0}C), and pH (▓ room temperature), (░ 35^{0}C) at OLR 530 mg COD/L/day and HRT of 40 days

Figure 2 shows benefit of controlling temperature of the reactor. At room temperature, the highest COD efficiency is only 84% whereas at controlled temperature at 35^{0}C, the highest COD efficiency is 92%. The trend of efficiency fluctuation is also better during temperature control than during room temperature. For the last two days reactor efficiencies are stable at 92% at controlled temperature of 35^{0}C. It is also shown in Figure 2 that pH during controlled temperature at 35^{0}C is more stable compared to pH during uncontrolled temperature. Even though the range of pH in both reactors is still in the normal pH range [4], however, pH actually shows the latest indication of unsuccessful degradation [8, 13]. The reactor pH could be still stable in the normal pH range when reactor buffer is capable to stabilize the reactor pH.

CH_4 produced during controlled temperature at 35^{0}C is also better compared to CH_4 produced during room temperature (Figure 3). The highest CH_4 production is 170 mL/day compared to 144 mL/day. Both curves show increased trends, however, the controlled temperature reactor produced increased trend steeper than the increased trend of uncontrolled temperature reactor.

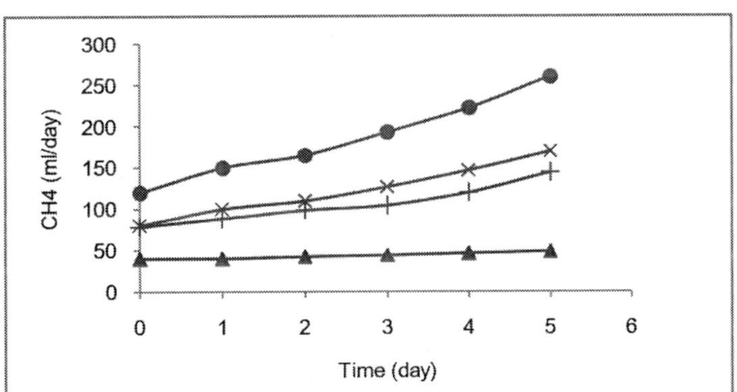

Figure 3. CH_4 production at OLR 530 mg COD/L/day (+, room temperature), (x, 35^{0}C), and at OLR 795 mg COD/L/day (▲, room temperature), (•, 35^{0}C)

3.3. Operations during organic loads of 795 mg COD/L/day

Such obtained during normal loads, the temperature controlled AFBR shows results much better than the results obtained during room temperature, especially in terms of reactor efficiencies (Figure 4) and CH_4 production (Figure 3). As shown in Figure 4, at room temperature, the highest COD efficiency is only 80% whereas at controlled temperature of 35^{0}C, the highest COD efficiency is 94%. Moreover, for the last two days, reactor efficiencies are stable at 94% at controlled temperature of 35^{0}C. During uncontrolled temperature, efficiencies of AFBR were almost the same at about 60% except at day 3 (Figure 4). This may result from the effect of temperature as well as the organic overload condition. Uncontrolled temperature reactor could not stand to be overloaded 1.5 times loads without maintaining reactor temperature at the optimum temperature [8, 13].

pH ranges measured during observation from both reactors were no distinctive differences (Figure 4). However, such have been mentioned above, pH ranges from both reactors which are in the normal pH range, might be resulted from the situation where pHs were not affected by variations of temperature. Effect towards pH is most influenced by buffer condition in the reactors. Buffer contained in the chemical compositions of synthetic wastewater may probably be capable to maintain the reactor pH at the normal pH range.

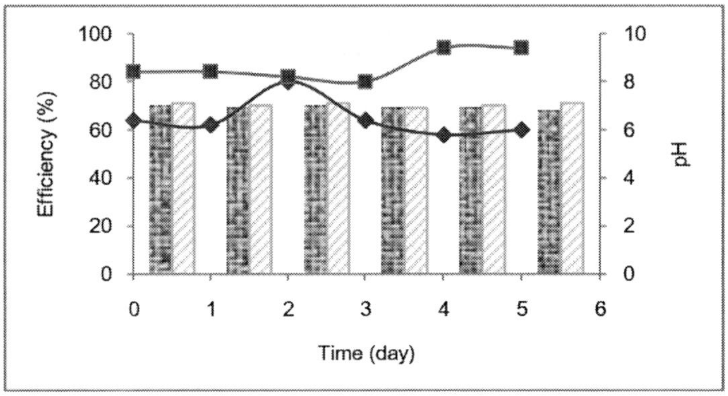

Figure 4. Reactor efficiency (♦ room temperature), (■ 35^{0}C), and pH (■ room temperature), (■ 35^{0}C) at OLR 795 mg COD/L/day and HRT of 40 days

Figure 3 shows increases of methane gas produced during temperature control at 35^{0}C at OLR 795 mg COD/L/day. Methane gas production increased from 120 to 260 mL/day. This is proportional with the increases of reactor efficiencies from 84 to 94%. On the other hand, methane gas produced during uncontrolled temperature only slightly increased from 40 mL/day to 48 mL/day during observation (Figure 3). The low increase of CH_4 production may have been resulted from the effect of uncontrolled temperature as well as the organic overload condition [1, 8, 14]. Uncontrolled temperature reactor could not stand to be overloaded 1.5 times loads without maintaining reactor temperature at the optimum temperature.

4. Conclusions

Temperature affects the performance of AFBR either during normal loads of 530 mg COD/L/day or OLR of 795 mg COD/L/day at the same HRT of 40 days. The effect is worse during 1.5 times normal OLR, which might be caused by the effect of temperature as well as the effect of organic loads. COD efficiencies and methane gas produced became lower but pH was not affected since reactor buffer could probably maintain the reactor pH at the normal pH range.

Acknowledgment

Financial support of this work by the Directorate Research and Community Services, Ministry of Research, Technology, and Higher Education, through the Program of Fundamental Research No. 0126.4/PL1. R7/LT/2016 is gratefully acknowledged.

References

[1] Budiastuti H 2011 Increased capabilities of W8 anaerobic reactor operated in sequencing batch in handling acclimatization and organic schock loads *Fluida J.* **7** 69-76

[2] Madigan M T, Martinko J M, et.al. 2015 *Brock Biology of Microorganisms 14th ed.* (Boston : Pearson)

[3] Yu H Q, Zheng X J, Hu Z H, Gu G W 2003 High-rate anaerobic hydrolysis and acidogenesis of sewage sludge in a modified upflow reactor *Water Science and Technology* **48** 69-75

[4] Shizas I and Bagley D M 2002 Improving anaerobic sequencing batch reactor performance by modifying operational parameters *Wat. Res.* **36** 363-367

[5] Budiastuti H, Widyabudiningsih D, Kurnia D R D 2015 Effect of seed sludge source on the performance of anaerobic fixed bed reactors *RSCE Proceeding, Bangkok, 24-25 September 2015*

[6] Mota V T, Santos S, Amaral M C S 2013 Two-stage anaerobic membrane bioreactor for the treatment of sugarcane Vinasse : Assessment on biological activity and filtration performance *Bioresource Technol* **146** 494-503

[7] Jerger W and Tsao G T 2006 *Anaerobic Bacterial in Ecological Aspects of Used-Water Treatment* (London : Academic Press)

[8] Budiastuti H 2004 Intensification of single state continuously stirred tank anaerobic digestion process using carriers *Dissertation, Murdoch University*

[9] Budiastuti H, Widyabudiningsih D, et al. 2016 Superiority of two stage anaerobic fixed bed reactors in handling organic shock loads *ICCE Proceeding, Nusa Dua, Bali, 12-13 October 2016*

[10] Bouallagui H, Torrijos M, et.al. 2004 Two-phases anaerobic digestion of fruit and vegetable wastes : Bioreactors performance *Biochemical Engineering J.* **21** 193-197

[11] Armfield 2006 Water treatment processes www.armfield.co.uk/images/w8_graphic1. gif.com Armfield Corporation 9 Agustus 2016

[12] Budiastuti H, Muhari E, Widyabudiningsih D, Ghozali M 2013 Degradation of leachate from municipal solid waste final disposal using two stage fixed bed anaerobic reactors, *SNTKK Proceeding, UPN Yogyakarta, 5 Maret 2013*

[13] Park M J, Jo J H, Park D, Lee D S, Park J M 2010 Comprehensive study on a two-stage anaerobic digestion process for the sequential production of hydrogen and methane from cost-effective molasses *International J. Hydrogen Energy* **35** 6194-6202

[14] Masse L and Masse D I 2016 Effect of soluble organic, particulate organic, and hydraulic shock loads on anaerobic sequencing batch reactors treating slaughterhouse wastewater at 20°C *Process Biochemistry* **40** 1225-1232

Performance of continuous stirred tank reactor (CSTR) on fermentative biohydrogen production from melon waste

K Cahyari[1,3,*], Sarto[2], S Syamsiah[2] and A Prasetya[2]
[1] Doctoral Student, Dept. Chemical Engineering, Faculty of Engineering, Gadjah Mada University, Yogyakarta, Indonesia
[2] Dept. Chemical Engineering, Faculty of Engineering, Gadjah Mada University, Yogyakarta, Indonesia
[3] Dept. Chemical Engineering, Faculty of Industrial Technology, Islamic University of Indonesia, Indonesia

E-mail : cahyari.khamdan@yahoo.com

Abstract. This research was meant to investigate performance of continuous stirred tank reactor (CSTR) as bioreactor for producing biohydrogen from melon waste through dark fermentation method. Melon waste are commonly generated from agricultural processing stages i.e. cultivation, post-harvesting, industrial processing, and transportation. It accounted for more than 50% of total harvested fruit. Feedstock of melon waste was fed regularly to CSTR according to organic loading rate at value 1.2 – 3.6 g VS/ (l.d). Optimum condition was achieved at OLR 2.4 g VS/ (l.d) with the highest total gas volume 196 ml STP. Implication of higher OLR value is reduction of total gas volume due to accumulation of acids (pH 4.0), and lower substrate volatile solid removal. In summary, application of this method might valorize melon waste and generates renewable energy sources.

1. Introduction

Hydrogen is the ultimate choice of energy carrier in future due to its superior properties. It emits less greenhouses gases, more energy efficient conversion up to 50%, higher energy densities, and it is renewable [1]. It can easily be produced from water electrolysis when there is sufficient free excess of electricity such as one from geothermal power plant. However, lack of excess of electricity leads to avoid the use of electrolysis method [2]. There is more promising method to produce hydrogen from cheap raw materials i.e. dark fermentation using organic waste as raw materials [3].

Fermentative production of hydrogen using the waste empowers microbial community to metabolize organic matter into volatile fatty acids. Simultaneously, it emits electron to hydrogenase converting proton into hydrogen (H_2). This method is influenced by many factors e.g. temperature, pH, substrate concentration, inhibitors, and reactor types such as continuous stirred tank reactor, plug flow reactor, fluidized bed reactor [4].

Continuous stirred tank reactor (CSTR) is widely used as means to convert reactants into valuable products in chemical industries. It is used commonly for liquid or suspension reaction yet rarely occupied as means for gas phase reaction. The fermentation of organic waste into renewable biohydrogen occurs in three phases i.e. liquid (diluted water), solid (total solid of melon waste) and gas phase (hydrogen and CO_2) [5]. Controlling optimum condition i.e. gas yield, substrate reduction, pH, inhibitory chemical concentration of the fermentation is challenging tasks on three phases CSTR. Moreover, change of organic loading rate (OLR) can cause significant interruption of CSTR stability performance. When OLR is high, there is tendency of accumulation of acids and inhibitory compounds leading to lower gas production and yield. On the other hand, lower OLR may reduce the unstable of CSTR due to sufficient ratio between source of microbial food (substrates) and certain population of microorganisms. Lower OLR also prevents accumulation of volatile fatty acids which maintain environmental condition at optimum pH level (pH 4 – 6). It is therefore important to study the performance of CSTR on biohydrogen production from organic waste.

Melon (*Cucumis melo*) is one of national fruit fresh trading commodities throughout Indonesia. According to National Statistics Bureau of Indonesia, there was more than 150,000 ton of harvested melon in 2014, which some portion turned into waste approximately 10 – 50%wt due to deterioration, physical damage or else [6]. This melon waste has been widely treated as organic waste through open dumping method. It causes environmental problems such as methane emission, water contamination, bad odor, and eventually global warming. To mitigate such effects, it was proposed to utilize this melon waste as raw material to produce biohydrogen as renewable energy.

Melon (*Cucumis melo*) contains 92 – 94%wt moisture, 6 – 8% dry matter (total solid), 6 g/ 100 g of fresh melon [7]. Melon seeds contains 52,3% of testa and 47.7% of kernel, oil 22.1 – 53.5%, 21.8% of crude protein [8]. Eugenol is major constituent in skin of pocket melon i.e.15,3% which is known also as anti-microbial agent [9], [10].

Therefore, this research was meant to investigate the performance of CSTR in fermentation of melon waste to produce hydrogen.

2. Materials and methods

2.1. Feedstocks

Orange and melon fruit waste was collected from local fruit market in Yogyakarta. The two were peeled and cut off, followed with grinding using kitchen blender, thereafter was called feedstock. It was stored at freezer (-5°C) prior to be used as feed to CSTR.

2.2. Inoculums

Hydrogen producing microorganisms was originally from residual sludge of cow dung biogas plant which had been treated through pH 5.0 (24 hours) and heat-shock treatment (95°C, 45 min).

2.3. Experimental Procedures

CSTR was initially filled with inoculums, substrates, and water for 3 days prior to feed with 1.2 g VS/ (l.d) organic loading rate. On the 4th day, quantity of 100 gr feedstock was fed to CSTR per day for 2 days followed with 1.8; 2.4; 3.0 and 3.6 g VS/ (l.d).

2.4. Analysis

Gas content was analyzed using Gas Chromatography Shimadzu 8A which equipped with MS 5A column and Thermal Conductivity Detector. It is run with Helium as gas carrier at 40 ml/min. temperature of oven, injector and detector was set at 60°C, 80°C and 80°C respectively. TS, VS and ash were measured based on standard method of APHA.

3. Results and discussion

Continuous stirred tank reactor (CSTR) was used as bioreactor for fermentative biohydrogen production from melon waste. The performance of CSTR was noted in term of total volume gas produced, volatile solid reduction (VS removal), and profile of pH evolution. These performance parameters are presented in Figure 1-3 respectively.

Figure 1 shows profile of total gas volume evolution during fermentation period with various value of organic loading rate. At initial stage, CSTR was loaded with organic loading rate value at 1.8 g VS/ (l.d) for period of 12 days. Gas production rate was significantly high at this stage. It was recorded that this fluctuate production achieved the highest total daily gas production 196 ml STP. At the beginning, total gas volume was high since there is sufficient microorganism to consume fresh feed of melon waste. On the next days, total gas volume decreased gradually due to accumulation of total organic solid which was not degraded during this period. On the following organic loading rate value at 2.4 g VS/ (l.d), the total gas volume was lower compare to the first period. The highest total gas production was 176 ml STP. It can be seen that there is no significant different between these two OLR values toward gas production. When OLR increase to 3.0 or 3.6 g VS/ (l.d), total gas volume decreased significant and consistent to value lower than 100 ml STP. On the day 40th – 42nd, there was no withdrawal of gas due to technical reactor problem, so that the produced gas was collected and measured on the day 43rd. This total cumulated gas was significantly high up to 447 ml STP/3 days (average 147 ml STP/day). It can be concluded that OLR value lower than 2.4 g VS/ (l.d) is better compare to higher value more than 3.0 g VS/ (l.d).

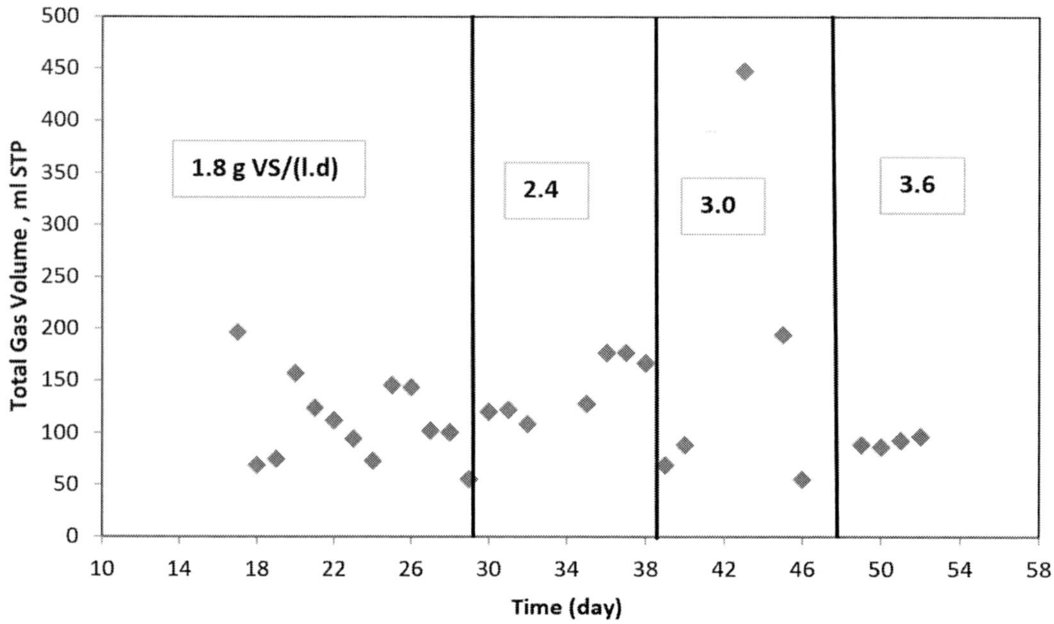

Figure 1. Effect of organic loading rate toward total volumetric gas production

Figure 2 shows value of volatile solid (VS) percentage of melon waste inlet and outlet of CSTR. At initial stage of OLR 1.8 g VS/ (l.d), the VS outlet of melon waste was 65 – 85%wt. The lowest value of VS outlet correspond to higher gas production indicating that more substrat was consumed and degraded by microorganism to produce volatile fatty acids and hydrogen. In the second period of OLR 2.4 g VS/ (l.d), the VS outlet slightly higher compare to the first one. Therefore, it implies that there was lower gas production as seen in Figure 1.

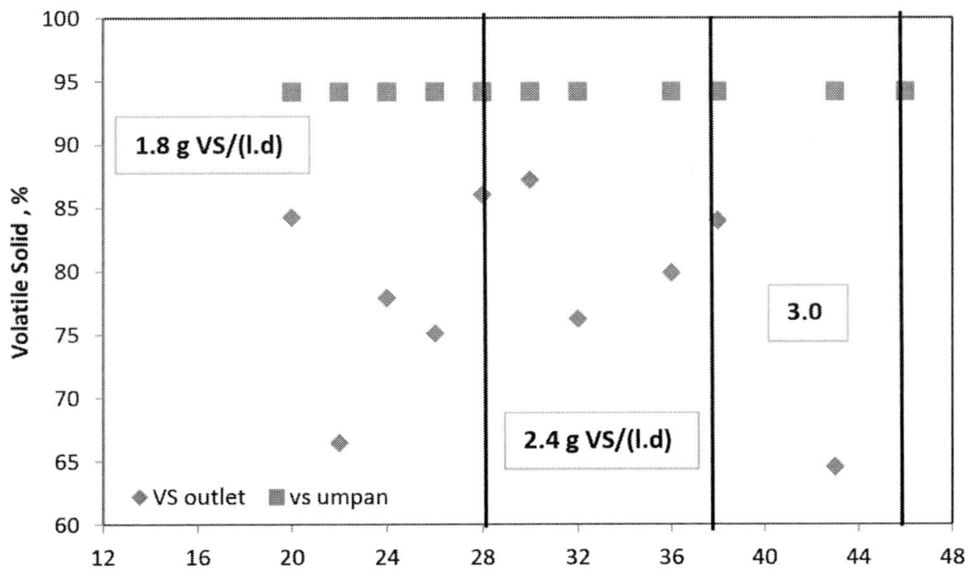

Figure 2. Characteristic volatile solid inlet and outlet of CSTR

Figure 3 shows the evolution of pH during each period of organic loading rate. At initial stage, pH of fermentation system was optimum at condition between 5 – 7. It indicates that there was no significant accumulation of volatile fatty acids during the first period OLR 1.8 g VS/ (l.d). However, increasing OLR to 2.4 g VS/ (l.d) was able to disturb the pH stability to value pH 4.0 or even further lower. This acidic condition occurred due to accumulation of volatile fatty acid which was generated from substrate degradation. Microorganisms was not able to sustain their metabolism of degradation of organic matter content in melon fruit waste.

Hydrogen production through dark fermentation is accomplished via complex microbial reaction. However, it is simplied in following reaction (equation 1 – 2). When acetate is the only volatile fatty acid products, maximum hydrogen gas production is achieved with yield 4 mol H_2/mol glucose or equivalent to 498 ml STP H_2/gram glucose. When butyric acids is the product, hydrogen yield become less equal to half of the first reaction [11].

$$C_6H_{12}O_6 + 2H_2O \rightarrow 2CH_3COOH + 4H_2 + 2CO_2 \quad (1)$$

$$C_6H_{12}O_6 + 2H_2O \rightarrow 2(CH_2)_3COOH + 2H_2 + 2CO_2 \quad (2)$$

Previous experiments on biohydrogen through dark fermentation using various type of substrates have been reported with wide range of practical hydrogen yield. For instance, utilization of peach pulp as substrate was reported with the highest yield 123.27 ml H_2/gTOC and production rate 35.6 ml H_2/h [12]. Mesophilic dark fermentation using novel strain Clostridium sp YM1 and glucose solution was conducted with hydrogen yield 1.7 mol H_2/mol glucose (equivalent to 211 ml STP H_2/gram glucose) [13]. Fermentation of food waste to produce hydrogen was also carried out in continuous stirred tank reactor with hydrogen yield 261 ml H_2/g VSadded and 379 ml H_2/(l.d) (38,6%v) [14]. In comparison, this experiment was carried out with the highest hydrogen yield and production rate 2.2 ml STP H_2/g VSadded (8%v H_2) and 49 ml/(l.d). This lower results was suspected due to pH descending evolution during fermentation (Figure 3). This pH decrease occurs because the CSTR was not designed with pH control and there was no addition of buffer solution, it was expected to maintain simplicity of reactor. In addition, accumulation of eugenol which acts as antimicrobial agents might also inhibits microbial metabolisms to produce high yield of hydrogen.

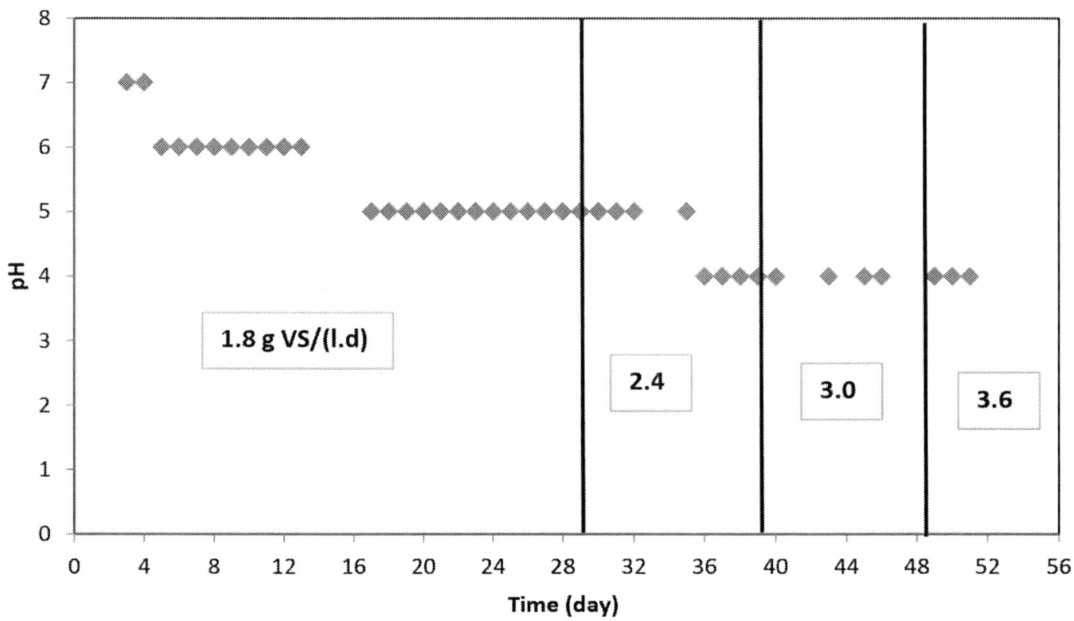

Figure 3. pH evolution during fermentation process in CSTR

4. Conclusion

Performance of biohydrogen production in CSTR was highly affected by the organic loading rate of melon waste. Optimum condition of OLR was achieved at value 2.4 g VS/ (l.d) with the highest total gas volume 196 ml STP. At higher OLR value, there is tendency of acid accumulation leading to lower gas production and substrate volatile solid removal. pH control system and addition of buffer solution might improve hydrogen production.

Acknowledgement

Authors acknowledge the financial support of Ministry of Researches and Higher Education of Republic Indonesia under Excellent Research of Higher Education Program 2016. Authors are also thankful to Indri Fajri Dewantry for technical assistance during the laboratory works.

References

[1] Balat M 2008 Potential importance of hydrogen as a future solution to environmental and transportation problems *Int. J. Hydrog. Energy* **33**(15) 4013–4029
[2] Cheng S and Logan B E 2007 Sustainable and efficient biohydrogen production via electrohydrogenesis *Proc. Natl. Acad. Sci.* **104**(47) 18871–18873
[3] Chong M-L, Sabaratnam V, Shirai Y and Hassan M A 2009 Biohydrogen production from biomass and industrial wastes by dark fermentation *Int. J. Hydrog. Energy* **34**(8) 3277–3287
[4] Lee K-S, Hsu Y-F, Lo Y-C, Lin P-J, Lin C-Y and Chang J-S 2008 Exploring optimal environmental factors for fermentative hydrogen production from starch using mixed anaerobic microflora *Int. J. Hydrog. Energy* **33**(5) 1565–1572
[5] Arooj M F, Han S-K, Kim S-H, Kim D-H and Shin H-S 2008 Continuous biohydrogen production in a CSTR using starch as a substrate *Int. J. Hydrog. Energy* **33**(13) 3289–3294
[6] Gustavsson J, Cederberg C, Sonesson U, Van Otterdijk R and Meybeck A 2011 Global food losses and food waste *Food Agric. Organ. U. N. Rom*

[7] Majkowska-Gadomska J 2010 The chemical composition of fruit in selected melon cultivars grown under flat covers with soil mulching *Acta Sci Pol Hortorum Cultus* **9**(2) 39 – 52

[8] Milovanović M and Picuric-Jovanovic K 2005 Characteristics and composition of melon seed oil *J. Agric. Sci.* **50**(1) 41–47, 2005

[9] Aubert C and Pitrat M 2006 Volatile compounds in the skin and pulp of Queen Anne's pocket melon *J. Agric. Food Chem.* **54**(21) 8177–8182

[10] Shah B, Davidson P M and Zhong Q 2013 Nanodispersed eugenol has improved antimicrobial activity against Escherichia coli O157: H7 and Listeria monocytogenes in bovine milk *Int. J. Food Microbiol.* **161**(1) 53–59

[11] Levin D B, Pitt L and Love M 2004 Biohydrogen production: prospects and limitations to practical application *Int. J. Hydrog. Energy* **29**(2) 173–185

[12] Argun H and Dao S 2016 Bio-hydrogen production from waste peach pulp by dark fermentation: Effect of inoculum addition *Int. J. Hydrog. Energy*

[13] Abdeshahian P, et al. 2014 The production of biohydrogen by a novel strain Clostridium sp. YM1 in dark fermentation process *Int. J. Hydrog. Energy* **39**(24) 12524–12531

[14] Reungsang A, Sreela-or C and Plangklang P 2013 Non-sterile bio-hydrogen fermentation from food waste in a continuous stirred tank reactor (CSTR): Performance and population analysis *Int. J. Hydrog. Energy* **38**(35) 15630–15637

Upgrading of bio-oil from the pyrolysis of biomass over the rice husk ash catalysts

B Sutrisno and A Hidayat
Chemical Engineering Department, University of Islam Indonesia
Jalan Kaliurang km 14,5 Yogyakarta 55584, Indonesia

E-mail : arif.hidayat@uii.ac.id

Abstract. The pyrolysis oils are complex mixtures of organic compounds that exhibit a wide spectrum of chemical functionality, and generally contain some water. Their direct use as fuels may present some difficulties due to their high viscosity, poor heating value, corrosiveness and instability. For possible future use as replacements for hydrocarbon chemical feedstocks and fuels, the liquids will require considerable upgrading to improve its characteristics. By esterification of the bio oil as the upgrading method, the properties of the bio-oil could be improved. In the paper, the upgrading of a bio-oil obtained by pyrolysis was studied over rice husk ash catalysts. The raw bio-oil was produced by pyrolysis of rice husk.From the experiment results, it can be concluded that the densities of upgraded bio-oil were reduced from 1.24 to 0.95 g.cm^{-3}, and the higherheating value increased from 16.0 to 27.2 MJ/kg and the acidity of upgraded bio-oil was also alleviated from 2.3 to 4.4. The results of gas chromatography-mass spectrometry (GC–MS) and FT-IR analysis showed that the ester compounds in the upgraded bio-oil increased. It is possible to improve the properties of bio-oil by esterifying the raw bio-oil.

1. Introduction

The global energy crisis and environmental concerns have boosted research initiatives on alternatives for fossil energy and has encouraged the exploration of renewable resources such as biomass for increased energy generation [1]. Thermochemical techniques for bio-oil production including fast pyrolysis liquefaction and byproduct oil collection from slow pyrolysis are gradually being developed and recommended. Fast pyrolysis has been developed considerably as a front technology for efficient conversion of biomass into a liquid product known as bio-oil, which can be used as a potential substitute for fossil fuels. This technology is industrially feasible and has advantage in terms of reduced cost[1–3]. Great progress in fast pyrolysis of biomass has been achieved, and liquid product yield is more than 70% on dry feed [4, 5]. However, bio-oil is not as good as fossil fuels and the drawbacks including high acidity, low heating value, high corrosiveness, high viscosity and poor stability limit its usage as high grade/transportation fuel. Consequently, upgrading of bio-oil before utilization is desirable to obtain a liquid product that can be used in a wider variety of applications. The current upgrading methods include catalytic hydrogenation, catalytic cracking and catalytic esterification [6–8]. The presence of a large amount of organic acids (formic acid, acetic acid and propionic acid) is the main reason for the strong acidity, high corrosiveness and poor stability of bio-oil [9–11].

Esterification reaction is an upgrading method that can convert carboxylic acids in bio-oil to esters. It is not only able to reduce corrosiveness of the bio-oil but can also prohibit some reactions catalyzed by these acids, for example, the reactions of oligomers or polymer formation (homopolymerization of aldehydes, phenol/aldehyde polymerizations, etc.) This treatment could be an attractive method for improving bio-oil properties due to its ease of operation, low investment cost and avoiding secondary reactions that may occur at high operating temperatures. Although limited studies on bio-oil upgrading by esterification reaction have been reported thus far, this upgrading technique is still the topic of recent interest. Doshi et al. (2005)upgraded the bio-oil derived from fast pyrolysis of sewage sludge via esterification using H_2SO_4 as a liquid acid catalyst. It was found that odor of the treated oil was improved from an extremely annoying level to a not annoying level (ASTM D1833-87) with about four times reduction in viscosity and up to 9% increasing in heating value. However, using solid catalyst in place of H_2SO_4 to reduce corrosion problem was also proposed. It is noted that increasing amount of alcohol for the esterification reaction can reduce heating value of the upgraded oil and therefore alcohol separation by fractional distillation would be needed in this process. Later, several researchers improved the process by using various types of solid acid catalyst and/or recovery of the alcohol by distillation [6-8]. Junming et al. (2008)reported that density and acidity were reduced (1.16 to about 0.93 g/cm^3 for density and pH from 2.82 to 6.21) and the gross calorific value increased from 14.3 to 23 MJ/kg using ethanol as a reactant and SO_4^{2-}/ZrO_2 as a catalyst at the reaction temperature of 55-77 °C.Mahfud et al. (2007) treated the bio-oil with a higher boiling point alcohol of n-butanol in the presence of a solid acid catalyst of Nafion SAC13 at 50-80 °C under reduced pressure of less than 10 kPa. The results indicated that the upgraded oil properties were improved, particularly the heating value and viscosity but the acidity was still close to that of the raw bio-oil. In studying the effect of catalyst type, Zhang et al. (2006) used solid acid catalyst $40SiO_2/TiO_2\text{-}SO_4^{2-}$ and solid base catalyst $30K_2CO_3/Al_2O_3\text{-}NaOH$ to compare their effectiveness. It was found that acidity of the bio-oil was lowered by the use of both catalysts but the solid acid catalyst was capable of converting volatile and nonvolatile organic acids into more esters. However, the strong acidification of solid acid catalyst tended to intensify the acidity of bio-oil.

Rice (*Oryza sativa L.*) is a primary source of food for billions of people and it covers 1% of the earth's surface. Globally, approximately 600 million tons of rice is produced each year. For every 1000 kg of paddy milled, about 220 kg (22%) of husk is produced. Rice husk (RH) is therefore an agricultural residue abundantly available in rice producing countries. Much of the husk produced from the processing of rice is either burnt or dumped as waste. RH is composed of 20% ash, 38% cellulose, 22% lignin, 18% pentose and 2% other organic components. Silica can be pyrolyzed at elevated temperature to form rice husk ash (RHA). RHA as biomass has promising role with high silica content (87–99%) and available abundantly, being a low cost waste source as amorphous silica precursors. Each tons of rice produces 200 kg of rice husk, and with the complete combustion will be generated 40 kg of RHA. Recently, about 146 million tons of RHA was produced annually worldwide and efforts are being made to overcome this environmental issue by utilization this material as catalyst support. In this research, the upgrading bio-oil from pyrolysis of biomass on rice husk ash as catalysts was studied.

2. Experimental

2.1. *Materials*

Rice husk ash (RHA) was collected from rice mills in Bantul Province, D.I. Yogyakarta, Indonesia. Methanol (analytical grade, Merck) was used for esterification reaction. Sulfuric acid p.a. (98 wt.%) was supplied from Merck.

2.2. *Catalyst Preparation*

RHA was ground to powder (particle size $\leqslant$ 0.28 nm) and sieved with a 200 mesh sieve. Then, 10 g of powdered ash was washed with 60 mL of 0.1 mol L^{-1}HCl for 1 h, and neutralized with deionized water. The washing step is to remove the trace of minerals/metal (Al, Ca, K, Mg, Mn and Na) contained in RHA and organic compounds. Finally, the clean ash was dried in an oven at 105 C for 2 h. The sulfonation of RHA was performed using the following procedure. RHA (5.0 grams) and a concentrated H_2SO_4 (98%) solution (100 mL) were placed into a 100 mL Teflon-tube autoclave. In this procedure, 5 grams of the RHA was immersed in 100 ml concentrated H_2SO_4 (98%) at 423 K for 12 hours. The product was then washed with hot distilled water until no sulphate ions were detected in the wash water and then dried.

2.3. Catalyst Characterization

The crystalline phase of the catalyst sample was confirmed using a powder X-ray diffractometer coupled with Cu Kα radiation. The surface areas (based on the BET equation), pore volumes and average pore diameter for the RHA and RHA catalyst were measured by N_2 adsorption–desorption isotherm apparatus. A FTIR spectrophotometer was used to characterize the chemical structure of alkali metal silicate at 400-4000 cm^{-1} range. The acid density of the catalysts was estimated by a titration analysis.

2.4. Esterification Reaction

Bio-oil required for this research was produced from rice husk ash with a size of 1-3 mm and moisture content of 8-10% on dry-weight basis. Bio-oil was produced by the fast pyrolysis process performed at a temperature of 450 °C with nitrogen carrier gas. The batch esterification was performed at a laboratory scale. A hot plate with a magnetic stirrer was used for heating the mixture in a three-neck 250 mL round bottom-flask fitted with a water-cooled reflux condenser. The round bottom-flask was held in a water bath to maintain the temperature. The esterification was performed by addition of 20 wt.% of methanol in the presence of RHA catalyst (4 wt.% of raw bio-oil) at a temperature of 60°C for 90 min. As a control raw bio-oil was also esterified without application of pretreatment by following the same esterification procedure as described above.In each experiment, once the reaction was complete, the liquid products were cooled to room temperature. The mixture was collected in centrifuge test tubes and which were centrifuged to separate the resulting aqueous and organic phases; the time for centrifuging to separation ranged from 2 to 4 h. Both phases were separated and weighed for mass balance computation. The composition of raw bio-oil and upgraded bio-oil was determined using the GC-MS (Gas Chromatography Mass Spectroscopy) instrument. The acid values were analyzed by titration procedure for the evaluation of free residual acidity. The physical properties of bio-oil such as density, kinematic viscosity, pH, and water content were characterized by commonly standar methods. Meanwhile, higher heating values (HHVs) were determined by bomb calorimeter.

3. Results and Discussion

3.1. Textural characterization

The catalyst characteristics, including the BET surface area, the average pore size, and the total pore volume of the RHA and RHA catalyst are presented in Table 1. The RHA had low surface areas. The acid treatment by sulfonation improved the surface area characteristics by opening the pores of the material. The acid treatment of the partially carbonized material improved the surface area from 1.510 to3.140 m^2/g. Meanwhile, the average pore size diameter decrease from 139.767 to 66.3547 nm. The larger pore size diameter would make the reactant molecules easily to access to the internal catalyst. The total acid density of the carbon solid acid catalyst was measured using a back titration method. The SO_3H density of the catalyst was significantly increasing after sulfonation. The catalyst exhibits high catalytic acid density.

Table 1. Physical properties of RHA and RHA catalyst.

Materials	**RHA**	**RHA/catalyst**
BET specific surface area (m^2/g)	1.510	3.140
Total pore volume (cm^3/g)	0.1055	0.1042
Average pore diameter (nm)	139.767	66.3547
Acid density (mmol/g)	0.05	2.24

The presence of amorphous silica in RHA and RHA catalystwere confirmed fromthe X-ray diffraction patterns shown in Figure 1. The RHA spectrum showed silica in the form of cristobalite (SiO_2 polymorph formed during RHA combustion process) and tridymite as the major crystalline phase. The SiO_2 (quartz) peaks could be observed for RHA catalysts.

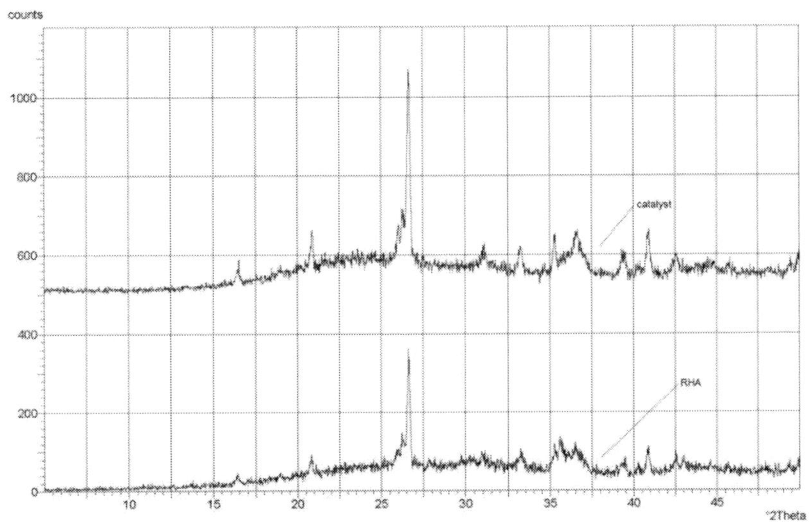

Figure 1. The XRD patterns of RHA and RHA catalyst.

In this study we have applied esterification to upgrade bio-oil from pyrolysis of biomass. Figure 2 compares acid values of raw bio-oil and the upgraded bio-oil. Figure1 indicates that after pretreatment of raw bio-oil with esterification reaction the acid values decreased from 89.54 of raw bio-oil to 56.82 mg KOH/g. The lowacid values of the upgraded bio-oil indicated the conversion of carbonyl compounds (mainly aldehydes and ketones) and alcohols to carboxylic acids.

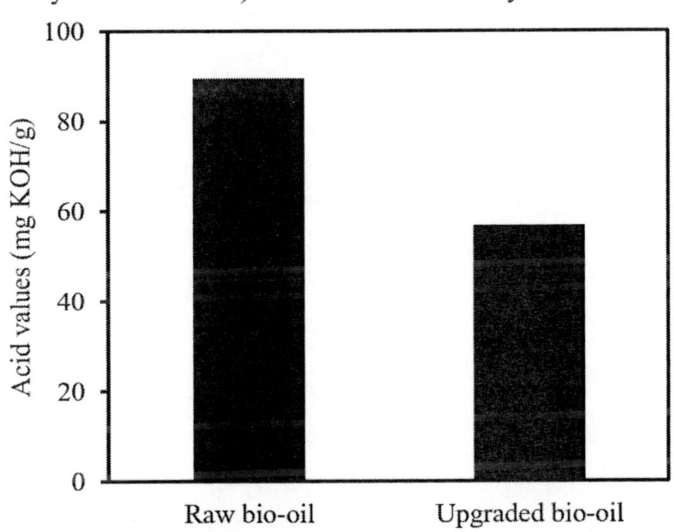

Figure 2. Acid values of raw bio-oil and the upgraded bio-oil.

Table 2 shows a comparison of physical and chemical properties of raw bio-oil and upgraded bio-oil after esterification reaction. As shown in Table 1, after pretreatment the acid value of upgraded bio-oil decrease to 89.54 mg KOH/g from 56.82 mg KOH/g the value of raw bio-oil. The decreasing of acid value indicates the oxidation of aldehydes/ketones and alcohols to acids. The viscosity decreased from 12.0 to 9.2 cSt, water content decrease from 31.4 to 4.7%. The HHV of the oxidized product increased from 16.0 to 27.2 MJ/kg. Density decreased from 1.2 to 0.95 g/ml and pH was reduced to 2.3 from 4.1. Oxygen content increased somewhat from 53.9 to 55.3 wt.%. This oxygen content was resulted from both increased acid and water content.

Table 2. Comparison of physical and chemical properties of raw bio-oil and upgraded bio-oil

Properties	Raw bio-oil	Upgraded bio-oil
HHV (MJ/Kg)	16	27.2
Acid Value (mg KOH/g)	89.54	56.82
Water content (%)	31.4	4.7
Density (g/ml)	1.24	0.95
Kinematic viscosity, 40 °C (cSt)	12	9.2
pH	2.3	4.4
Elemental analysis (%)		
C	39.1	64.7
H	7.7	9.8
N	0.3	0.4
O	52.8	25.1

Carbon content of upgraded bio-oil nearly doubled from 39.1 to 64.7 that of raw bio-oil. Hydrogen content was higher for upgraded bio-oil at 9.8% compared to the 7.7% of raw bio-oil. Nitrogen increased from 0.3% for raw bio-oil to 0.4% for upgraded bio-oil. Upgraded bio-oil oxygen content was reduced by the esterification of raw bio-oil with a decrease from 52.8 to 25.1%. Therefore, the esterification of raw bio-oil to produce upgraded bio-oil substantially increased the quality of its most important characteristics (HHV, acid value, water, hydrogen and oxygen content, and pH value); density and viscosity remain essentially the same. The only negative change was the small increase in nitrogen content.

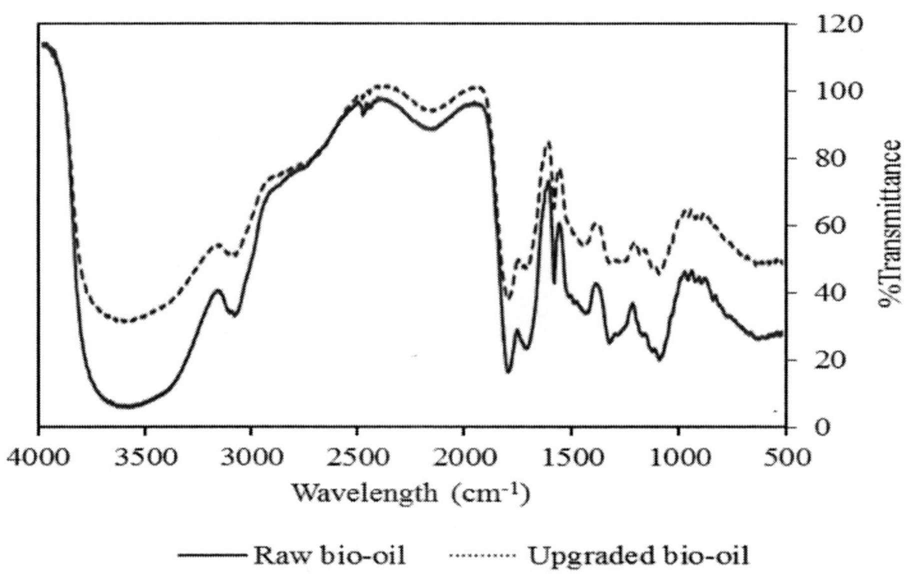

Figure 3. FTIR spectrum comparison of raw bio-oil and upgraded bio-oil.

Figure 3 compares the FTIR spectra of raw bio-oil and upgraded bio-oil. Characteristic vibrational modes are observed at 1650-1710 cm^{-1} (C=O stretching) and 3600-3000 cm^{-1} (-OH stretching). It was observed that C=O stretching was decreased and OH stretching was intensified, indicating that carbonyl compounds (mainly aldehydes and ketones) were converted to carboxylic acids. The bands at

1100-1300 cm^{-1} was corresponding to C-O stretching. An adsorption peak at 1710 cm^{-1} which attributed to C=O stretching. The band around 2850-2980 cm^{-1} is due to aliphatic CH stretching. The bands at 3600-3000 cm^{-1} can be assigned to the stretching vibration of OH stretching. From Figure 3, it was evident that after esterification reaction C=O stretching peak was decreased and OH stretching was intensified. The decreasing of C=O stretching peak and increasing of OH stretching indicate increase in carboxylic acids after esterification reaction. Furthermore, on esterification the intensity of CH stretching band dramatically increased and OH absorption band is significantly decreased due to conversion of carboxylic acids to esters and decrease in water content. As shown in Figure 2, it is clear that the decreasing of OH stretching peak of upgraded bio-oil is significantly higher than raw bio-oil; which implies that conversion of acids to esters and separation of water after esterification of upgraded bio-oil was improved compare to the esterification of raw bio-oil.

Table 3. Raw bio-oil and upgraded bio-oil chemical composition analysis by GC-MS with area percentages.

Raw bio-oil		**Upgraded bio-oil**	
Compound name	**Area (%)**	**Compound name**	**Area (%)**
Ethanone,1-(4-hydroxy-3-methoxyphenol)	19.55	1-Butanol	34.12
Phenol,2-methoxy-4-methyl-	10.81	Acetic acid,butyl ester	8.99
Acetic acid	5.1	Propanoicacid,butyl ester	3.04
Phenol,2-methoxy-	4.15	Butanoicacid,octyl ester	2.69
2-Propanone,1-hydroxy	3.63	Phenol,4-methyl	2.59
Phenol,2-methoxy-3-(2-propenyl)-	3.28	Cyclopentanone,2-methyl	2.51
Phenol,2-methoxy-4-propyl-	3.16	Phenol,5-methoxy-2,3-dimethyl	2.06
Phenol,4-ethyl-2-methoxy-	2.61	Benzaldehyde,3,4-dihydroxy	1.99
3,4-Anhydro-d-galactosan	2.1	Hexanoicacid,butyl ester	1.96
Benzeneaceticacid,alpha-hydroxy-	1.96	Pentanoicacid,butyl ester	1.82
Furfural	1.94	3-Undecene,6-methyl-	1.47
Acetic acid,fluoro-,ethyl ester	1.93	Oxalic acid,isobutylnonyl ester	1.36
Hexanoic acid,1-methylhexyl ester	1.85	Cyclohexanone,2-butyl	1.08
2-Cyclopenten-1-one,2-hydroxy-3-methyl	1.84	Ethanol,2-(4-(1,1-dimethylpropyl)))	1.07
Acetic acid,1-methylethyl ester	1.82	Phenol,2-ethyl-6-methyl	1.04
2-Furancarboxaldehyde,5-methyl-	1.78	3-Methyl cyclopentanone	0.89
Eugenol	1.78	2-Heptanone	0.88
Phenol,2-methoxy-4-(1-propenyl)-	1.65	Cyclohexanone,3-dimethyl	0.87
Heptanoic acid	1.42	Cyclopropane,trimethyl methylene	0.86
2-Cyclopenten-1-one,2-hydroxyl-	1.42	Cyclopentane,1,2,3,3,5-pentamethyl	0.86
Vanillin	1.39	Phenol,4-propyl	0.84
1,2-Benzenediol,4-methyl-	1.31	4-(N-Dimethylamino) phenol,acetate	0.83
4-Hydroxy-2-methylacetophenone	1.27	Cyclohexanone,2,3-dimethyl	0.83
3-Methoxy-4-methylaniline	1.25		
Glutaraldehyde	1.16		
Phenol,2-methyl-	1.14		

Table 3 shows the chemical composition of raw bio-oil and upgraded bio-oil analyzed by GC-MS. Approximately 50 chemical compounds were identified by GC-MS in both samples. As shown in this section the major chemical compounds present in the upgraded bio-oil are esters, ethers, ketones, phenols and alcohols

4. Conclusions

From the experiment results, it can be concluded that the densities of upgraded bio-oil were reduced from 1.24 to 0.95 and g cm^{-3}, and the higherheating value increased from 16.0 to 27.2 MJ/kg and the acidity of upgraded bio-oil was also alleviated from 2.3 to 4.4. The results of gas chromatography-mass spectrometry (GC–MS) and FT-IR analysis showed that the ester compounds in the upgraded bio-oil increased. It is possible to improve the properties of bio-oil by esterifying the raw bio-oil.

Acknowledgement

The author would like thank to Direktorat Riset dan Pengabdian Masyarakat, Direktorat Jenderal Penguatan Riset dan Pengembangan, Kementerian Riset, Teknologi, dan Pendidikan Tinggifor financial support on this research.

References

[1] Bridgwater A V, Meier D, Radlein D 1999 *Org.Geochem.* **30**(12) 1479–1493
[2] Fatimah I, Hidayat A, Setiawan K H 2010 *Asian J. Chem.* **22** (5) 3793
[3] Hidayat A, Rochmadi, Wijaya K, Nurdiawati A, Kurniawan W, Hinode H, Yoshikawa K and Budiman A 2015 *Energy Procedia* **75** 969
[4] Hidayat A, Rochmadi, Wijaya K, Hinode H, Budiman A 2013 *Asian J. Chem.* **25** (3) 1569-1579
[5] Bridgwater A V, Peacocke G V C 2000 *Sustainable Renewable Energy Rev* **4** (1)1–73
[6] Qin T F 2006 *Biomass Chemical Engineering* **40(suppl)** s78–s85
[7] Czernik S, Bridgwater A V 2004 *Energy Fuels* **18**(2) 590–598
[8] Xu J M, Jiang J C, Lu Y J 2007 *Modern Chemical Industry* **27**(7) 13–17
[9] Zhang Q, Chang J, Wang T J, Wu C Z, Xu Y, Zhu X F 2006 *Journal of Fuel Chemistry and Technology* **34**(6) 680–683
[10] Branca C, Giudicianni P, Diblasi C 2003 *Ind. Eng. Chem. Res.* **42**(14) 3190–3202
[11] Mohan D, Pittman C U, Steele P H 2006 *Energ. Fuels* **20**(3) 848–889
[12] Oasmaa A, Kuoppal A, Solantausta Y 2003 *Energ.Fuels* **17** (2) 433–443
[13] Doshi V A, Vuthaluru H B and Bastow T 2005 *Fuel Process. Technol.* **86**(8) 885
[14] Junming X, Jianchun J, Yunjuan S, Yanju L 2008 *Biomass Bioenerg.* **32**(11) 1056
[15] Mahfud F H, Melian-Cabrera I, Manurung R, Heeres H J 2007 *Process Saf. Environ. Prot.* **85**(B5) 466
[16] ZhangQ., ChangJ, Wang Tand XuY(2006)*Energ. Fuels***20**(6)2717.

Vacuum infusion equipment design and the influence of reinforcement layers addition to the resin infusion time

A H Saputra[1,a] *and **G Setyarso**[2,b]
[1,2] Chemical Engineering Department, Universitas Indonesia, Indonesia

E-mail: [a]sasep@che.ui.ac.id, [b]gifa_setya@ymail.com

Abstract. The characteristic of composite material is greatly influenced by the manufacture method of composite. The conventional method that has been used such as hand lay-up and spray up are simple and easy to apply but the composite tend to have a void in it because of the air trapped during the manufacture process. Vacuum infusion is one of the modern composite manufacture process which can replace the conventional method. The problem of this method happens when the resin infusion time become longer due to the addition of reinforcement layers. When the resin infusion time is longer than the resin's gel time, the resin will become gel and not able to flow into the mold. In order to overcome this problem, a study that observe the influence of reinforcement layers addition to the resin infusion time is needed. In this study, vacuum infusion equipment for composite materials manufacturing process that are designed consists of: 1x1m glass as the mold, 1L PVC tube for the resin container, 1L glass tube for the resin trap, and ½ HP vacuum pump with 7 CFM vacuum speed. The resin that is used in this study is unsaturated polyester resin (UPR) and the fiber used as reinforcement is fiber glass. It is observed that the more number of reinforcement layers the longer resin infusion time will be. The resin infusion time (in seconds) from two until six layers respectively for the area of 15x20cm are: 88, 115, 145, 174, 196; for the area of 15x25cm are: 119, 142, 168, 198, 235; and for the area of 15x35cm are: 181, 203, 235, 263, 303. The maximum reinforcement layers that can be accommodated for each 15x20cm, 15x25cm, and 15x35cm area are respectively 31 layers, 29 layers, and 25 layers.

1. Introduction

Composite material is a material that consist of two or more material that has a different shape and chemical composition from its origin material [1]. Composite material is a material consists of two or more materials that has a different characteristics physically and chemically which form a material with new characteristic [2]. Generally, composite material consist of two main components which are matrix and reinforcement. The manufacture process of the composite plays an important role to the characteristic of the composite. Hand lay-up and spray-up method are the examples of composite manufacture method that has been widely used. These two methods are simple, cheap, and doesn't need many equipment to be applied. On the other hand, this method has some limitation in which it really depends on the ability of the operator. Some of the time, the quality of the repetition products don't have the same quality. Furthermore, this method also have a big possibility for the composite to have a trapped air in it. This condition will lead to the decrease of quality to the composite.

Vacuum infusion is one of the modern composite manufacture method that can be the solution of the problems that exist in hand lay-up and spray-up method. By using the vacuum infusion method, it minimize the possibility of the air trapped in the composite. This method also allows the composite to be made in repetition with the same quality because it depends on the equipment design, in spite of depend on the operator's ability. Vacuum infusion is performed by put in the reinforcement material into the mold before the resin is infused to the mold with the help of vacuum pump. Vacuum infusion method can distribute resin resulting well which resulting in the composite with low void content and increase the mechanical properties of composite [3].

Beside the manufacture method, the manufacture duration is also important to be noticed because resin has gel time that needs to be considered. The manufacture duration should not pass the gel time because when the gel time passed, the resin will change form into gel. This condition can be a problem because the resin gel cannot be infused into the mold and will interrupt the manufacture process. Based on this condition, a study that relates the influence of reinforcement layers on the resin infusion time in vacuum infusion equipment is needed. The vacuum infusion equipment that is used in this study is designed and used to observe the resin infusion time. The variations that used are the mold area and the number of reinforcement layers. The resin used in this study is unsaturated polyester resin and fiberglass is used as the reinforcement. In addition, the maximum number of reinforcement that can be accommodate by the mold in the designed vacuum infusion equipment is also observed in this study.

2. Experimental

2.1 Materials

The resin used in this study is unsaturated polyester resin with EVERPOL 324 AR-2 brand as specified in Table 1. The fiber used as reinforcement in this study is fiberglass that is cut into layers with area of 15x20cm, 15x25cm, and 15x35cm.

Table 1. Specification sheet EVERPOL 324 AR-2

Product Name	**EVERPOL 324 AR-2**	
Type	Unsaturated Polyester Resin (Ortho Type)	
	SPECIFICATION	
ITEMS	ACTUAL	STANDARD
N.V. (%)	56.5	Check
Viscosity	540 cps	400-600
Acid Value (Solution)	24.9	15-25
Color	Pink	Pink
Transparency	Turbid	Turbid
Gel Time	21'04"	20'-25'
TI	2.5	1.5-2.0

2.2 Vacuum infusion process

The principle of vacuum infusion is using the negative pressure to suck the air inside the mold and at the same time the resin is also sucked into the mold so that the reinforcement is impregnated in the resin. After the layer of resin and matrix are formed, the remaining resin will be sucked out of the mold. As a result, the layer with same thickness can be obtained by using this process. The material manufactured by using vacuum infusion process has a higher fiber to resin ratio than the hand lay-up and spray-up method resulting in the composite material that is stronger and lighter. The flow of resin going through the reinforcement can be determined by using Darcy Law that explain the flow of fluid going through porous medium. Darcy Law can be seen in Eq. 1.

$$Q = \frac{KA\Delta P}{\mu L} \quad (1)$$

Where Q = resin flow (m^3/s), K = material permeability (m^2), A = area where resin flows into (m^2), ΔP = pressure difference in lamination ($kg/m.s^2$), μ = dynamic viscosity of resin (kg/m.s), and L = distance that needed to be traveled by resin (m). Eq. 1 will be used further in this study to analyze the effect of the reinforcement layer addition to the resin infusion time.

2.3 Vacuum infusion equipment design

The first step of the equipment design is deciding the size of desired composite material. In this laboratory scale study, the composite is made in form of rectangle layers with the size of 15x20cm, 15x25cm, and 15x35cm. Vacuum infusion equipment designed for this study consist of: 1x1m glass as the mold, 1L PVC tube for the resin container, 1L glass tube for the resin trap, and ½ HP vacuum pump with 7 CFM vacuum speed. These equipment are then assembled like as seen in Figure 1.

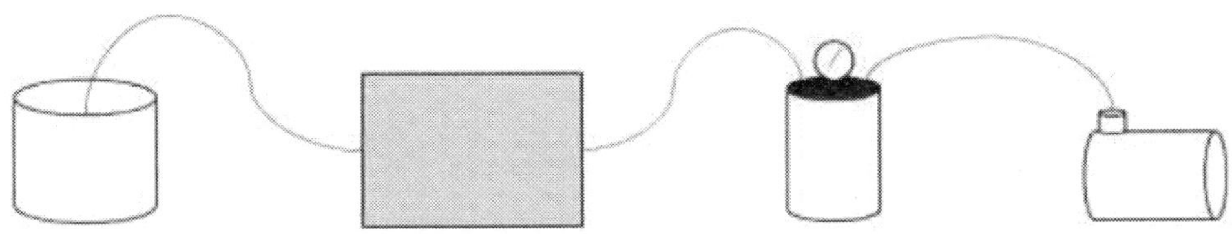

Figure 1. Vacuum infusion equipment

This vacuum infusion equipment is then used to discover the effect of reinforcement layer addition to the resin infusion time. Beside the different area of mold, which are 15x20cm, 15x25cm, and 15x35cm, the number of reinforcement layer are also varied. The number of reinforcement used in this study are 2 until 6 layers. The data obtained from this study is the resin infusion time that is measured by using the stopwatch which is turned on when the vacuum pump starts to operate and resin inlet is opened and is turned off when the resin wet the reinforcement completely.

3. Results and Discussion

3.1 The effect of reinforcement layer addition to the resin infusion time

The infusion time data obtained from this study can be seen in Table 2. Table 2 shows that the more number of reinforcement resulting in the longer infusion time. The more number of reinforcement provide a longer distance for the resin to wet the reinforcement. According to Darcy Law from Eq. 1, when the distance needed to be traveled (L) by the resin is further, the flow rate (Q) of resin becomes lower. As a result, the time needed for resin to wet the reinforcement become longer.

From the view of fluid mechanic, as seen in Figure 2a, resin flows from resin inlet through the gap in spiral tubing to the flow media and the lamination beneath. Resin flows horizontally to the other lamination and resin outlet while flows vertically to the lamination under the flow media. Resin that has reached the spiral tubing on the other side is then flows through the gaps in the spiral tubing and flows out to the resin outlet. As seen in Figure 2b, there are a position difference between the resin in the top side and the bottom side. Fluid or in this case resin, tend to flow through the media that has a high permeability. As a result, on the top side which is closest to the flow media, the resin wet the reinforcement faster than the one in the bottom side. After flowing through flow media, resin is then flows down through the pores in reinforcement layer like fluid that flows into capillary tube. This condition resulting in the longer infusion time with the addition of the reinforcement layer because the distance that is needed to be traveled by the resin to wet the reinforcement become longer.

Table 2. Resin infusion time for different reinforcement area and number of reinforcement layer

Number of Reinforcement Layer	Resin Infusion Time (s)		
	Reinforcement Area	Reinforcement Area	Reinforcement Area
	15 x 20 cm	15 x 25 cm	15 x 35 cm
2	88	119	181
3	115	142	203
4	145	168	235
5	174	198	263
6	196	235	303

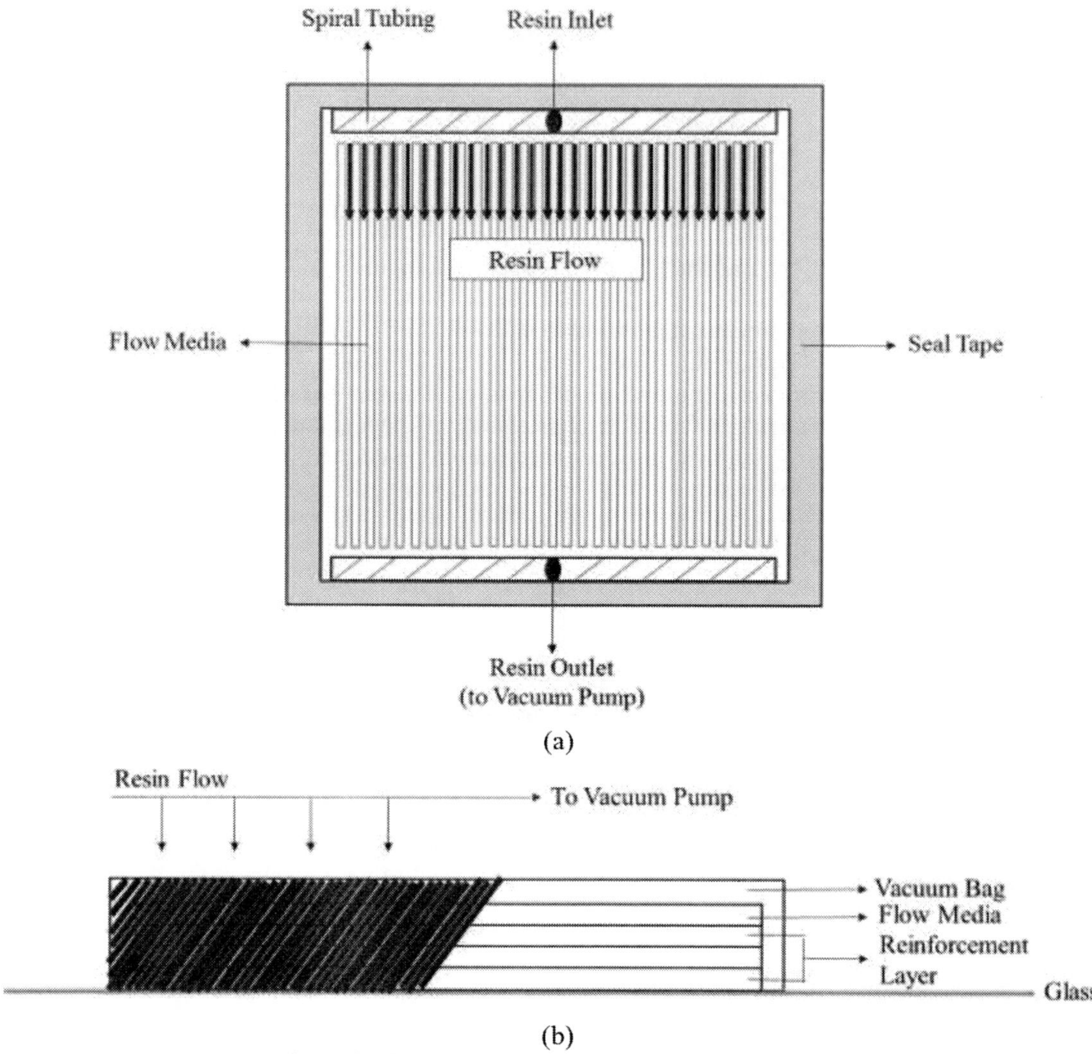

Figure 2. The resin fluid mechanic from top view (a) and side view (b)

3.2 The calculation of maximum reinforcement layer

Beside the effect of reinforcement layer addition to the resin infusion time, the number of reinforcement layer that can be accommodate by the mold is also needed to be observed for further use of the vacuum infusion equipment. By using the vacuum infusion equipment that has been designed, the infusion time data for different number of reinforcement layer and area are obtained and can be seen in Table 2. The specification of the resin used in this study, including the gel time, can be seen in Table 1. The calculation of maximum reinforcement layer is performed by using the linear equation of the extrapolated data from Table 2. The extrapolated graph for the different reinforcement area can be seen in Figure 2. By using the linear regression, the linear equation of infusion time for the reinforcement area of 15x20cm, 15x25cm, and 15x35cm respectively are $y = 27.5x + 33.6$; $y = 28.8x + 57.2$; and $y = 30.4x + 115.4$ with y as the infusion time and x as the number of reinforcement layers.

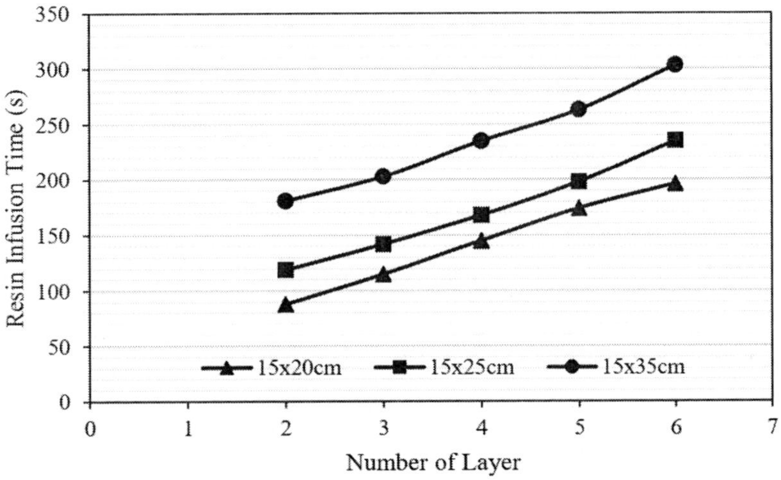

Figure 3. The resin infusion time for different number of layer and different area of composite

From Table 1, it can be seen that the gel time of EVERPOL 324 AR-2 resin is 21 minutes 4 seconds so the maximum time for extrapolation is set to 15 minutes with the assumptions that the rest 6 minutes is used for mixing and degasing process of the resin before infused into the mold. The number of maximum reinforcement layer can be obtained by substituting 15 minutes or 900 seconds as y parameter for each linear equation that has been obtained before. It is then observed that the maximum number of reinforcement layers that can be accommodated by the designed equipment are 31 layers for the reinforcement area of 15x20cm, 29 layers for the reinforcement area of 15x25cm, and 25 layers for the reinforcement area of 15x35cm. This condition is because for the wider reinforcement area, the resin infusion time as seen in Table 2 is longer due to the longer distance needed to be traveled by resin.

4. Conclusion

Vacuum infusion equipment was designed and used to discover the influence of reinforcement layers addition to the resin infusion time. Unsaturated polyester resin was used as the reinforcement. The designed equipment consists of: 1x1m glass as the mold, 1L PVC tube for the resin container, 1L glass tube for the resin trap, and ½ HP vacuum pump with 7 CFM vacuum speed. The wider area and the more number of reinforcement layers resulting in the longer infusion time. Overall, vacuum infusion equipment that has been designed in this study can be used to make composite material with unsaturated polyester resin as the reinforcement with the maximum reinforcement layers that can be accommodated for each 15x20cm, 15x25cm, and 15x35cm area are respectively 31 layers, 29 layers, and 25 layers.

5. References

[1] Smith W F 1996 *Principles of Materials Science and Engineering: 2nd Edition* (Singapore : McGraw-Hill)

[2] Cahya E 2004 *Materi Pelatihan Komposit (Composite)* (Bogor : PT. Arindo Pacific Chemicals)

[3] Callister Jr, William D 2002 *Fundamentals of Material Science and Engineering : An Introduction* (New Jersey : Wiley)

[4] Thomas S, Joseph K, Malhotra S K, Goda K and Sreekala M S 2012 Advances in polymer composites: Macro- and microcomposites – State of the art, new challenges, and opportunities *J. Polym Composite* **1** 1-16

[5] Wisojodharmo L A, Roseno S 2012 The use of vacuum assisted resin infusion process on the manufacturing of wind blade composites *J. Materials and Science Engineering* **A2** 72-76
[6] Refiadi G, Koswara A L 2005 Pembuatan material komposit polimer dengan metoda Vacuum Assisted Resin Infusion (VARI) *Jurnal Berita Teknologi Bahan dan Barang Teknik,* **19** 40-45
[7] Goren A, Atas C 2008 Manufacturing of polymer matrix composites using vacuum assisted resin infusion molding *Archives of Materials Science and Engineering Journal* **24** 117-120
[8] Schuster J, Govignon Q, Bickerton S 2014 Processability of biobased thermoset resins and flax fibres reinforcements using vacuum Assisted resin transfer moulding *Journal of Composite Materials* **4** 1-11
[9] Hull D 1981 *An Introduction to Composite Materials* (Cambridge: Cambridge University Press)
[10] Ulrich, Kainer K 2006 *Metal Matrix Composites: Custom-made Materials dor Automotive and Aerospace Engineering* (Weinheim: Wiley)
[11] McCathie D 2012 *Darcy's Law and its Relevance to Infusion* (Silverdale:Vacmobiles)
[12] Dholakiya B 2012 Unsaturated polyester resin for specialty applications *Intechopen* **10** 167-201
[13] Wang R M, Zheng S R, Zheng Y P 2011*Polymer Matrix Composites and Technology* (Cambridge: Woodhead Publishing)
[14] Grimsley B W, Hubert P, Song X, Cano R J, Loos A C and Pipes R B 2001 *Flow and Compaction During The Vacuum Assisted Resin Transfer Molding Process* (Virginia : NASA Langley Research Center)

IOP Conf. Series: Materials Science and Engineering **162** (2017) 012016 doi:10.1088/1757-899X/162/1/012016

Review of neural network modelling of cracking process

M N Rosli and N Aziz*
School of Chemical Engineering, Engineering Campus, Universiti Sains Malaysia, Seri Ampangan, 14300 Nibong Tebal, Seberang Perai Selatan, Penang, Malaysia

E-mail : chnaziz@usm.my

Abstract. Cracking process is a very important process that converts low value products into high value products such as conversion of naphtha into ethylene and propylene. The process is nonlinear with extensive reaction network. Thus, nonlinear technique such as artificial neural network is explored to develop the model of the system. The paper will review and discuss the research works done on the technique in modelling cracking process using artificial neural network starting from early 1990s until recent development in 2015. Timeline is provided to show progression of work done throughout the years, the main issues addressed, and the proposed techniques for each. In the next section, the main objective of each work and each techniques explored by previous researchers is discussed in more detail. A table that summarizes previous works is provided to show common works done throughout the years. Lastly, potential gap for future works in the area is highlighted.

1. Introduction

Cracking is a very important process as it has the ability to convert low value heavies material such as heavy vacuum gas oil and atmospheric gas oil into high value ethylene and propylene. There are three prominent methods for hydrocarbon cracking i.e. steam cracking, thermal cracking, and catalytic cracking. Cracking process is nonlinear and semi-continuous, contributed by its extensive reaction networks, continuous deposition of by-product, and constantly changing operating condition. Throughout the cracking period, coke will form as by-product and deposit in the coil which lowering the heat transfer coefficient of the coil, and composition of the feed will also start to change as the fresh feed will be mixed with recycled feed. Due to that, operating condition at the beginning of the process will continue to change until coke deposit layer has become so thick that the heat input required will cause the coil metal to reach maximum tube metal temperature. By that time, operator will stop the cracking process and initiate decoking process to remove the coke.

Looking at the nonlinear and continuously changing nature of the cracking process, researchers and industrial practitioners in the area of system identification have started to explore non-linear identification technique such as Artificial Neural Network (ANN), which are known to be able to handle nonlinearities. They also incorporate multimode model to develop inferential model that can best represent the process under different operating condition. This paper will review the previous research work on modelling of cracking process using ANN. Finally, from all the works reviewed, the research gaps will be highlighted for future research.

2. Progression of ANN modelling of cracking process

Modelling cracking process initially starts with developing a first principle model (FPM) from mass, momentum, and energy balances couple with cracking reaction mechanism. The attempt to develop FPM proved to be very extensive plus the amount of assumptions that need to be made to simplify the reaction network and determining many other unknowns [1]. Since then, empirical techniques such as ANN have been explored instead of FPM.

2.1. Overview of ANN research progress

ANN is one of the most attractive nonlinear techniques used by researchers for modelling the cracking process. In 1994, researchers start to explore ANN to model cracking process. The works starts with Feng et al. which used Multilayer Perceptron (MLP) with functional input to model thermal cracking furnace [2] and McGreavy et al. which used conventional MLP to model Fluid Catalytic Cracking (FCC) reactor [3] . Since then, the efforts to improve the method continue until today. As shown in Figure 1, earlier effort focused on utilizing single hidden layer feedforward network. From year 2000 onwards, contributed by more research in the area, researchers started to take notice of the limitations of ANN and focused their work on improving the techniques.

In early 2000, two important aspects of modelling have been explored – online implementation and model validity under different condition. Realizing the importance of maintaining the ANN model during actual plant operation, Nagy et al. has introduced adaptive ANN to enable online training of the model [4]. Another researcher take notice of changing operating conditions of cracking at the feed, after mixing with recycle feed, and at furnace run length that is caused by deposition of coke. In the work of Zhuang and Yu, a method using dynamic biases has been proposed to cater for the changing operating condition [5]. Normally during actual plant operation, plant dynamics consistently change due to many factors and thus, the training dataset used during offline training might not cover all operating conditions. This type of data is considered as unseen data. Bootstrap Aggregated Neural Network (BANN) has been introduced to enhance the generalization ability of the network to work with unseen data [6,7].

During the development of the model, many researchers work on improving the prediction errors of the model as well as improving the computation time. Huge amount of training data leads to long computation time and it can be difficult for online training. To reduce computation time and training load, an effort to reduce dimensionality of the data has been done by integrating Principle Component Analysis (PCA) into the network input as demonstrated in [8,9]. Starting from year 2006, researchers have started to combine ANN with other methods creating hybrid such as combination of FPM with ANN [10], incorporate Genetic Algorithm (GA) into ANN to optimize network parameters, which is known as GA-NN technique [10], and integrate Fuzzy Inference System (FIS) into the ANN layers and called as ANN-FIS model [1]. Another research work on this area is using the conventional Feedforward MLP and also Radial Basis Function NN for the modelling. The improvement proposed is more towards optimizing the cracking furnace operation parameters by using evolutionary methods such as Cultural Algorithm (CA), Differential Evolution algorithm (DEA) [11], Particle Swarm Optimization (PSO) [12,13] and GA [14,15] to vary input parameters in order to maximize output.

Within the span of 20 years, the research works have varied from exploring different structures and network configurations to improve prediction error until incorporating evolutionary methods to optimize cracking furnace operation.

2.2. Summary of research works from 1994 until 2015.

In this section, summary is provided for each paper that is related to the application of ANN in modelling cracking process, covering thermal cracking, steam cracking, and catalytic cracking. In the year 1994, Feng et al. [2] has ventured into using ANN to model thermal cracking of naptha, with addition of using functional units at input layer. A model has been developed to predict the yield of ethylene, the yield of propylene, and Kinetic severity function (KSF). McGreavy et al. has explored the use of ANN in modelling FCC to predict yield distribution of main products of light hydrocarbons and byproducts such as coke [3]. These two are the first attempt to model cracking process using ANN.

Elkamel et al. has explored the use of feed forward network to model hydrocracking unit for light vacuum gas oil [16]. A multiple inputs multiple outputs model has been used, with different model has been developed for different modes of operation, which are the Aviation Turbine Kerosene (ATK) mode and middle distillate mode. Result shows that different output has different errors produced.

Nagy et al has developed two ANN models to be used with Model Predictive Control (MPC) for FCC unit [4]. One is offline-trained and the other is adaptive. Adaptive ANN is trained online using past values of manipulated and controlled variables, to cater for unmeasured process disturbances. Both models shows superior performance compared with PI controller, and adaptive is showing better result than offline ANN. GA method is proposed to improved convergence of learning. Data used in the research is obtained from simulation through MATLAB FCCU package.

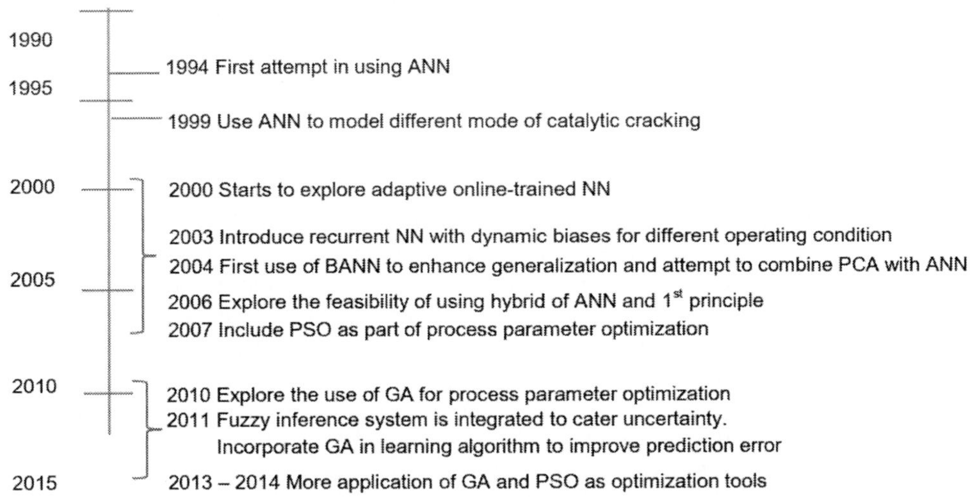

Figure 1. Evolution of ANN modelling of cracking process

Research work by Michalopoulos et al. focused on finding the optimum structure for good generalization abilities [17]. Through trial and error, the optimum structure is obtained. The MLP model of FCC to predict percent conversion is compared with other non-linear statistical model derived from previous research. It proved to shows better performance than the statistical model.

Through previous experience, it is acknowledged that cracking process exhibit changing process condition after the start of the process. Researchers start to focus on that aspect during modelling. The first attempt has been done Zhuang and Yu to explore the use of recurrent neural network and introduced dynamic biases to cater for different feed conditions and running phases of the steam cracking furnace of ethane-naphtha mixture [5]. Comparison has been made between Back-propagation (BP) and Elman network. The result shows that the use of dynamic biases significantly improves the accuracy of the model and convergence time. The proposed structure achieved convergence in lower epoch value than BP network. Another similar research by Zhu and Li model the cracking furnace using two-layer network, with the first layer function is to determine the operating condition and the second layer is a layer of BANN with sequential training specific for each operating condition [9]. Both methods avoid generalization and develop a specific method to identify the operating condition and introduce a solution for each condition.

Another research group is working on reducing computation time of network learning. In order to do that, the input data need to be reduced. Zhuang and Yu [8] have proposed a PCA in combination with dynamic neural network to model naphtha cracking furnace and predict the cracking severity. The procedure to conduct the PCA has been presented clearly in the paper for reference. Output of PCA becomes an input to the Elman network. Combination of PCA with Elman network shows better prediction than BP network and Elman network. Zhu and Li have combined PCA with Input Training Neural Network (ITNN) to develop a nonlinear PCA [9]. The output from the PCA-ITNN is input into the Radial Basis Neural Network (RBNN) to produce a model for naphtha cracking furnace to predict ethylene and propylene yield. The integration of PCA not only increases learning speed, it also reduces prediction error primarily due to the noise removal and dimensionality reduction by PCA.

The research by Zhou et al. has introduced a new method of ANN. Two ANN models of cracking furnace is developed to predict product yields and cracking degree [6]. The first model is MLP trained by LM algorithm and the second one is using bootstrap aggregated with sequential training. However, the MLP is found to be lacking generalization abilities and prone to error when presented with unseen data. The proposed Boostrap Aggregated Network (BANN) enables several models of ANN to be developed with different objective. Then, the outputs of individual networks are combined to give overall output and capable of producing better performance than single MLP. Zhou et al. has also utilized BANN with additional of one layer to determine process operating condition prior to proceeding to BANN layer [7]. In the work, BANN model is developed for each operating condition.

To improve the performance of ANN and overcome its limitation, several researchers have proposed hybrid technique which is a combination with first principle model, genetic algorithm, or fuzzy logic. The first one to propose is Bhutani et al. whereby three methods has been explored to model industrial hydrocracking unit [18]. The method explored are First principle model (FPM), ANN model, and a combination of both (hybrid model). The three models have been compared and evaluated, the result shows that ANN model alone provide much more robust and less absolute percent error compared to hybrid and FPM. Wang and Xu in their research developed ANN model of cracking furnace and introduces GA in the network learning algorithm to utilize its stochastic properties and improve likelihood of finding error surface global minima [10]. To handle uncertainty in the process, Sedighi et al. has introduced Fuzzy inference system into the scope and coupled it with neural network [1]. The NN-FIS has been compared with regular ANN and First principle model. Evaluation between models has been done using Analysis of Variance (ANOVA) to see prediction capability of each. NN-FIS shows superior performance of all three. The downside of NN-FIS is that it is very specific to feed type and furnace type. New models need to be developed for new feed and reactor.

In the recent years, more researchers start to focus on developing method to search for optimum process parameters in order to maximize product yields. Thus, ANN combined with evolutionary methods such as PSO, GA, CA, and DEA comes into the scene. PSO starts to be explored by Li et al. where naphtha cracking furnace has been modelled using single hidden layer network, and Multi-objective PSO has been incorporated to find the optimum operating parameter of the furnace that will maximize yield of ethylene and propylene [11]. Several other researches also using PSO, such as Geng et al., has explored the combination of Multi-modes PSO (MMPSO) and Radial Basis Function Neural Network (RBFNN) to model cracking furnace [13]. The model will be integrated with advanced process control system of COT to optimize cracking depth and maximizing ethylene and propylene yields. MMPSO has been incorporated to further optimize the ANN model.

In their work, Pu and Liu have introduced a new type of optimization algorithm. The work focuses on developing a model for monitoring process state. Elman network has been used to model cracking furnace severity state, instead of the typical product yield. The model has been integrated with Cultural algorithm (CA) and Differential Evolution algorithm (DEA) focuses on the training. The network output will be used as tool to monitor cracking severity [12].

Liu et al., in their research, has focused on the aspect of cycle oil as recycle feed to FCC after considering that cracking behavior of cycle oil is largely unknown and it affect cracking severity [14]. RBFNN has been used to develop the model to product light oil yield, and GA and PSO has been used to find optimum operating parameters to maximum yield of light oil. Bispo et al. has modelled an FCC unit using Feedforward ANN and expected to integrate the inferential model in Model Predictive Control system [15]. GA and PSO method has been used in their work to obtain optimum operating condition.

Table 1. Summary of techniques employed in cracking process ANN modelling

Technique employed	Issues to be solved	Year	Reference
MLP	Modelling	1994, 1999, 2000, 2001, 2006, 2007, 2014	[2] [16] [4] [17] [18] [19] [20] [11] [21] [10] [15] [3]
Adaptive NN	Changing process condition online	2000	[4]
Elman NN	Modelling	2004, 2010	[8] [12]
Radial Basis Function NN	Modelling	2006, 2010, 2013	[9] [13] [22] [14]
Recurrent NN	Modelling	2003	[5]
Bootstrap Aggregated NN	Model generalization	2004, 2006	[6] [7]
Hybrid FPM-NN	Explore combination of method	2006	[18]
Hybrid GA-NN	Finding error global minima	2011	[10]
Hybrid NN-FIS	Uncertainty in the process	2013	[1]
Optimization with GA	Reactor optimum parameters	2010, 2011, 2013, 2014	[22] [14] [15]
Optimization with PSO	Reactor optimum parameters	2007, 2010, 2014	[11] [13] [14] [15]
Optimization with CA and DEA	Reactor optimum parameters	2010	[12]
Integration with PCA	Reduce data dimensionality and computation time	2004, 2006	[8] [9]

3. Future works

There are still many research gaps that can be addressed in future works. First, the gap is in the development of a complete severity control system package that consist of fault diagnosis system, data pre-processing, online ANN model training, post-processing, as well as implementing the model online in actual plant. Currently, no such work has been reported in journal. Secondly, it has been observed that different practitioners used different methods and different procedures to develop the ANN model. This has resulted in difficulties to compare the result of each work and also resulted in slow progression in the area. Thus, there is a need to establish standardized methodology and procedures that can be applied to build ANN model for different cracking process and different cracking feed. Lastly, throughout the years, it has been observed that very few works have been done to improve the learning algorithm. As highlighted in one of the works, one of the challenges of ANN is long computation time and the learning to stop at local minima in the error surface [10]. Thus, it proposed that more work needs to be done to develop a method to improve the learning speed as well as the likelihood of finding global minima.

References

[1] Sedighi M, Keyvanloo K, and Towfighi J 2011 Modeling of thermal cracking of heavy liquid hydrocarbon- application of kinetic modeling, artificial neural network, and neuro-fuzzy models *Ind. Eng. Chem. Res.* **50** 1536–1547

[2] Feng Q, Jinshou Y, and Weisun J 1994 Modelling of industrial thermal cracking furnaces via functional-link artificial neural network *Industrial Technology. Proceedings of the IEEE*

[3] McGreavy C, Lu M L, Wang X Z and Kam E K T 1994 Characterisation of the behaviour and product distribution in fluid catalytic cracking using neural networks, *Chemical Engineering Science* **49** (24 A) 4717-4724

[4] Nagy Z, Agachi S, Bodizs L, 2000, Adaptive neural network model based nonlinear predictive control of a fluid catalytic cracking unit European Symposium on Computer Aided Process Engineering *Elsevier Science B.V* **10**
[5] Zhuang X F, Yu J Z 2003 Modelling ethylene and propylene yield for cracking furnace based on a new kind of recurrent neural network. *The Fourth International Conference On Control And Automation (ICCA'03), 10-12 June 2003, Montreal, Canada*
[6] Zhou Q, Zhang J, Xu Y 2004 Predicting product yield profile and cracking degree in ethylene furnace *8th International Conference on Control, Automation, Robotics and Vision, Kunming, China. 6-9th December 2004*
[7] Zhou Q, Xiong Z, Zhang J, and Xu Y 2006 Hierarchical neural network based on product quality prediction of industrial ethylene process *Springer-Verlag Berlin Heidelberg*
[8] Zhuang X F and Yu J S 2004 A novel approach for modelling cracking furnace severity, *Proceedings of the 5th World Congress on Intelligent Control and Automation, June 15-19, 2004*
[9] Zhu Q X and Li C F 2006 Dimensionality reduction with input training neural network and its application in chemical process modelling *Chinese J. Chem. Eng.* **14**(5) 597-603.
[10] Wang J F, Xu H Y 2013 Ethylene plant genetic network modelling and optimization *IEEE*.
[11] Li C F, Zhu Q X and Geng Z Q 2007 Multi-objective particle swarm optimization hybrid algorithm with neural network: An application on industrial cracking furnace, *Ind. Eng. Chem. Res.* **46** 3602-3609
[12] Yunxia Pu Y X and Liu M D 2010 Early warning technology of cracking severity based on improved cultural differential algorithm, *Proceedings of the 8th World Congress on Intelligent Control and Automation, July 6-9 2010, Jinan, China*
[13] Geng Z Q, Shang T F, Li F 2010 Integrated MMPSO and RBFNN for optimal control of cracking depth in ethylene cracking furnace *2010 Sixth International Conference on Natural Computation (ICNC 2010)*
[14] Liu Y B, Tu Y S, Li C Y, Yang C. H. 2013 Catalytic cracking and PSO-RBF neural network model of FCC cycle oil *China P. E. T. Process P. E.* **15** (4) 63-69
[15] Bispo V D S, Silva E S R L, Meleiro L A C 2014 Modelling, Optimization, and control of a FCC unit using neural networks and evolutionary method, *Engevista*, **16** (1) 70-90
[16] Elkamel A, Al-Ajmi A and Fahim M 1999 Modelling hydrocracking process using artificial neural network *Petrol. Sci. Technol.* **17** (9&10) 931-954
[17] Michalopoulos J, Papadokonstadakis S, Arampatzis G and Lygeros A 2001 Modelling of an industrial fluid catalytic cracking unit using neural networks *Trans IChemE* **79** (A) March 2001
[18] Bhutani N, Rangaiah G P and Ray A K 2006 First principles, data-based, and hybrid modelling and optimization of an industrial hydrocracking unit *Ind. Eng. Chem. Res.*, **45** 7807-7816
[19] Eckert E, Belohlav Z, Vanek T, Zamostny P, Herink T, 2007 ANN modelling of pyrolysis utilizing the characterisation of atmospheric gas oil based on incomplete data *Chem. Eng. Sci.* **62** 5021-5025
[20] Nabavi R, Niaei A, Salari D, Towfighi J 2007 Modelling thermal cracking of LPG - Application of artificial neural network in prediction of main product yield *J. Anal. Appl. Pyrol.* **80** 175–181
[21] Niaei A, Towfighi J, Khataee A R, Rostamizadeh K. 2007 The use of ANN and mathematical model for prediction of the main product yield in the thermal cracking of naphtha *Petrol. Sci. Technol.* **25** 967-981
[22] Shang T F, Geng Z Q 2010 Online operation optimization based on GA-RBF network in ethylene cracking furnace 978-1-4244-7330-4/10 *IEEE*

IOP Conf. Series: Materials Science and Engineering **162** (2017) 012017 doi:10.1088/1757-899X/162/1/012017

Simulation of ethane steam cracking with severity evaluation

M N Rosli and N Aziz[a*]
School of Chemical Engineering, Engineering Campus, Universiti Sains Malaysia, Seri Ampangan, 14300 Nibong Tebal, Seberang Perai Selatan, Penang, Malaysia

E-mail : chnaziz@usm.my

Abstract. Understanding the influence of operating parameters towards cracking severity is paramount in ensuring optimum operation of an ethylene plant. However, changing the parameters in an actual plant for data collection can be dangerous. Thus, a simulation model for ethane steam cracking furnace is developed using ASPEN Plus for the assessment. The process performance is evaluated with cracking severity factors and main product yields. Three severity factors are used for evaluation due to their ease of measurement, which are methane yield (Ymet), Ethylene-Ethane Ratio (EER) and Propylene-Ethylene Ratio (PER). The result shows that cracking severity is primarily influenced by reactor temperature. Operating the furnace with coil outlet temperature ranging between 850°C to 950°C and steam-to-hydrocarbon ratio of 0.3 to 0.5 has led to optimum main product yield.

1. Introduction

Ethylene is one of the major petrochemical products in the world as it is a primary building block in the production of many other chemicals and products. Common feedstocks of steam cracking are ethane, ranges of naphtha, and other range of hydrocarbons. Steam cracking process took place in a furnace with tubes of plug flow reactor passing through the furnace firebox. Prior to entering the reactor, the hydrocarbon feed is preheated in the convection section of the furnace, and then mixed with superheated steam at a specific ratio. After that, the mixed feed enters the radiant section, which is the furnace firebox, where it is heated to a range of targeted reaction temperature by radiation and reaction rapidly take place. Cracked gas exiting the radiant section is cooled rapidly by Transfer Line Exchanger (TLE) to below reaction temperature in order to stop primary reactions and minimize side reactions. The cracked gas is fed to cracked gas compressor and delivered for further purification and recovery, including recovery of unreacted feed, methane, and C3+ product. The unreacted feed is typically recycled and mixed with fresh feed while methane and C3+ is typically used as fuel gas. From the description, it can be concluded that steam cracking is an energy intensive process. Figure 1 shows simple schematic of steam cracking furnace.

Ethane, being the simplest feedstock, requires the highest temperature (800°C – 900°C) to crack and thus, it utilizes huge amount of energy to operate. Other than temperature, critical process parameters that affect steam cracking performance are reactor pressure and steam-hydrocarbon ratio [1]. Operating the cracking furnace at optimum operating envelope is paramount in order to maintain optimum equipment efficiency, maximizing ethylene yield, while minimizing operating cost. Evaluation of cracking process also requires understanding of the cracking reaction kinetic. It is mainly accepted that cracking is most accurately represented by free-radical reaction mechanism [2]. Since free-radical reaction scheme consist of several hundreds of reactions, most researchers use molecular reaction scheme to simplify the computation in order to determine cracking reaction products [3, 4].

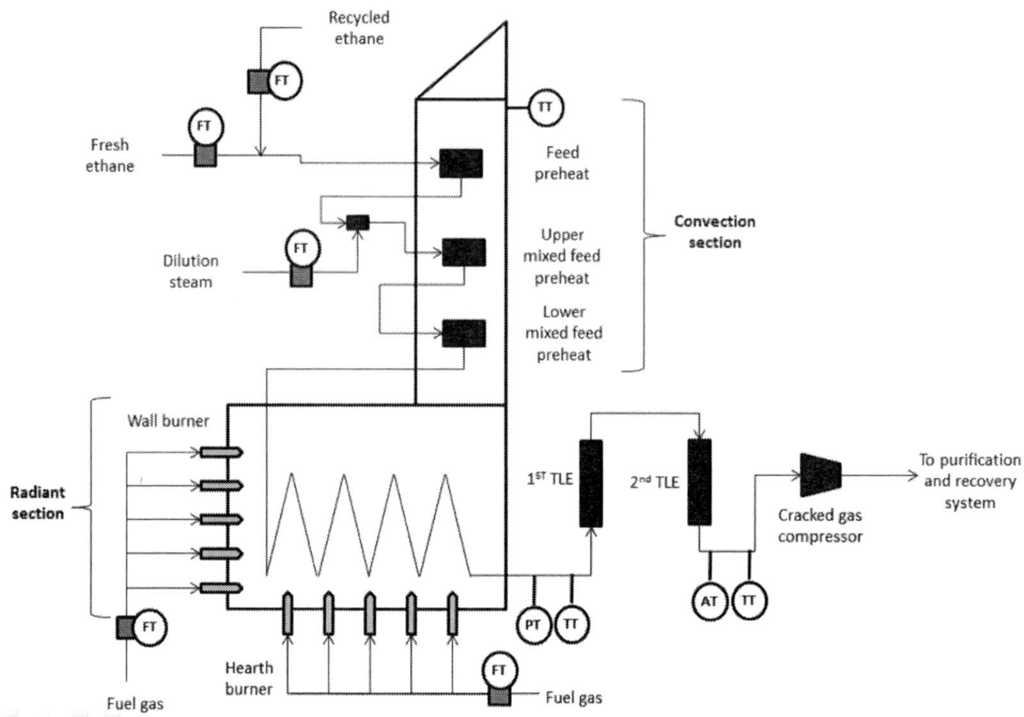

Figure 1. Schematic of ethane steam cracking furnace

Another important factor to consider is the cracking severity, which determines the extent of cracking reaction and cracking product distribution. Maintaining cracking severity at target is important to avoid over-cracking as it yields unwanted products (heavier hydrocarbons and coke formation) and to avoid under-cracking as it will not crack enough hydrocarbon feed to produce valuable product. To control the cracking severity, industry typically uses Coil Outlet Temperature (COT) as indicator. Maintaining the COT at a target temperature region that favour ethylene yield is correlated with cracking severity control. However, it has been reported that referring solely on COT does not accurately indicate the cracking severity [5, 6]. It needs to be used in tandem with other factors. Thus, the study incorporate methane yield (Ymet), Ethylene-Ethane Ratio (EER), and Propylene-Ethylene Ratio (PER) in the assessment of cracking severity.

2. Model Development

Having a process model will enable offline analysis of plant under various operating conditions and operating strategies in order to understand the impact of making parameter changes prior to implementation at actual plant. It is also useful as a tool to familiarize the plant personnel with the process behaviour during training. Using ASPEN Plus, the steam cracking furnace is modelled as plug flow reactor with ethane and superheated steam as feed streams. No recycle feed is included in the model to simplify process analysis. Using the reactor geometry and similar to the one used by P. Ranjan et al [7], the reactor has a length of 10.5 m with 48 tubes and an inner diameter of 0.085 m. Maximum reactor operating pressure is 2.2 barg and steam-to-hydrocarbon ratio between 0.3 – 0.5.

Using the molecular reaction scheme proposed by P. Ranjan et al. [7], the reactions and kinetic parameters are incorporated in the model, as per Table 1. Coke formation reactions are not included in the model.

Table 1. Ethane cracking reaction scheme

No	Reactions : Order	Order	Forward reaction A (s^{-1}) or (L $(mol.s)^{-1}$) E (kJ mol^{-1})		Reverse reaction A (L $(mol.s)^{-1}$) E (kJ mol^{-1})	
1	$C_2H_6 \leftrightarrow C_2H_4 + H_2$	1	A=4.6 x10^{13}	E=272.8	A=8.49 x10^{8}	E=136.5
2	$C_3H_6 \leftrightarrow C_2H_2 + CH_4$	1	A=7.2 x10^{12}	E=274.2	A=3.81 x10^{8}	E=147.2
3	$C_2H_2 + C_2H_4 \rightarrow C_4H_6$	2	A=1.0 x10^{15}	E=172.6		
4	$C_2H_4 + C_4H_6 \rightarrow C_6H_6 + 2H_2$	2	A=8.3 x10^{12}	E=144.6		
5	$C_3H_8 \leftrightarrow C_3H_6 + H_2$	1	A=5.8 x10^{10}	E=214.6	A=9.03 x10^{5}	E=93.5
6	$C_3H_8 + C_2H_4 \rightarrow C_2H_6 + C_3H_6$	2	A=2.5 x10^{16}	E=247.1		
7	$2C_3H_6 \rightarrow 3C_2H_4$	1	A=7.3 x10^{12}	E=268.5		
8	$2C_2H_6 \rightarrow C_3H_8 + CH_4$	1	A=3.8 x10^{11}	E=273.0		
9	$C_4H_{10} \leftrightarrow C_4H_8 + H_2$	1	A=1.6 x10^{12}	E=260.9	A=1.78 x10^{7}	E=135.1
10	$C_2H_4 + C_2H_6 \rightarrow C_3H_6 + CH_4$	2	A=7.0 x10^{16}	E=252.8		
11	$C_3H_6 + C_2H_6 \rightarrow C_4H_8 + CH_4$	2	A=1.0 x10^{17}	E=251.1		

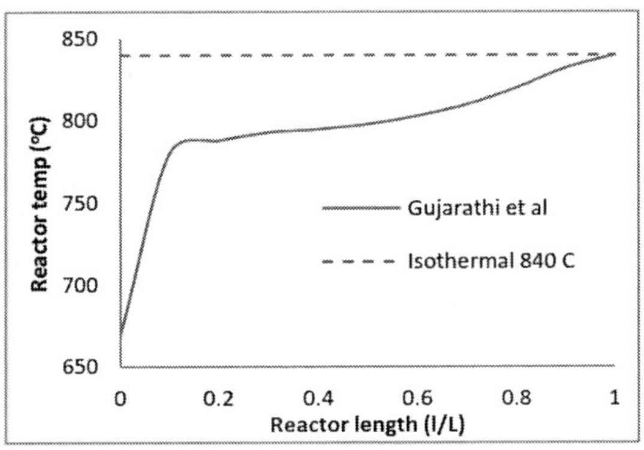

Figure 2: Optimal temperature profile proposed by Gujarathi et al

In the literature, Gujarathi et al. [4] proposed a reactor temperature profile that shows a nonlinear temperature distribution along the reactor with outlet temperature of 840°C. Using the proposed profile (as shown in Figure 2) and a constant reactor temperature profile of 840°C, evaluation of thermodynamic property methods are done. There are six property methods that are suitable to be used for modelling ethylene production system, which are Peng-Robinson, SYSOP0, SRK, UNIFAC, UNIQUAC, and GRAYSON. Selection of property method is done by comparing the resulting product compositions calculated from each method. The result, as shown in Figure 3, shows that all property methods estimated similar percentage of hydrogen, ethylene, ethane, and methane in the reactor product, at both temperature profiles. Since the result using all six property methods is the same, , any property method can be used in the simulation. Thus, , GRAYSON property method is selected as it is also recommended by ASPEN for ethylene production system. As for temperature profile, constant reactor temperature profile is selected to be used in the study.

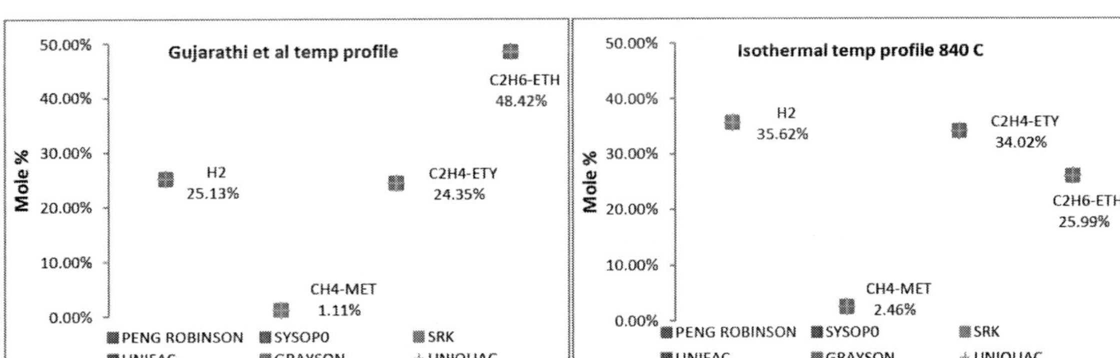

Figure 3: Prediction of cracking products using different thermodynamic property methods
Figure 3a (left) – Comparison using Gujarathi et al profile
Figure 3b (right) – Comparison using isothermal profile at 840 C

3. Process parameter analysis

3.1. Effect of reactor temperature.

Cracking is an endothermic reaction, and thus, require continuous energy input for the reaction to be activated. Stopping the heat supply to the system will cease the reaction almost immediately. Figure 4 shows the effect of reactor temperature to the product composition and different measure of severity applied to it. From Figure 4, it is observed that at 650°C, minimal ethane conversion occurs. Only at 700°C, the reactions start to take place as ethane molecules gain more energy from the heat. Ethane conversion rapidly occur producing ethylene and other products. Ethylene production maximizes between 850°C – 950°C. Beyond that range, ethylene production starts to drop and methane is seen to increase, even though ethane conversion continues. This is primarily due to side reactions activated at higher temperature causing ethane and ethylene to react with other components producing more methane and other by-products [6]. Thus, valuable product is loss while fresh feed and energy is still consumed, reducing the profitability of the plant. To maximize ethylene yield, normal operating strategy is to bring the COT towards higher limit of the optimum temperature range. To minimize further ethylene conversion into other products, immediately after exiting the reactor, cracked gas is cooled down below 650°C.

As shown in Figure 4b, COT is linearly proportional to EER and methane yield, and as shown in Figure 4c, with respect to ethylene yield, COT also exhibit similar patterns as other severity factors. As reactor operating temperature increases, ethylene yield and methane yield rapidly increases. PER, EER, and methane yield indicate high activity at high temperature region, although after certain temperature, ethylene yield start to decrease even with increasing trend of EER and methane yield. This is mainly due to ethane rapid conversion and slow conversion of ethylene into methane and other by-products [6], leading to increasing EER. Reduction in PER also indicate that propylene disappear at higher temperature together with ethylene. Analysing all four severity factors provide an idea of the product distributions of the cracking at any stage and also the optimum severity region that needs to be maintain to ensure optimum ethylene yield.

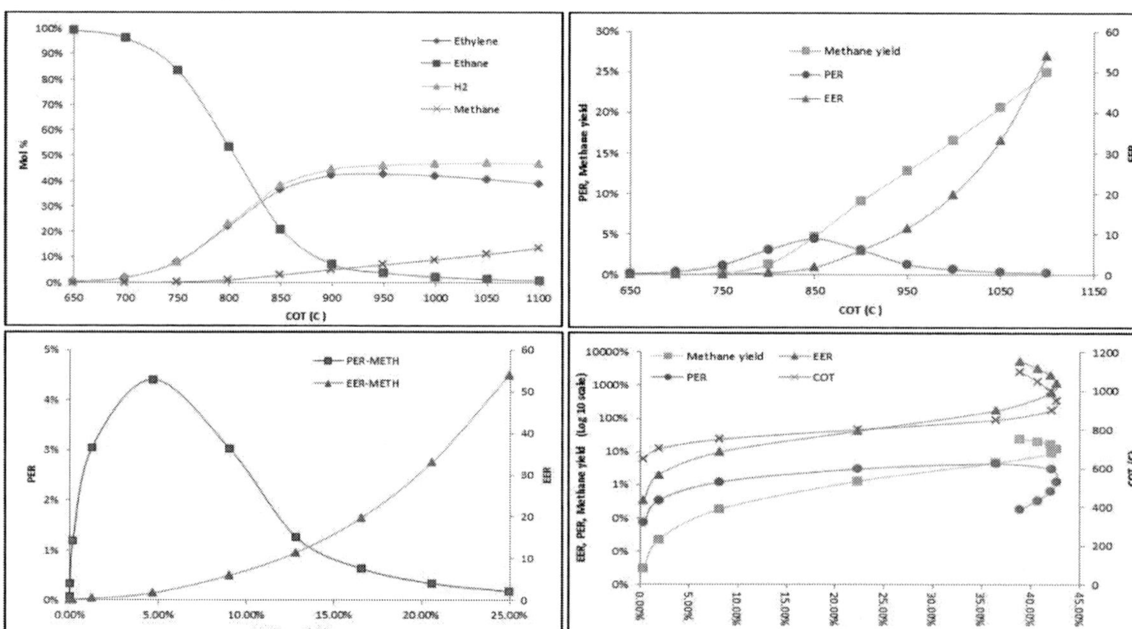

Figure 4. Assessment at different COT. Figure 4a (top left) – Effect of COT to product yield. Figure 4b (top right) – Methane yield, EER, and PER at varying COT. Figure 4c (bottom left) – Relationship between PER and EER with Methane yield. Figure 4d (bottom right) – Comparison of different severity factors with respect to ethylene yield.

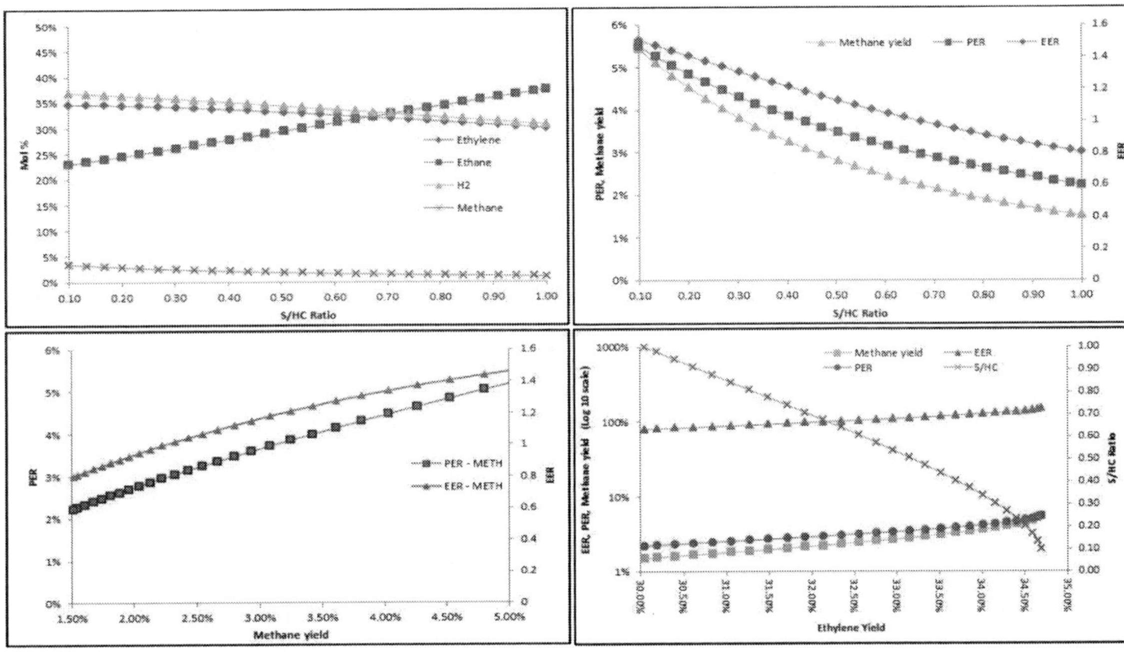

Figure 5: Assessment at different S/HC ratio.
Figure 5a (top left) - Effect of S/HC to product yield. Figure 5b (top right) - Methane yield, EER, and PER at varying S/HC.. Figure 5c (bottom left) - Relationship between PER and EER with Methane yield.
Figure 5d (bottom right) - Comparison of different severity factors with respect to ethylene yield

3.2. *Effect of steam to hydrocarbon ratio.*

Cracking of ethane into ethylene and hydrogen is a reversible reaction. In order to keep the reaction favourable towards ethylene, partial pressure of ethane need to be reduced. Thus, steam is introduced into the feed stream to reduce the partial pressure of ethane. Since steam is inert (no reaction with ethane or other components), its presence does not produce by-products. Figure 5 shows the effect of steam to hydrocarbon ratio to the product composition and different measure of severity applied to it. From the results in Figure 5, increasing the steam to hydrocarbon ratio (S/HC) beyond 0.3-0.5 does not contribute to economic benefits in terms of ethylene yield. All severity factors exhibit linear proportional pattern with each other at varying S/HC. Further increasing S/HC is shown to reduce cracking activities. Thus, maintaining steam only between 0.3-0.5 is sufficient to reduce partial pressure in order to keep the reaction equilibrium towards ethylene and reduces reversal to ethane.

4. Summary

Three severity factors were evaluated together with COT and the results showed that temperature plays the most important role in the process. Reactor temperature heavily influence the product distribution from the cracking process compared to steam to hydrocarbon ratio. It is also found that operating at higher temperature region led to high ethane consumption. Moreover, by-products and methane have been produced due to ethylene consumption. Analysis shows that operating at a temperature range of 850°C – 950°C and steam to hydrocarbon ratio of 0.3-0.5 produces maximum ethylene yield while minimizing byproducts. Analysis of multiple severity factors enables engineers to evaluate the cracking performance at every angle and the effect on several products simultaneously. To further improve the accuracy of the model, coke formation can be incorporated together with addition of recycle feed. This will enable analysis of cracking process under the presence of impurities from the recycle stream and the effect to cracking severity.

Acknowledgement
The authors gratefully acknowledge the financial support given by Universiti Sains Malaysia through Research University Grant No. 1001/PJKIMIA/814237.

References

[1] Gao G Y et al 2009 Optimal operation of tubular reactors for naphtha cracking by numerical simulation *Asia Pac. J. Chem. Eng.* **4** (6)
[2] Sundaram K M and Froment G F 1978 Modelling of thermal cracking kinetics - III Radical mechanisms for the pyrolysis of simple paraffins, olefins, and their mixtures *Ind. Eng. Chem. Fundam.* **7** (3)
[3] Sundaram K M and Froment G F 1977 Modelling of thermal cracking kinetics - I Thermal cracking of ethane, propane, and their mixtures *Chem. Eng. Sci.* **32**
[4] Gujarathi A M et al 2009 Simulation and analysis of ethane cracking process
[5] Ghashghaee M, Karimzadeh R 201, Multivariable optimization of thermal cracking severity, *Chem. Eng. Res. Des.* **89** 1067 – 1077
[6] Geem K M v, Reyniers M, Marin G B 2005 Two severity indices for scale-up of steam cracking coils *Ind. Eng. Chem. Res.* **44** 3402 – 3411
[7] Ranjan P et. al. 2012 Modeling of ethane thermal cracking kinetics in a pyrocracker *Chem. Eng. Technol.* **35(**6) 1093–1097
[8] Yancheshmeh M S S et. al. 2013, Modeling of ethane pyrolysis process: A study on effects of steam and carbon dioxide on ethylene and hydrogen productions, *Chem. Eng. J.,* **215–216** 550–560

Evaluation of shrinking core model in leaching process of Pomalaa nickel laterite using citric acid as leachant at atmospheric conditions

K C Wanta[1,2], I Perdana[1] and H T B M Petrus[1,*]

[1] Center of Advanced Material and Mineral Processing, Department of Chemical Engineering, Universitas Gadjah Mada
Jl. Grafika 2, Bulaksumur, Yogyakarta 55281, INDONESIA

[2] Department of Chemical Engineering, Universitas Katolik Parahyangan
Jl. Ciumbuleuit 94, Bandung 40141, INDONESIA

E-mail : kcwanta@unpar.ac.id / bayupetrus@ugm.ac.id

Abstract. Most of kinetics studies related to leaching process used shrinking core model to describe physical phenomena of the process. Generally, the model was developed in connection with transport and/or reaction of reactant components. In this study, commonly used internal diffusion controlled shrinking core model was evaluated for leaching process of Pomalaa nickel laterite using citric acid as leachant. Particle size was varied at 60-70, 100-120, -200 meshes, while the operating temperature was kept constant at 358 K, citric acid concentration at 0.1 M, pulp density at 20% w/v and the leaching time was for 120 minutes. Simulation results showed that the shrinking core model was inadequate to closely approach the experimental data. Meanwhile, the experimental data indicated that the leaching process was determined by the mobility of product molecules in the ash layer pores. In case of leaching resulting large product molecules, a mathematical model involving steps of reaction and product diffusion might be appropriate to develop.

1. Introduction

In this era, the use of nickel laterite in mineral processing has to be done, though the nickel content in laterite ore is small. It needs to be done as a result of depletion of nickel sulphide resources in the world [1-5]. During this time, smelting method dominates in mineral processing which the ore has low nickel content [2, 3]. However, high energy consumption and high operational cost in smelting process make the other process, like leaching process at atmospheric condition should be done [6,7]. The utilization of this process is expected to be useful and applicable in industry.

One of the important factor in the application of leaching process at atmospheric conditions in industrial scale is information about process mechanism, mathematical model, and constants' value related on that process. This information will be useful in scale-up process. In the study about leaching process, shrinking core model is the most popular model among researchers. Several studies, like the studies conducted by Astuti, et.al., Thubakgale et.al., Agacayak and Zedef, used this model and were able to show that shrinking core model were the most suitable to describe the physical phenomenon of leaching process [7-9].

Shrinking core model was introduced at the first time by Yagi and Kunii (1955) and is described completely in the text book of Chemical Reaction Engineering written by Octave Levenspiel. This model describes that the reaction occurs first at the outer skin of particle and then the reaction zone moves into the deeper unreacted zone. The zone where reaction has occurred will become ash layer and the layer is assumed to be inert [10].

Leaching process of nickel laterite is a heterogeneous process which involves more than one phase i.e. the solid-liquid phase. Thus, this process will be controlled by several steps, such as diffusion through liquid film layer, diffusion through ash layer, and chemical reaction. The overall rate of the process is determined by the slowest step [7]. Types of the shrinking core models in leaching process can be summarized in the following equations [7,10] :

Diffusion through liquid film layer control	: $k_f . t = x$	(1)
Diffusion through ash layer control	: $k_d . t = 1 - 3(1\text{-}x)^{0.67} + 2(1\text{-}x)$	(2)
Chemical reaction control	: $k_r . t = 1 - (1\text{-}x)^{0.33}$	(3)

where x is the reacted fraction; t is time; k_f, k_d, and k_r are reaction rate constants. Most of models used for leaching process are of diffusion through ash layer control model.

In this present study, the use of ash layer diffusion controlled shrinking core model was evaluated against the experimental results from leaching process of Pomalaa nickel laterite using citric acid as leachant.

2. Materials and Methods

2.1. Materials

In this study, nickel laterite was limonite type and was mined from Pomalaa, Southeast Sulawesi Province, Indonesia. Based on the result of x-ray fluorescence (XRF) test, this sample contained 2.73% wt of nickel. As leachant, citric acid (Merck) was used in the form of solution with a concentration of 0.1 M.

2.2. Experimental design

300 ml of 0.1 M citric acid solution was poured into the three-neck flask and was heated until 85°C. After the temperature was reached, 60 grams of nickel laterite sample was added into the reactor. The leaching process was done for 2 hours isothermally. In this experiment, the particle size was varying at fractions of 60-70 mesh, 100-120 mesh, and -200 mesh. Periodically, samples were taken and then the solid phase was separated from the liquid phase using centrifuge at 1.000 rpm for 10 minutes. The filtrate (liquid phase) was diluted with distilled water as much as 20 times. Then, the elements in the sample was analyzed using Inductively Coupled Plasma-Optical Emission Spectroscopy (ICP-OES).

3. Results and Discussion

The effect of particle size on recovery nickel is shown in Figure 1. Based on that figure, it can be seen that the smaller the particle size was used, the higher the nickel recovery was obtained. For the small particles, the ash layer formed by the reaction is relatively thin so that the molecule transport through the layer is faster. However, Figure 1 also shows that the particle size with a fraction of -200 mesh has no significant effect on the nickel recovery.

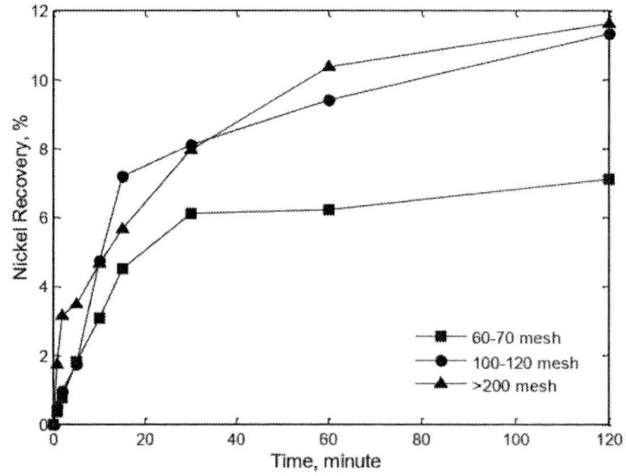

Figure 1. The effect of particle size on nickel recovery

In this study, shrinking core model determined by diffusion through ash layer is evaluated and the simulation results can be seen in Figure 2. Figure 2 shows that shrinking core model is not suitable to describe the phenomenon of the leaching process. The results suggests that the leaching process of Pomalaa nickel laterite using citric acid as leachant is not determined by only one-step controlling process as commonly used in shrinking core models for leaching processes.

Simulation result from this study is not in agreement with the the results reported by some previous researchers. Astuti, et.al. studied leaching process of Pomalaa saprolitic nickel laterite using citric acid as leachant. Their results showed that shrinking core model with diffusion through ash layer controlling process was the most suitable model [7]. Similar conclusion was also reported by other researchers [8,9] that showed that the shrinking core model with diffusion controlling process was much better than other models. However, since the evaluation of the models was merely based on single component data, the conclusion might be misled.

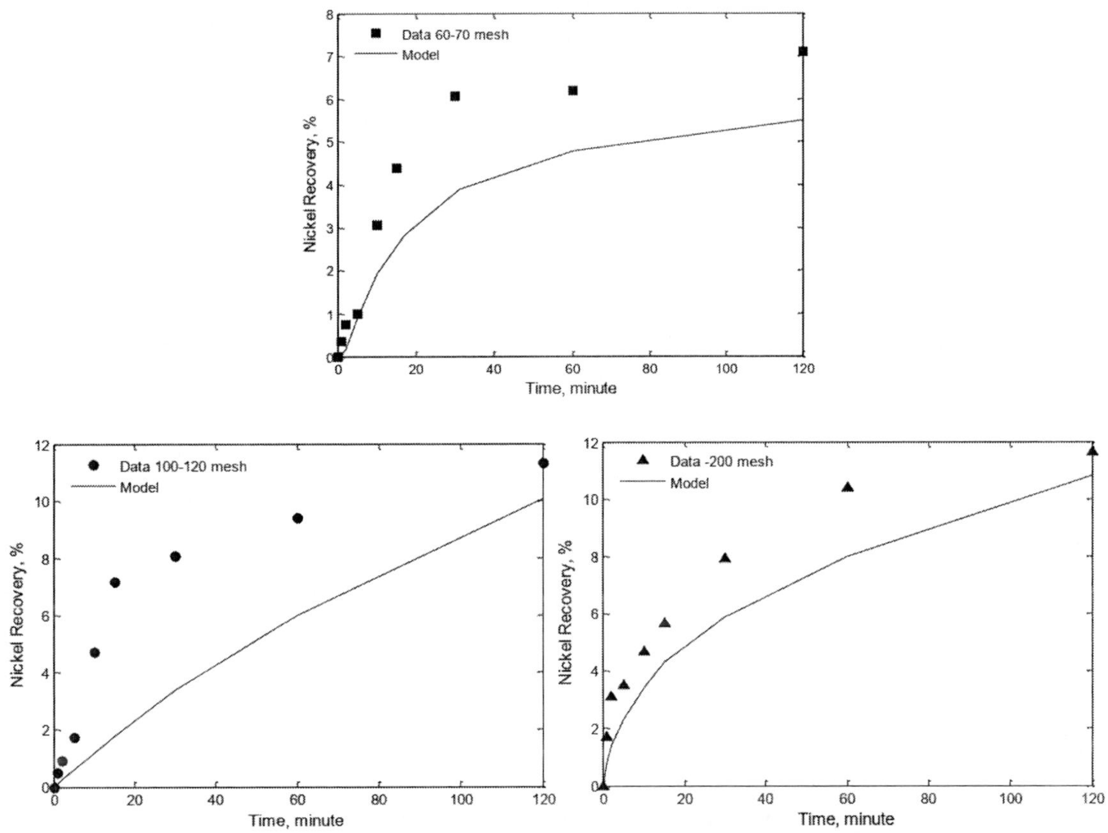

Figure 2. Simulation results using diffusion controlled shrinking core model

The experimental data as shown in Figure 1 and 3 indicates that there is another mechanism, besides internal diffusion that might determine the overall leaching process.

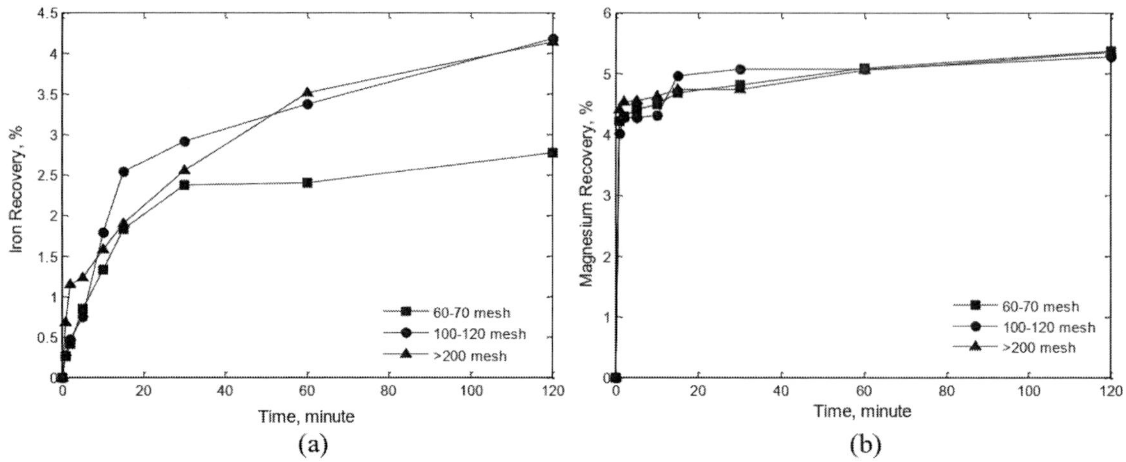

Figure 3. The effect of particle size on iron and magnesium recovery

As shown in Figure 1 and 3, particle size plays a role in the recovery of nickel and iron. The recovery of nickel and iron was found higher with the decrease of the particle size. However, even though the same leachant was used in all experiments, recovery of magnesium was not influenced by the particle size. Reaction between citric acid and nickel or iron compounds might result in large product molecules that experience steric hindrance of internal diffusion. This result suggests that the leaching process is not controlled by internal diffusion of the reactant molecules. On the other hand, diffusion of product molecules in the ash layer seems to have more important role.

Since the conventional shrinking core model is insufficient to approach the experimental data in this work, development of new mathematical model is necessary. A model that considers steps of chemical reaction and product diffusion through ash layer might be appropriate to describe the physical phenomenon of leaching process of Pomalaa nickel laterite using citric acid as leachant.

4. Conclusion

Shrinking core model was not suitable to describe the phenomenon of leaching process Pomalaa nickel laterite using citric acid as leachant. Experimental data indicated that diffusion of product molecules played a role in the leaching process. Therefore, development of mathematical model involving steps of reaction and product diffusion through ash layer might be necessary.

Acknowledgements
The authors acknowledge analysis equipment support provided by Department of Chemical Engineering, Universitas Gadjah Mada. The authors also acknowledge materials support from Indonesian Institute of Sciences (LIPI).

References

[1] Dalvi A D, Bacon W G and Osborne R C 2004 The past and the future of nickel laterites *PDAC 2004 International Convention, Trade Show, & Investors Exchange*
[2] Kyle J 2010 Nickel laterite processing technologies – Where to next? *ALTA 2010 Nickel/Cobalt/Copper Conference, Perth, Australia*
[3] Wanta K C 2016 Kinetika proses leaching nikel laterit Pomalaa dengan menggunakan asam sitrat sebagai leachant *Thesis, Universitas Gadjah Mada*
[4] McDonald R G and Whittington B I 2008 Atmospheric acid leaching of nickel laterites review – Part I. Sulphuric acid technologies *Hydrometallurgy* **91** 35-55

[5] McDonald R G and Whittington B I 2008 Atmospheric acid leaching of nickel laterites review – Part II. Chloride and bio-technologies *Hydrometallurgy* **91** 56-69

[6] Astuti W, Hirajima T, Sasaki K and Okibe N 2016 Comparison of effectiveness of citric acid and other acids in leaching of low-grade Indonesian saprolitic ores *Minerals Engineering* **85** 1-16

[7] Astuti W, Hirajima T, Sasaki T and Okibe N 2015 Kinetics of nickel extraction from Indonesian saprolitic ore by citric acid leaching under atmospheric pressure *Minerals & Metallurgical Processing* **32** (3) 176-185

[8] Thubakgale C K, Mbaya R K K and Kabongo K 2012 Leaching behavior of a low-grade South African nickel laterite *International Journal of Chemical, Molecular, Nuclear, Materials, and Metallurgical Engineering* **6** (8) 228-232

[9] Agacayak T and Zedef V 2012 Dissolution kinetics of a lateritic nickel ore in sulphuric acid medium *Rocnik* **17** 33-41

[10] Levenspiel O 1999 *Chemical Reaction Engineering* (New York : John Wiley & Sons)

Kinetics study of carbon raiser on the reduction of nickel laterite from Pomalaa, Southeast Sulawesi

H T B M Petrus[1], A R Rhamdani[1], A D P Putera[1], I W Warmada[2], A T Yuliansyad[1], and I Perdana[1]

[1] Center of Advanced Material and Mineral Processing, Department of Chemical Engineering, Universitas Gadjah Mada
Jl. Grafika 2, Bulaksumur, Yogyakarta 55281, INDONESIA

[2] Department of Geological Engineering, Universitas Gadjah Mada
Jl. Grafika 2, Bulaksumur, Yogyakarta 55281, INDONESIA

E-mail : bayupetrus@ugm.ac.id

Abstract. As one of the top ten on nickel laterite ore resources in the world, Indonesia must have been initiating the nickel processing in total amount of about 1.5 million tonnes. In regard to the low nickel laterite processing, one of the possible product is nickel pig iron (NPI) needed for the stainless steel industries. In this study carbon raiser that is waste from oil industries was used to replace metalurgical coke. The kinetic of nickel laterite reduction using carbon raiser was studied and compared with anthrasite coal. In this work, the author conducted the reduction of nickel laterite ores by both carbon raiser and anthrasite coal as reductant, in air and CO_2 atmosphere, within the temperature ranged from 800°C and 1000°C. Two models were applied, sphere particle geometry model and Ginstling-Brounhstein diffusion model, to study the kinetic parameters. The results indicated that type of reductants and reduction atmosphere greatly influence the kinetic parameters. The obtained values of activation energy were varied between 17.44-18.12 kcal/mol.

1. Introduction

Nickel is one of the most important materials in the world. 66% of nickel consumption used for stainless steel making, while the others 18% used in other alloy making, and the rest used for other purposes. There are two main nickel sources, sulfide ore and laterite ore. At least, 60% of nickel demand is supply by sulfide ore besides 80% of nickel deposit lies on laterite ore [1]. Laterite is a highly weathered material rich in secondary oxides of iron, aluminium, or both. In addition, laterite often contains minor amounts of nickel, cobalt, and chromium [2]. One of the reasons that laterite ore is not used as the main nickel source is because of its low nickel content but recently China as the biggest stainless steel producer had found a way of utilizing laterite ore. They used it to produce nickel pig iron (NPI), as a substitute for pig iron in the stainless steel making process, using blast furnace or rotary kiln – electric furnace (RKEF). One of the important processes in NPI making is the reduction process. In the conventional NPI production, coal was used as a reductant.

There have been several studies of nickel laterite reduction. But to the best of our knowledge, very few works have been presented on studying the kinetic of nickel laterite reduction. The kinetic study of nickel laterite reduction using gaseous reductant shows that the reduction mechanism was diffusion controlled [3] but the kinetic study of nickel laterite reduction using solid reductant has not been studied yet. Therefore, this present work investigated the kinetic of nickel laterite reduction using solid reductant. The effect of using various kind of reductant to the kinetic parameter was also studied.

2. Materials and Methods

Anthracite coal, and carbon raiser was used as a reducing agent. Laterite, anthracite coal, and carbon raiser were provided by Indonesian Institute of Sciences. All of the materials above were crushed using mortar and then screened to get the desired size (-100+120 mesh). Laterite was characterized by

XRD (Shimadzu 7000), XRF (Thermo Scientific Niton XL3t). The proximate analysis of reductants was done by PAU UGM.

2.1. *Pelletization*

Laterite, starch, and anthracite charcoal were mixed together with laterite/reductant/binding agent ratio of 4:1:0.3. Then 5 mL of distilled water was added to the mixed sample and shaped into a ball pellet using hand. The pellet was dried at 110°C for 5 h. Using the same procedure, another pellets were made with carbon raiser as a reductant.

2.2. *Reduction of the pellets*

The pellets were roasted using two different atmospheric conditions, air and CO_2. Muffle furnace was used for air atmospheric roasting and tube furnace was used for CO_2 atmospheric roasting with CO_2 flow rate of 3 L/min. The pellets were roasted at temperature of 800, 900, and 1000°C with roasting time varied from 15 to 240 minutes. After the roasting process, the pellets were cooled and placed in desiccator. Thereafter, the pellets were crushed using mortar and were characterized using XRD (Shimadzu 7000).

2.3. *Fraction of reduction reaction*

To evaluate the kinetics parameters, the term fraction of reaction (*f*) was used. Eq (1) represents the formula for fraction of reaction. The equation was modified from the term used in iron ore reduction process [4]. The maximum weight loss of a pellet in Eq (1) was obtained from the maximum weight loss of laterite and reductant. The maximum weight loss of laterite was obtained from the total of removable oxygen from laterite. The total of removable oxygen from laterite is the oxygen binds to the iron and nickel.

$$f(t) = 100\frac{\Delta W_t}{\Delta W_{max}} \tag{1}$$

where:

f = fraction of reaction.
ΔW_t = weight loss of pellet (g).
ΔW_{max} = maximum weight loss of pellet (g).

2.4. *Kinetic models*

The experimental data (*f*) were applied to the following mathematical models :

Sphere Particle Geometry Model [5]

$$1-(1-f)^{1/3} = kt \tag{2}$$

where:

t = time (s).
k = kinetic constant (1/s).

$$k = k_0 exp\left(-\frac{E_a}{RT}\right)$$

where :

k_0 = pre-exponential factor (1/s).
E_a = activation energy (kcal/mol).
T = reduction temperature (K).
R = universal gas constant (kcal/mol/K).

Ginstling-Brounhstein Diffusion Model [5]

$$1-\frac{2}{3}f-(1-f)^{2/3}=kt \quad (3)$$

3. Results and Discussion

3.1. Material characterization

The chemical compositions of nickel laterite used in this research was showed in the Table 1. The laterite was dominated by Fe and Si element and based on the analysis of the XRD pattern using Match! 2 software (Figure 1), the main mineral composition of the laterite was goethite (FeOOH), lizardite ($Mg_3(Si_2O_5)(OH)_4$), clinochlorite ($(Mg,Fe^{2+})_5Al(AlSi_3O_{10})(OH)_8$), and quartz ($SiO_2$). The proximate analysis of the reductants was done by PAU UGM and the results was showed in the Table 2. The results showed that all of the reductants used had different characteristics. Anthracite coal had the highest fixed carbon content and the lowest ash content while carbon riser had the lowest fixed carbon and volatile matter content.

Table 1. Main Components of Laterite from Pomalaa, Southeast Sulawesi

Components	SiO_2	Fe_2O_3/Fe*)	NiO/Ni*)	MgO	Al_2O_3	Na_2O	Cr_2O_3
(wt.%)	26,94	43,83/30,60	3,19/1,73	12,67	5,13	2,73	1,63

*) Analyzed using XRF Metal

Table 2. Proximate Analysis of Reductans (%wt)

Reductor Type	Moisture	Volatile matter	Ash	Fixed Carbon
Anthracite coal	2,3	7,4	2,5	87,9
Carbon riser	4,7	12,6	13,5	69,2

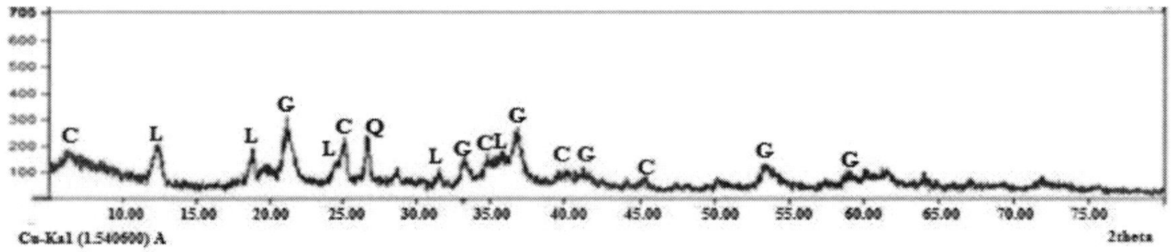

Figure 1. XRD pattern of initial laterite ore [Q: quartz; L: lizardite; G: goethite]

3.2. Fraction of reaction

Experimental data which were the sample mass from time to time converted into fraction of reduction reaction (*f*). The maximum weight loss from a pellet made with one type of reductant is different from a pellet made with the other type of reductant because the ash content of a reductant is different from the other reductant. The maximum weight loss of the pellet is 80.4% for the sample made with anthracite coal, and 73.5% for the sample made with carbon raiser. Figure 2 shows the graphical representation of fraction of reduction vs time.

From the experimental data, fraction of reaction value was already reached a constant value before reached its maximum value at certain time. The maximum value of fraction of reaction was 48%. It could happen because the reduction reaction has stopped. The reason of it could be two things. The first reason is the ratio between laterite and reductant was not high enough which makes the reductant has already used up before could reduce the minerals completely. The second reason is there are a reaction between iron oxide and quartz which form fayalite ($FeSiO_4$) [6]. Fayalite has relatively low melting point (1100°C) [7] and already melt during the reduction process which makes the porosity of the pellet decreased. The decreasing of porosity of pellet could prevent reducing gas diffusion. On the other hand, fayalite could also act as a nucleating agent on iron metal forming process. During the reduction process, iron metal layer formed around the surface of fayalite and prevent the contact of reducing gas. The fraction of reaction reached a constant value faster at higher temperature. It could happen because the reaction rate is faster at higher temperature. The reaction has stopped even when CO gas is used as a reductant instead of using solid reductant like in this study [3]. It concluded that the second reason was better at explaining why the reaction has stopped. The reduction reaction was fast at the early stage of the reduction process because at the early stage, the devolatilization of the reductant occurs alongside the dehydroxylation of the minerals in the laterite which makes the weight loss of the sample is very high.

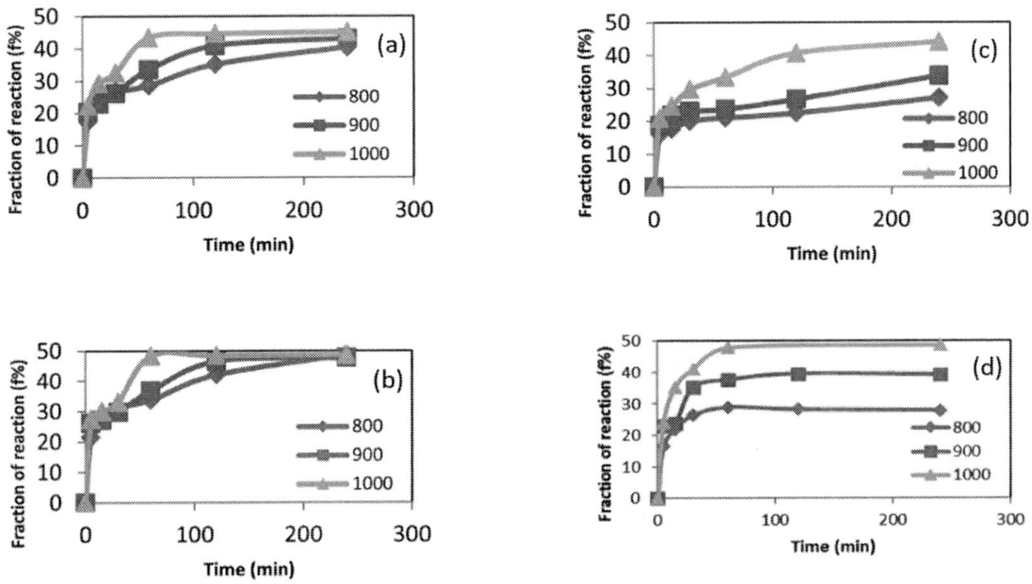

Figure 2. Fraction of reaction as a function of time of the laterite with anthracite (a), and carbon raiser (b) in air atmospheric condition and with anthracite (c), and carbon raiser (d) in CO_2 atmospheric condition.

There was a slope change on the fraction of reaction vs time data on 0-5 minutes and 5-60 minutes reduction duration. This could happen because there were changes of reduction mechanism between these two periods. In the early stage of reduction, the process will follow the reaction controlled mechanism and after that the process will follow diffusion controlled mechanism [3]. The changes of slope in the previous study happened at 20 minutes reduction duration [3], but in this study, the changes happened at 5 minutes duration time. It could happen because the reaction happened in this two study is different. The changes of slope happened in the very early stage of the reduction. These early changes made the evaluation of kinetic parameters could only be done starting at 5 minutes reduction time.

The maximum value of fraction of reduction was almost the same (around 47%) for both reductions happened in air and CO_2 atmosphere, but based on the XRD analysis of the reduced sample, the mineral phase of both sample was different. The comparison of XRD pattern of the reduced samples was shown in Figure 3. The reduction in CO_2 atmosphere proceeds more complete compared with reduction in air atmosphere based on the loss of hematite and magnetite peaks. It could happen because there is oxygen in the air atmosphere reduction which could oxidized the reduced mineral in the sample.

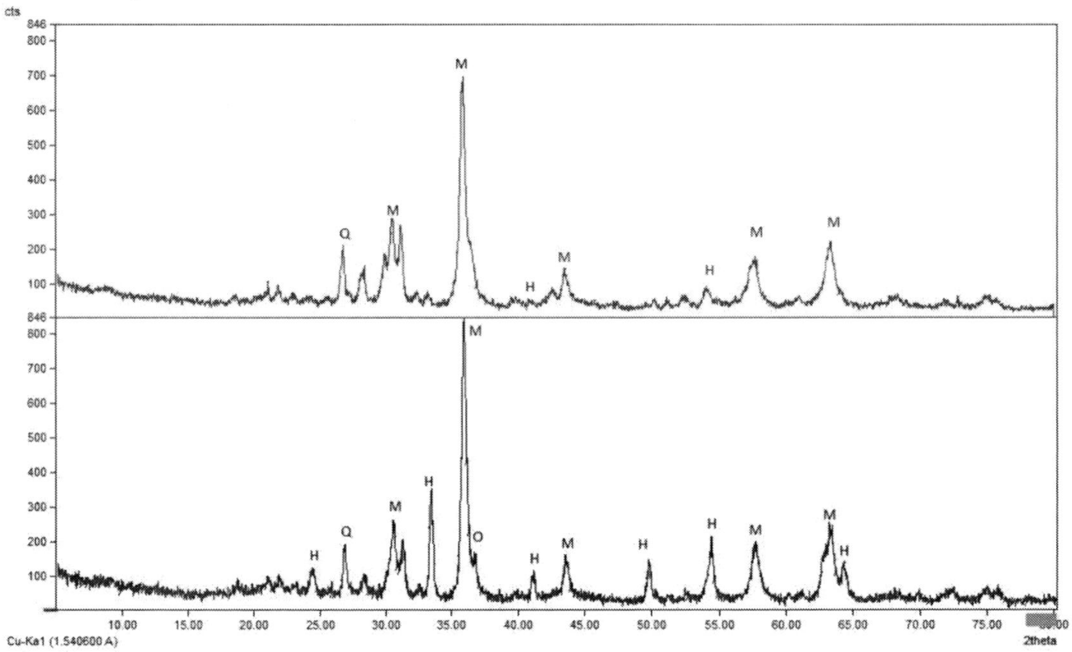

Figure 3. XRD pattern laterite roasted in air condition (a) anthrasit (b) carbon raiser at 1000°C for 240 minutes. [Q: quartz; O: olivine; H: hematite; M: magnetite]

3.3. *Kinetic analysis*

For the kinetic analysis of the experimental results, the kinetic model equations (2) – (3) have been applied to the experimental data for the temperatures of 800, 900, and 1000°C. The graphical representation of the fitting of experimental data to the kinetic model were shown in Figure 3 and Figure 4.

Based on the simulation, both models could represent the reduction process well enough because the error in the data fitting process was small enough for both model applied. However, based on literature study, the diffusion model was considered as the model that could represent the process. It also fitted with the reason why the reaction stopped at a certain time. The kinetics parameters obtained from evaluation of the data using diffusion model was listed in Table 3. The variation of activation energy means that the reaction mechanism happened is different for each reductant and atmosphere condition. The reaction mechanism differences happened because the reactions which took place consist of various step such as carbon oxidation, reducing gas diffusion, and reduction reaction hence the reaction was very complex.

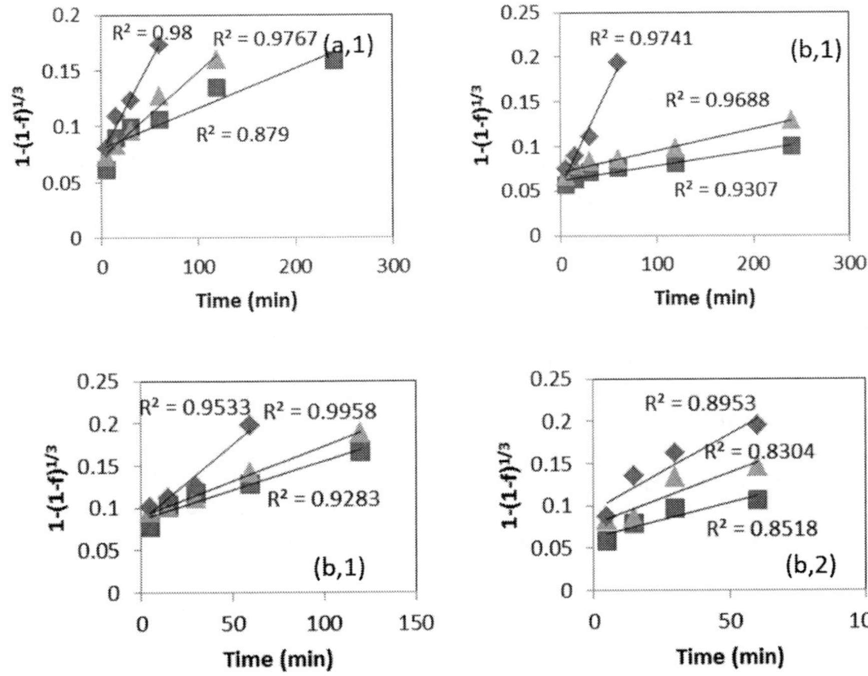

Figure 4. Applications of the sphere particle geometry model to the experimental data of sample reduced with anthracite coal (a), and carbon raiser (b) in air atmosphere (1) and CO_2 atmosphere (2) at temperature of reduction of 800 (■), 900 (▲), and 1000°C (♦).

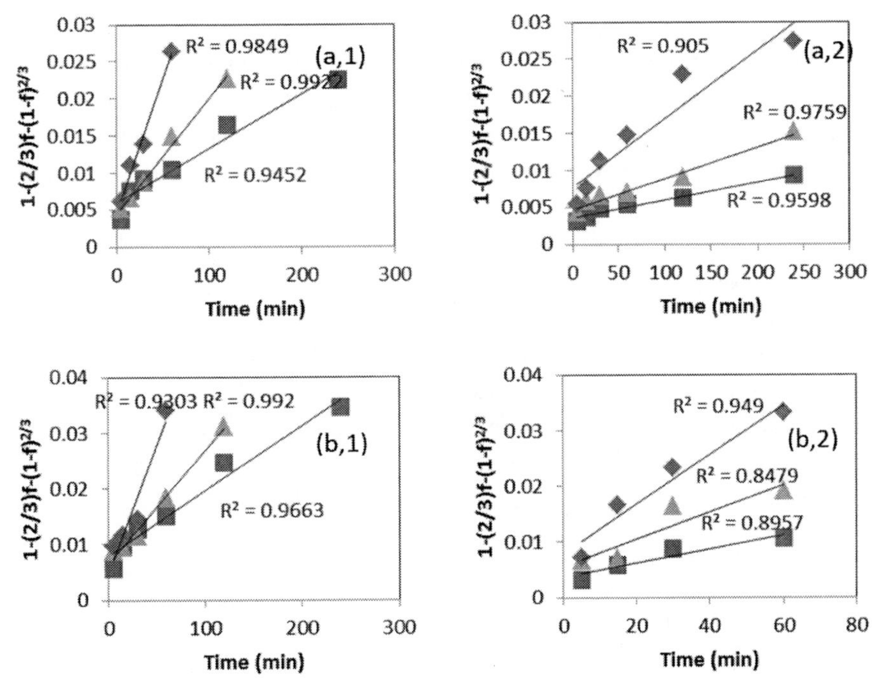

Figure 5. Applications of the Ginstling-Brounhstein diffusion model to the experimental data of sample reduced with anthracite coal (a), and carbon raiser (b) in air atmosphere (1) and CO_2 atmosphere (2) at temperature of reduction of 800 (■), 900 (▲), and 1000°C (♦).

Table 3. The value of Activation Energy (E_a) dan k_0 from Simulation of Ginstling – Brounhstein Diffusion

Reductant Type	**Atmosphere**			
	Air		**CO_2**	
	k_0 (1/min)	Ea (kcal/mol)	k_0 (1/ min)	Ea (kcal/mol)
Anthracite coal	0.3182	17.4445	0.1083	18.1217
Carbon riser	0.5366	18.1169	0.5714	18.0832

4. Conclusion

From the results the following conclusions were drawn:

1) Carbon raiser is potential to replace coking coal in the NPI production.
2) Both model applied were good at representing the experimental data, but Ginstling-Brounhstein diffusion model was considered better at representing the process because it has relatively smaller error and was supported by literature study.
3) From applying Ginstling-Brounhstein diffusion model, the obtained value of activation energy was varied around 17.44 kcal/mol to 18.12 kcal/mol depends on the type of reductant and the reduction atmosphere used.

Acknowledgements
The authors acknowledge the financial support provided by Directorate General of Higher Education (DIKTI), Ministry of Technology Research and Higher Education. The authors also acknowledge materials and analysis equipment support from Indonesian Institute of Sciences (LIPI).

References

[1] Pickles C A 2004 Microwave Heating Behaviour of Nickeliferous Limonitic Laterite Ores *Minerals Engineering* **17** 775-784
[2] Olapinekun E O 2000 Kinetics of Leaching Laterite *Int. J. Miner. Process.* **60** 9-14
[3] Zevgolis E, Zogradifis C, Halikia I, Perrarki M 2009 Roasting Reduction Kinetics of an Indonesian Nickeliferous Laterite Ore *3rd AMIREG International Conference (2009) : Assessing the Footprint of Resource Utilization and Hazardous Waste Management, Athens, Greece* 459-465
[4] Agrawal B B, Prasad K K, Ray H S 1990 The Relationship between the Degree of Reduction (α) and the Fraction of Reaction (*f*) during Reduction of Ore-Coal Mixture *ISIJ International* **30**(8) 997-999
[5] Khawam A, Flanagan D R 2006 Solid-State Kinetic Models : Basic and Mathematical Fundamentals *J. Phys. Chem. B.* **110** 17315-17328
[6] Jiangan C, Tichang S, Qi L 2014 Study of the Mechanism of Direct Reduction Roasting of the Limonite in Jiangxi *Journal of Chemical and Pharmaceutical Research* **6**(3) 671-678
[7] Vogt J H L 1922 The Physical Chemistry of the Crystallization and Magnetic Differentiation of Igneous Rock *The Journal of Geology* **7** 611-630

Effect of catalyst concentration and reaction time on the extraction of glucomannan from porang (*Amorphophallus oncophyllus*) flour via acid hydrolysis

A C Kumoro[1, 2], T H A Yuganta[2], R Ratnawati[1] and D S Retnowati[1]
[1] Department of Chemical Engineering, Faculty of Engineering, Diponegoro University. Prof. H. Soedarto, SH Road-Tembalang, Semarang-INDONESIA 50275
[2] Master of Chemical Engineering Program, Department of Chemical Engineering, Faculty of Engineering, Diponegoro University. Prof. H. Soedarto, SH Road-Tembalang, Semarang-INDONESIA 50275

E-mail: andrewkomoro@undip.ac.id

Abstract. High demand of glucomannan for various applications has attracted the attention of researchers to look for efficient extraction method from its botanical sources. The aim of this study is to investigate the effect of catalyst concentration and reaction time on the yield and purity of glucomannan, and profile of reducing sugar during glucomannan extraction from crude porang flour via acid hydrolysis. The effect of catalyst concentration was found to be more pronounced over the effect of reaction time. When catalyst concentration was varied from 0.03125 to 1 M, extraction of glucomannan from porang flour for 1 hour at 60°C yielded 40.5 to 70% glucomannan with purity of 47.35 to 90.18% (w/w). The yield and purity of glucomannan obtained from extraction using catalyst concentration of 1 M for 0.25 to 3 hour ranged between 49 to 66.67% and 58.32 to 90.18% (w/w), respectively. Reduction in glucomannan yield and purity observed at high catalyst concentration and prolong reaction time was likely to be due to over-decomposition. Glucomannan with highest purity (90.18% w/w) was obtained at 66.67% yield from acid hydrolysis of porang flour using 0.5 M hydrochloric acid solution with flour:water ratio of 1:50 at 60°C for 1 hour.

1. Introduction

Glucomannan is a neutral polysaccharide composed mainly of β, 1-4 linked D-mannose and D-glucose with molar ratio of 1.6:1. The backbone is lightly branched, which short side branches at the C-3 position of the mannoses. Some acetyl groups randomly present at the C-6 position of the sugar units [1]. This acetyl groups, which is believed to contribute to the solubility and gelling properties are located every 9 to 19 sugar units. Those properties are related to high water absorbance (100 g of water per g of glucomannan), and its high molecular weight, $10^5 - 10^6$ g/mol. Glucomannan has been used as traditional Chinese medicine for thousands of years for treatment of asthma, cough, hernia, breast pain, burns and skin disorders [2]. Recently, clinical studies have proven the ability of glucomannan to lower blood sugar, lower blood cholesterol, helping weight loss and promote healthy micro flora in intestinal [3]. With regard to biodegradability and good gel-forming ability, glucomannan is also used in pharmaceutical field for drug delivery system [4]. In addition, due to unique rheological and gelling properties glucomannan is also used in food industries as emulsifier and stabilizer for foods, drinks, and cosmetics products [2].

Glucomannan can be obtained from various botanical sources. However, the tuber of porang (*Amorphophallus oncophyllus*), which belongs to the family of Araceae, has been reported to be the most potential source due to highest glucomannan content and largest availability [5]. Because porang plant is sensitive to high temperatures, dry atmosphere and high wind [6], it is cultivated in Indonesia as secondary crops under shady environment such as the teak, mahogany or sonokeling plantation [7]. Generally, the glucomnnan content in porang flour is about 50% to 70% [8]. The impurities trapped in glucomannan molecules are starch, protein, and insoluble sugar, which are usually derived from tissue space in the tuber [2]. Starch is the main impurity that seriously affects the purity and quality of glucomannan, especially reducing viscosity and increasing turbidity [9]. In addition, starch molecules usually interact with glucomannan and make them even more difficult to be separated for its low solubility under mild condition. However, gelatinization behaviour efficiently boosts its solubility. In fact, glucomannan molecules possess better swelling ability and become insoluble in aqueous ethanol, whereas the impurities do not. Gelatinization temperature of porang flour is between 68.5-77.6 °C [10]. These properties are beneficial for extraction of glucomannan from porang flour by removal of the starch via hydrolysis [11] and followed by precipitation of glucomannan by addition of alcohol [12]. Based on the swelling property in alcohol solution, ethanol precipitation is the most common and practical method to purify glucomannan and to remove the impurities on the surface or trapped inside of porang flour particles [13]. Glocumannan has lowest solubility when ethanol concentration in the solution of about 45% v/v by which its separation from the solution may occur via precipitation [14]. Alcohol contributes to unfold glucomannan molecular chain and expose the impurities trapped in the macromolecules to the solution for easier removal.

The traditional processing methods to obtain purified porang flour are by dry or wet processing which can separate starch granules from larger glucomannan granules. The dry processing method includes milling of dried porang tuber chips into crude porang flour, which is subsequently purified via wind-sifting [15]. Porang flour obtained using such method is of low purity and is usually marketed as food commodity at a low price [14]. Wet methods, which include enzymatic hydrolysis, ethanol washing, centrifugation may results high purity of glucomannan, but at higher operating cost [2]. Therefore, it is clear that more efficient technologies are required to be developed to replace the existing porang flour glucomannan purification technology.

In the case of glucomannan extraction via starch hydrolysis, the use acid hydrolysis has never been broadly explored. Operating parameters such as temperature, reaction time, acid concentration and solid-to-liquid ratio play critical roles in obtaining optimum acid hydrolysis of starch [16]. In general, the catalytic activity of hydrochloric acid during hydrolysis of starch is higher than that of sulfuric acid [17] and phosphoric acid [18]. This high catalytic activity, non disturbing residue after neutralization (sodium chloride) on glucomannan precipitation using alcohol and its simplicity are the advantages of using hydrochloric acid as catalyst. The drawback of acid hydrolysis is merely slower reaction rate as compared to enzymatic hydrolysis [2]. The governing factor of the homogeneous acid catalysis turned out to be the acidity (pH) of the solution, which is directly proportional to the concentrations of the hydroxonium ions (H_3O^+) exist in the solution [19]. On the other hand, the temperature and the reaction time have less influence on the hydrolysis reaction [20].

The aim of this study is to investigate the effect of catalyst concentration and reaction time on the yield and purity of glucomannan, and profile of reducing sugar during glucomannan extraction from porang flour via acid hydrolysis.

2. Materials and Method

2.1. Materials

The crude porang flour used in this research was purchased from PT. Prima Agung Sejahtera-Surabaya. Other chemicals used for extraction and analysis were of analytical grade (purity ≥ 98% w/w) and purchased from Sigma Aldrich via authorized chemicals distributor in Semarang-Indonesia. All the chemicals were used directly without further treatments.

2.2. Method

The glucomannan extraction from porang flour was performed using acid hydrolysis employing hydrochloric acid as catalyst. The details of the experiments are as follows:

2.2.1. Glucomannan extraction

Ten grams of crude porang flour was suspended in 500 mL hydrochloric acid solution of predetermined concentration in a 1000 mL three-neck round bottom flask at room temperature. The mixture was heated in a water bath heater to achieve reaction temperature at 60°C under continuous stirring. Reaction temperature was kept at 60°C and continuous stirring was performed to allow acid hydrolysis of starch. When a desired reaction time (0.25, 0.5, 1, 2 and 3 hours) was achieved, the reaction was terminated by quenching of the reaction mixture to room temperature using chilled water. Fifty millilitres sample was withdrawn for reducing sugar analysis. The rest of the reaction mixture was then diluted to 1000 mL with demineralised water, prior to centrifugation (9000 ×g, 30 min, 25°C) to remove the insoluble materials. Subsequently, rotary evaporation is performed to reduce the volume of the filtrate to ~1/3 the original volume. Glucomannan presents in the solution is precipitated overnight with addition of 95% (v/v) ethanol at 4°C, followed by centrifugation (9000 ×g, 40 min, 25°C). The resultant pellet was washed twice with anhydrous ethanol and subsequently isolated by vacuum filtration, before being freeze-dried for 48 h. The dried material was ground and sieved to produce purified glucomannan and was then subjected to yield and purity analysis.

2.2.2. Analysis

Starch content of porang flour was determined by acid hydrolysis method [21]. Total nitrogen analysis was conducted using the Kjeldahl method for protein determination [22]. The protein content was then calculated by applying the nitrogen conversion factor of 5.7, as proposed by the U.S. Food Chemicals Codex (FCC) and European Commission [15]. The reducing sugar of the sample was determined using phenol-sulfuric acid colorimetry [23]. While the glucomannan purity was analysed based on the reducing sugar content of the hydrolysate and porang flour sample solution as previously used by Chua *et al.* [24].

3. Results and Discussion

As the reaction proceeded, it was obvious that the dispersed porang flour granules absorbed water and swelled. As the heat was continuously supplied to the reaction mixture, the starch in the flour underwent gelatinization. Therefore, the shearing effect of stirring is expected to intensively get rid of the gelatinized outer part of the starch molecules and dissolves into the solution [25]. Thus, hydrolysis will soon take place.

3.1.1. Chemical composition of crude porang flour.

Prior to extraction study, porang flour was subjected to composition analysis. The results are presented in Table 1.

Table 1. Chemical composition of porang flour (% w/w).

Component	This work	Literature [26]
Glucomannan	57.56	59-60
Starch	24.64	10-30
Protein	7.43	5-14
Soluble sugar	4.86	3-5
Ash	5.51	3.4-5.3
Inorganics	-	2.6-7

It is clear in Table 1 that glucomannan and soluble sugar contents of the porang flour used in this study were lower than those of porang flour obtained from the corm of porang plant grown in China [26], but still within the range of glucomannan content of porang flour produced in Thailand [8]. The protein content fell within the range of that mentioned in the literature. However, the starch content was slightly on the upper level of that given in the literature [26]. This suggests more extensive starch removal via acid hydrolysis is required to obtain high purity glucomannan. The differences in chemical composition of porang flour may be affected by ecological environment, growing condition and harvest time [27]. In addition, the utilization of fertilizers, shades, planting densities and flour manufacturing process also may significantly influence the chemical composition of porang flour [13, 28, 29].

3.1.2. Effect of catalyst concentration

The effect of catalyst concentration ranging from 0.03125 M to 1 M was studied by hydrolysis at 60°C using flour: water ratio of 1:50 for 1 hour. The results are shown in Figure 1 and Figure 2.

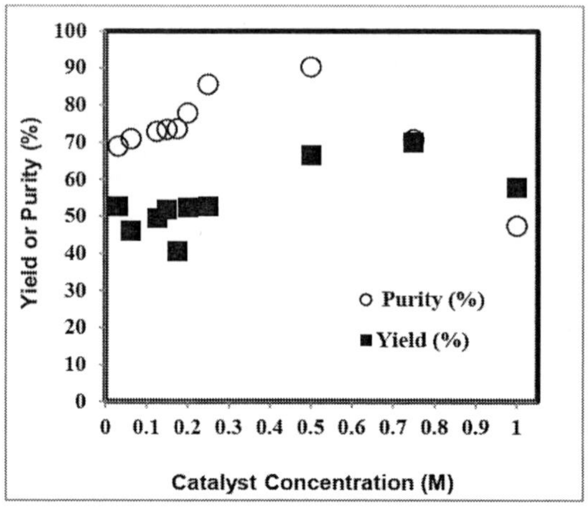

Figure 1. Yield and purity of glucomannan at various catalyst concentrations

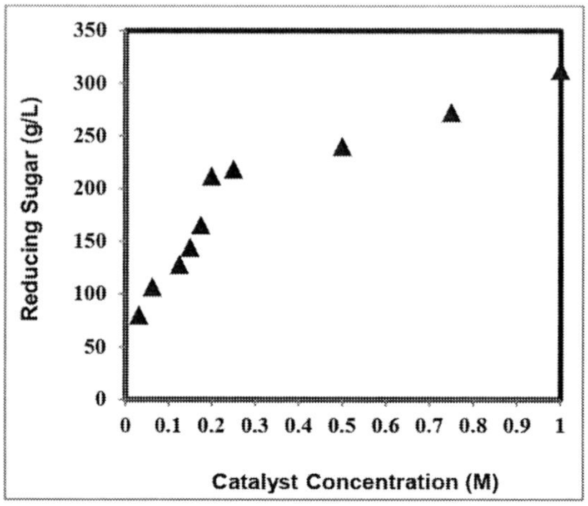

Figure 2. Reducing sugar at various catalyst concentrations

As seen in Figure 1, glucomannan with highest purity (90.18% w/w) was obtained from hydrolysis of porang flour using 0.5 M hydrochloric acid solution. Since homogeneous acid hydrolysis of starch is governed by the acidity (pH) of the solution, which is directly proportional to the concentrations of the hydroxonium ions (H_3O^+) exist in the solution [19], then the concentration of hydroxonium ions at that condition is 0.5 M. Based on this principle, similar result was reported by Tanaka *et al.* for recovery of glucomannan from porang flour using sulphuric acid [30]. The highest recovery was achieved using 0.25 M sulphuric acid, which corresponded to 0.5 M hydroxonium ions.

Glucomannan yields varied slightly with catalyst concentrations used. However, the average glucomannan yield was 54 %. Since glucomannan was obtained from a series of processing step including hydrolysis, filtration, centrifugation, vaporization, drying, grinding and sieving, then loss of glucomannan within this process was inevitable consequence.

Generally, the rate of hydrolysis and the reducing sugar formation rise with increasing acid concentration (Figure 2), possibly due to the increase in the activity of hydrogen ions taking part in the reaction as catalyst [17]. Theoretically, purity of glucomannan should increase steadily as more starch as the main impurity being hydrolyzed into reducing sugar. However, purity of glucomannan decreased when the catalyst concentration was higher than 0.5 M. Usually, the treatment with concentrated acid and/or high temperature causes the degradation of polysaccharides to form byproducts such as furfural and hydroxymethylfurfural, and thus reducing glucomannan purity [31]. Over decomposition of glucomannan into other products may have been the cause of this phenomenon [30].

3.1.3. Effect of reaction time.

The effect of reaction time ranging from 0.25 hour to 3 hour was studied by hydrolysis at 60°C using flour: water ratio of 1:50 employing 0.5 M hydrochloric acid solution as catalyst. The results are depicted in Figure 3 and Figure 4.

Figure 3 shows that yield and purity of glucomannan increase as hydrolysis goes by to 1 hour, but level off afterward. Highest yield (66.60%) and purity (90.18% w/w) of glucomannan was achieved at 1 hour reaction. Acid hydrolysis leads to partial degradation of polysaccharides into soluble lower molecular weight polymers, oligosaccharides and monosaccharides. The proportion of the soluble compounds depends on the operation conditions. Temperature, acid concentration, and reaction time are the most crucial parameters in the hydrolysis of hemicelluloses, since they affect hydrolysis rate and selectivity [32]. Degradation of polysaccharide, such as ι-carrageenan under mild acid conditions required a longer time (3 hour) than that for κ-carrageenan (1.5 hour) [33].

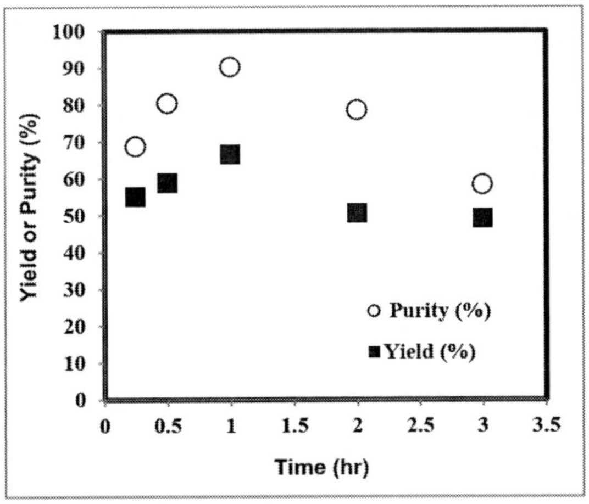

Figure 3. Yield and purity of glucomannan at various reaction times

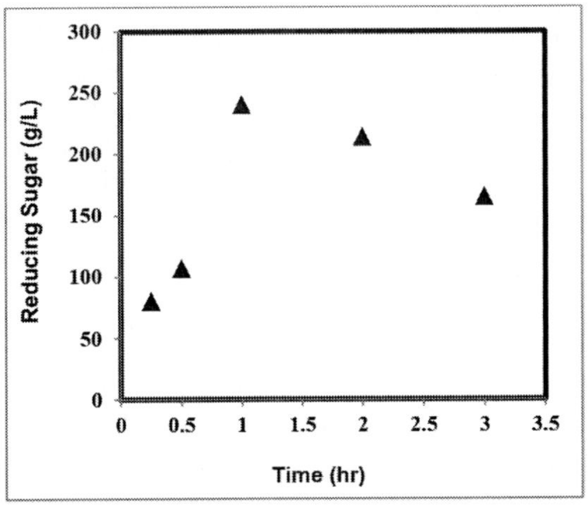

Figure 4. Reducing sugar at various reaction times

It is observed in Figure 4 that reducing sugar concentration increased proportionally to the reaction time in the beginning of the hydrolysis process, but level off after 1 hour. Theoretically, hydrolysis of starch will produce 60% glucose [34]. Prolong reaction time decreased the reducing sugar concentration indicating possible decomposition of glucose to degradation products. In the acid hydrolysis of polysaccharides, the acid catalyzes not only the conversion of polysaccharides into monosaccharides, but also there are further degradation reactions of monosaccharides into furan and carboxylic acids [35]. Decreased of monosaccharides from acid hydrolysis of polysaccharides due to degradation under prolong time and increased acid concentrations was also reported previously [36].

This proposed extraction method resulted in a comparable purity of glucomannan with that previously reported by Xu *et al.* [37] and Chua *et al.* [38], which is about 90% (w/w). However, the method proposed in this work is simpler, lower temperature and requires shorter extraction time. The ethanol washing technique used by Xu *et al.* [37] employed 40% (v/v) ethanol solution, feed:liquid ratio of 1:6, temperature 68°C for 4 hour to obtain 90% (w/w) purity. Chua *et al.* [38] also used ethanol washing to purify glucomannan from porang flour utilizing 50% (v/v) ethanol solution for multi stage washing at room temperature to achieve 92% (w/w) purity. The differences between the method of Xu *et al.* [37] and Chua *et al.* [38] were the temperature, ethanol concentration and the number of stages. Unfortunately, both research groups did not report the yield obtained from their experiments.

4. Conclusion

Based on the experimental results, it can be concluded that acid hydrolysis offers attractive potential as an efficient method for extraction of glucomannan from crude porang flour. Both catalyst concentration and reaction time affect the yield and purity of glucomannan significantly. Possible degradation of glucomannan and reducing sugar under high catalyst concentration and prolong reaction time was observed. Glucomannan with highest purity (90.18% w/w) was obtained from acid hydrolysis of porang flour using 0.5 M hydrochloric acid at 60°C for 1 hour.

Acknowledgement
The authors would like to express their gratitude to Ministry of Education and Culture the Republic of Indonesia for its financial support through Non Tax National Revenue-Diponegoro University Budget Execution List Year 2016 under contract No.: 1052-43/UN7.5.1/PG/2016.

References
[1] Katsuraya S, Okuyama K, Hatanaka K, Oshima R, Sato T and Matsuzaki K 2003 *Carbohydr. Polym.* **53(2)** 183
[2] Chua M, Baldwin T C, Hocking T J and Chan K 2010 *J. Ethnopharmacol.* **128(2)** 268
[3] Yao-ling L, Rong-hua D, Ni C, Juan P and Jie P 2013 *J. Single Mol. Res.* **1(1)** 7
[4] Alonso-Sande M, Teijeiro-Osorio D, Remunan-Lopez C and Alonso M J 2009 *Eur. J. Pharm. Biopharm.* **72(2)** 453
[5] Ohashi S, Shelso G L, Moirano A L and Drinkwater W L 2000 Clarified konjac glucomannan. United States Patent 6,162,906.
[6] Wootton A N, Luker-Brown M, Westcott R J, and Cheetham P S J 1993 *J. Sci. Food Agric.* **61** 429
[7] Flach M and Rumawas F 1996 *Plants Yielding Non-Seed Carbohydrates* (Leiden: Prosea Foundation and Backhuys Publishers).
[8] Tatirat O and Charoenrein S 2011 *LWT – Food Sci. Technol.* **44(10)** 2059
[9] Yoshimura M, Takaya T and Nishinari K 1998 *Carbohydr. Polym.* **35(1)** 71
[10] Aprianita A, Vasiljevic T, Bannikova A and Kasapis A 2014 *J. Food Sci. Technol.* **51(9)** 1784
[11] Kumoro A C, Ngoh G C, Hasan M, Ong C H and Teoh E C 2008 *Asian J. Sci. Res.* **1(4)** 412
[12] Retnowati DS and Kumoro AC 2012 *Reaktor* **14 (1)** 46
[13] Jianrong Z, Donghua Z, Srzednicki G, Kanlayanarat S and Borompichaichartkul C 2008 Proc. Proc. Asia-Pac. Symp. on assuring quality and safety of agri-foods Vol. 837 (Bangkok: International Society for Horticultural Science) pp. 345
[14] Zhao J, Zhang D, Srzednicki G, Kanlayanarat S and Borompichaichartkul C 2010 *Int. Food Res. J.* **17** 1113
[15] Parry J M 2011 Konjac glucomannan *Food Stabilisers, Thickeners and Gelling Agents* ed A Imeson (United Kingdom: John Willey & Sons, Ltd.) pp 198 – 216
[16] Taherzadeh M J and Karimi K 2007 *BioRes.* **2** 472
[17] Tasi´c M B, Konstantinovi´c B V, Lazi´c M L and Veljkovi´c V B 2009 *Biochem. Eng. J.* **43** 208
[18] Fontana J D, Mitchell D A, Molina O E, Gaitan A, Bonfim T M B, Adelmann J, Grzybowksi A and Passos M 2008 *Food Technol. Biotechnol.* **46 (3)** 305
[19] Salmi T, Murzin D Y, Maki-Arvela P, Kusema B, Holmbom B and Willfor S 2014 *AIChE J.* **60(3)** 1066
[20] Betancur A D and Chel G L 1997 *J. Agric. Food Chem.* **45** 4237
[21] Puswatien P, Siong T E, Kantasubrata J, Craven G, Feliciano R R and Judprasong K 2011 *The ASEAN Manual of Food Analysis* (Phutthamonthon: Institute of Nutrition, Mahidol University)
[22] AOAC 1999 *Official Methods Of Analysis Method 988.05* (Gaithersburg: AOAC International)
[23] Cuesta G, Suarez N, Bessio M I, Ferreira F and Massalsi H 2003 *J. Microbiol. Methods* **52** 69
[24] Chua M, Chan K, Hocking T J, Williams P A, Perry C J, Baldwin T C 2012 *Carbohydr. Polym.* **87** 2202
[25] Kolusheva T and Marinova A 2007 *J. Univ. Chem. Technol. Metall.* **42(1)** 93
[26] Li B, Xia J, Wang Y and Xie B J 2005 *J. Agric. Food Chem.* **53** 7404
[27] Brown D 2000. *Aroids, Plants of the Arum Family* (Oregon: Timber Press)
[28] Douglas J A, Follett J M and Waller J E 2005 *Acta Hortic.* **670** 173

[29] Douglas J A, Follett J M and Waller J E 2006 *N. Z. J. Crop Hortic. Sci.* **34** 139
[30] Tanaka Y, Okamoto K, Matsushima A, Ota T, Matsumoto Y and Akasaki T 2013 *Anal.Sci.* **29** 1049
[31] Otieno D O and Ahring B K 2012 *Biores.Technol.* **112** 285
[32] Bian J, Peng P, Peng F, Xiao X, Xu F and Sun R C 2014 *Food Chem.* **156** 7
[33] Quemener B and Lahaye M 1998 *J. Appl. Phycol.* **10** 75
[34] Woiciechowski A L, Nitsche S, Pandey A and Soccol C R 2002 *Braz. Arch. Biol. Technol.* **45(3)** 393
[35] Chandel A K, Chandrasekhar G, Radhika K, Ravinder R and Ravindra P 2011 *Biotechnol. Mol. Biol. Rev.* **6(1)** 8
[36] Kumar R and Wyman C E 2008 *Carbohydr. Res.* **343(2)** 290
[37] Xu W, Wang S, Ye T, Jin W, Liu J, Lei J, Li B and Wang C 2014 *Food Chem.* **158** 171
[38] Chua M, Chan K, Hocking T J, Williams P A, Perry C J and Baldwin T C 2012 *Carbohydr. Polym.* **87** 2202

IOP Conf. Series: Materials Science and Engineering **162** (2017) 012021 doi:10.1088/1757-899X/162/1/012021

The capacity of modified nickel catalysts derived from discharged catalyst of fertilizer plants for NO_x treatment

T M P Ha[1, 2], N T Luong[2], P N Le[2].
[1] University of Social Science and Humanity, Vietnam National University in Ho Chi Minh City
[2] Vietnam Petroleum Institute

E-mail : phuc.htm@hcmussh.edu.vn; nguyenlp.pvpro@vpi.pvn.vn.

Abstract. In Vietnam for recent years, a large amount of hazardous waste containing nickel (Ni) derived from discharged catalyst of fertilizer plants has caused environmental problems in landfill overloading and the risk of soil or surface water sources pollution. Taking advantage of discharged catalyst, recycling Ni components and then synthesizing new catalysts apply for mono-nitrogen oxides (NO_x) treatments is an approach to bring about both economic and environmental benefits.

This study was carried out with the main objective: Evaluate the performance of modified catalysts (using recovered Ni from the discharged RKS-2-7H catalyst of Phu My Fertilizer Plant) on NOx treatment. The catalysts was synthesized and modified with active phases consist of recovered Ni and commercial Barium oxide (BaO), Manganese dioxide (MnO_2) / Cerium (IV) oxide (CeO_2) on the support Aluminium oxide (γ-Al_2O_3). The results show that the modified catalysts with Ni, Ba, Ce was not more beneficial for NO_x removal than which with Ni, Ba, Mn. 98% NO_x removal at 350°C with the start temperature at 115°C and the T_{60} value at 307°C can be obtained with 10Ni10Ba10Mn/Al catalyst.

1. Introduction

Fertilizer is a small industry in Vietnam, whereas an average of 105,7 m^3 of waste catalyst discharged every year from reforming, methanogenesis and NH_3 synthesis processes of one fertilizer plant. It becomes a source of solid waste pollution because of containing a lot of metal and toxic contaminants (Al_2O_3, ZnO, Fe_2O_3, Fe_3O_4, NiO, Cr_2O_3, CuO, $CuCO_3$, CaO, CoO, MoO_3 and so on). In particular, the catalysts in the reforming and methanogenesis processes often contain Nickel (Ni) which is a heavy metal with the potential to pollute soil and water sources. Whereas, the Ni recycling from spent catalysts is a co-benefits approach including both environmental and economical solutions. According to Yang, the commercial Ni compounds cost is considerably high (21,46 USD/1kg in 2010 compared to 3,3 - 5,5 USD/1kg spent Ni) [1]. Regarding to the environmental application, Ni is known as a transition metal valuable which is emerging in term of research to be the precursor for polluted air treatment by catalysis, replaces for expensive precious metals [2-4]. Besides, in the field of industrial and civil polluted air treatment, using catalyst attached directly to the treatment device helps to transform unwanted substances into non-toxic form. In addition, mono-nitrogen oxides treatments (NO_x, produced during combustion, included nitric oxide -NO and nitrogen dioxide - NO_2) commonly exists in the emission of fossil-fuel or natural-gas-fuel processes which can lead to photochemical smog phenomenon. In regard with NO_x emissions from internal combustion engines, three-way NO_x decomposition (de-NO_x) catalysts with the combination of several metal compounds had been commercialized. These types of catalyst based on precious metal precursors such as Pt (Platinum), W (Wolfram) or economic transition metals such as Ni, Ti (Titanium), Co (Cobalt), Ce (Cerium) and so forth. Acknowledge the situation, this research mainly questioned on the de-NO_x performance of modified Ni catalysts derived from discharged catalyst of fertilizer plants.

2. Materials and methods

2.1. Components of the modified Ni catalysts

In the modified-catalyst preparation, Al_2O_3 is used as the catalyst support, recycled Ni solution from spent catalyst is used as the catalytic active sites along with Ba, Mn, Ce as catalytic supporting components. A proposition of Huong et al explain the mechanism of the NO_x decomposition by reducing reaction on Ni sites could be as the following protocol: NO_x molecules are adsorbed on Ni and decomposed into N_2; oxygen atoms separated from NO_x is retained on the oxygen vacancies of catalyst, and then ionized and pumped out through the redox reactions [3]. In a number of related research [3-8], the de-NO_x catalysts contain 0.1 – 30 wt% active phases by weight, whereas the Ni, Ce, Co active phase catalysts contain 3 – 20 wt%. Hence, this research prepared 5 – 30 wt% recycled-Ni/Al catalysts to test the NO_x conversion catalytic capability of recovered-Ni components. Referring to the catalytic support, γ-Al_2O_3 is selected due to the high surface areas (about 130 - 280 m^2g^{-1}), uniform porous structure, greatly capable of dispersing active phases, extremely temperature resistant and very high mechanical strength [9]. Thus, γ-Al_2O_3 can increases catalytic durability in harsh reaction conditions [10]. Besides, be readily available and reasonably priced, it is no doubt that fine grain alumina is the most appropriate support for the high temperature NO_x catalytic conversion.

Regarding to the Ba component in de-NO_x catalysts which probably enhances catalytic reactions through types of mechanism as giving electrons for catalytically active metals promotion to a higher potential energy at common surface, which means they are less tightly bound [11]; releasing O_2 to form oxygen vacancies on the catalytic active phases; increasing catalyst defects on which NO easily adsorbs and dissociates [11]; increasing the NO_x retention ability and time on catalyst surface [12]; basic sites play an important role in the NO_2^-formation and adsorption - the intermediate products of NO decomposition to N_2 [13]. The amount of 5 – 20 wt% BaO are prepared on Ni/Al catalysts to investigate the NO_x catalytic decomposition by basic Ba elements.

Be a representative of d-block transition metals, Mn component in de-NO_x catalysts can generate anion-oxide defects which also improves the heat resistance for catalysts [14]. In the prepared catalyst, Mn could exist as Manganese dioxide (MnO_2 - black powders) and an inconsiderable amount of Manganese (III) oxide (Mn_2O_3) with the weight ratio is 5-20 wt% MnO_2. Besides, another type of transition metal is f-block elements with Ce as a significant component in de-NO_x catalytic activity. It is reported that Ce catalytic performance in NO_x treating reactions is only less favorable than alkaline and earth-alkaline metals. Studies by Zhu [15], Nguyen [6], Delmon [16], Pomonis [17, 18], Nitadori [19, 20] investigated on this component as favorable active or supporting sites in $LaCoO_3$, La-Sr-Ce-Fe-O, $LaMO_3$ (M = Mn, Fe, Co) catalysts for the decomposition of NO_x. In de-NO_x catalysts, Ce creates O_2 storages which reduces the "poisoned by oxygen" syndrome for active sites [15]; increase the metals dispersion on support thereby reduces the metal cluster problem and improves temperature resistance of supports [15]; demotes NH_3 product in NO_x reduced reactions [6] (increases N_2 product in another words). Ce exists in the catalyst as the form of Cerium (IV) oxide (CeO_2 - light yellow crystals) and a small amount of Ce_2O_3, with the weight ratio is 5-20 CeO_2 wt%.

2.2. Experimental and analysis methods

2.2.1. Recovery Ni from discharged catalyst

The studied waste catalyst has the trade name RKS-2-7H which is used in the secondary reforming process at Phu My Fertilizer Plant. The Ni content in this waste catalyst is approximately 12.01%. The Ni recovery process from waste catalyst in laboratory scale is composed by three steps: material handling (crushed and sifted, then decoking); preparation (using 1.5M HNO_3 acid solution, 1:20 g/ml solid:liquid ratio, the temperature and time of digestion are 100°C in 1 hour) [21].

2.2.2. Catalyst preparation

Two main methods are used widely to prepare catalytic active phases are co-precipitation and metal impregnation. The co-precipitation method (precipitation deposition, **Figure 1**) is chosen due to the research orientation for making up to 40 wt% soaked-up metal components in catalysts. Nitrate solutions are intentionally precipitated due to their good solubility in water, popularity and low-price (compared with oxalate or formate), easy removal after been furnaced (better than halides, sulfates). The pH value-stabilization at 10 is taken by NH_3 solution. At pH 10, the surface of alumina support will be negatively charged (so as the electrostatic interaction with Ni positively charged component could be better) [22]. The reducing pretreatment is performed to make up the metallic active phase (by the mixture of 30 vol% H_2 : 70 vol% N_2 at 550°C in 4 hours). For the $10Ni_{nc}10Ba10Mn/Al$ catalyst, pure $Ni(NO_3)_2$. $6H_2O$ is used instead of recovered Ni solution and there is no pH adjustment required.

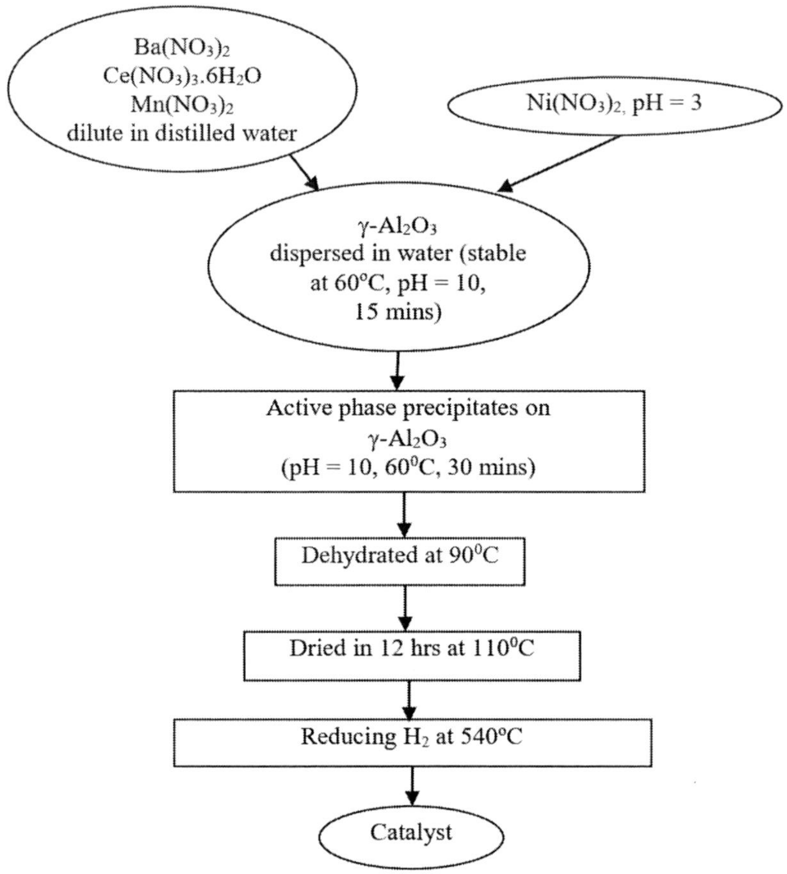

Figure 1. Catalyst synthesis process from the recovered Ni

2.2.3. Model-based testing

After synthesis and pretreatment processes, catalysts would be tested for catalytic activity in NO_x conversion through the micro-flow control system. Each catalyst loading is 0.1 g. The testing temperature is within range from the room temperature to 500°C. Each of CO/CO_2 and NO/NO_x concentrations will be electrical recorded every 30 seconds by the Ultramat 6 and CLD 700 AL equipment, then expressed as graphs by the Virtual Bench program.

The testing air mixture includes: 500 ppm NO, 10000 ppm CO as a co-existing reducing agent, and N_2 as carrier gas. The gas volume flow rate is 10 $L.h^{-1}$ (~166 $mL.min^{-1}$). The gas hourly space velocity value (GHSV) is 110000 $hours^{-1}$.

2.2.4. Physicochemical properties analyzing methods

The specific surface area of samples are measured – including the pore size distribution by the BET (liquid nitrogen adsorption method). Chemical properties of catalysts are determined by the XRD method (X-ray diffraction) in regard to existing crystal phases. The BET and XRD instrument are TriStar 3020 and Bruker AXS D8 respectively.

2.2.5. Data processing and analysis

The original data of gaseous concentrations are demonstrated as light-off graphs (conversion efficiency vs temperature) to identify the specific values: T_{60}, the conversion beginning temperature, the maximum conversion efficiency in the tested temperature range, the conversion efficiency at 350°C. Here are the much detail definition:

T_{60}: the temperature at which the conversion efficiency is 60%. For the NO concentration input is 500 ppm (613 mg/Nm^3 NO), 60% conversion efficiency is appropriate to adapt local environmental standards (250 mg/Nm^3).

Conversion efficiency at 350°C: according to related research showing the peak performance of NO_x at about 350°C [4, 7, 8]. This value can express the low-temperature de-NO_x performance of catalysts as well as be comparative to other related studies.

3. Results and discussion

3.1. Effect of the Ni active sites to catalytic activity

The tested Ni content changed from 5% to 10%, 20% and 30% by weight and denoted as xNi/Al (x = 5, 10, 20, 30 respectively). It can be seen that when Ni content increases from 5% to 10% by weight, the catalytic activity increased significantly (16% up to 63% at 350°C). However, it is not favorable for NO_x conversion as increasing the Ni content up to 20 or 30 wt%.

Results by XRD of the Ni/Al catalyst show that the Ni peaks appear at positions of $2\theta = 45°$ and 52°, which indicates the metallic Ni existing on the surface (**Figure 2**). When Ni content in the samples gradually increase, the intensity of Ni° peaks also gradually increase along with a decrease in the intensity of the Al_2O_3 peaks. Obviously, recycled Ni component is well-distributed on the catalyst surface. In fact, Ni° plays a role in Ni/Al catalyst for NO_x conversion. The more Ni sites exists on the surface, the earlier NO_x conversion starts. It was evidenced that the catalyst 20Ni/Al and 30Ni/Al continuously show the earlier NO_x conversion beginning in accordance with 5Ni/Al and 10Ni/Al samples. However in cases of 20Ni/Al and 30Ni/Al, increasing too much active sites on the support can cause obscuring and agglomeration phenomena, thereby reducing available active sites of the catalyst. The Ni contribution induce decreases in the specific surface area and pore volume. It appears that samples with x = 5 and 10 decreases 8 and 14 m^{2}g^{-1} in specific surface area and 0,01 and 0,04 cm^{3}g^{-1} in pore volume, compared to non-Ni/Al_2O_3 sample. And they are up to 23 and 39 m^{2}g^{-1} specific surface along with 0,05 and 0,09 cm^{3}g^{-1} pore volume respective decreasing happened to samples with x = 20 and 30, compared to non-Ni/Al_2O_3 sample.

Among the catalysts prepared and tested, 10Ni/Al which content 10% Ni by weight has the only 116 m^{2}g^{-1} surface area and 0.21 cm^{3}g^{-1} pore volume (Table **1**), but it has shown the maximum catalytic effectiveness of Ni sites in NO_x dissociation and removal (reaches T_{60} at 332°C, reaches 63% at 350°C and shows the highest efficiency with 98,5% at 471°C).

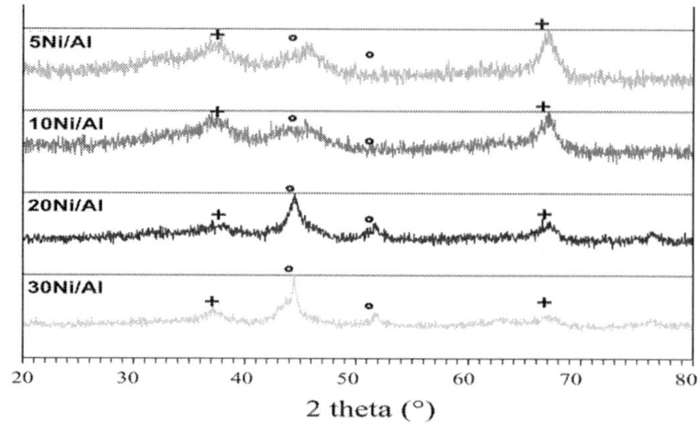

Figure 2. X-ray diffraction pattern of the xNi/Al catalyst - (°) Ni, (+) γ-Al_2O_3

3.2. Effect of supporting sites to catalytic activity

3.2.1. Effect of Ba content to catalytic activity

The Ba component was aimed at increasing the defect as well as the adsorption affinity for NO_x of catalysts. The Ba content is prepared on 10Ni/Al are 5%, 10%, 15% and 20% by BaO weight and denoted as 10NiyBa/Al (y = 5, 10, 15, 20 respectively). Results from the experiments shows that the catalyst with 10 wt% Ba content exhibits the best NO_x conversion.

Compare to the non-Ba catalyst, the NO_x performance finished promptly with the optimal efficiency is 98% at 407°C (64°C sooner than 10Ni/Al); although the Ba addition neither start the conversion and reach T_{60} value earlier nor achieve a higher performance at 350°C (63%, equal to 10Ni/Al). In fact, the NO_x performance is only enhanced after 350°C (the conversion rate is 0,61%/°C compared to 0,29%/°C in similar temperature period of 10Ni/Al). The BaO addition reduces the surface area massively (about 25% of 10Ni10Ba/Al compared to 10Ni/Al); this can be the reason for non-improved catalytic activity in early stages of Ba modified samples. However, the XRD results show that 10Ni10Ba/Al has clearer Al_2O_3 and Ni° peaks than non-Ba samples. Hence, despite there is the surface area and pore volume smaller than non-Ba modified samples (decreased 29.1 m^2g^{-1} and 0,04 cm^3g^{-1} respectively) , the 10Ni10Ba/Al sample has a better distribution of Ba and Ni on surface.

Noticeably, this result was approximate to the drawing of Choi et al in which their 2Pt20Ba/Al_2O_3 catalyst shows 65% NO_x conversion at the same temperature (350°C) [8]. Therefore, it is no doubt to say that with merely two catalytic components Ni and Ba coated, the modified Ni catalyst shows their competition to traditional precious ones.

3.2.2. Effect of Mn content to catalytic activity

Mn is added to the catalyst 10Ni10Ba/Al with the content of 5%, 10%, 15% and 20% by MnO_2 weight and denoted as 10Ni10BazMn/Al (with z = 5, 10, 15, 20 respectively). With the amount of 10 wt% Mn, the catalyst is improved marvelously compared to non-Mn modified catalysts (lower beginning temperature, T_{60} reached sooner, the maximum efficiency is 98% at only 350°C). The XRD results (**Figure 3**) show that the MnO_2 peaks of 10Ni10Ba10Mn/Al sample have the highest intensity that can express a large amount of Mn supporting sites on the catalyst surface.

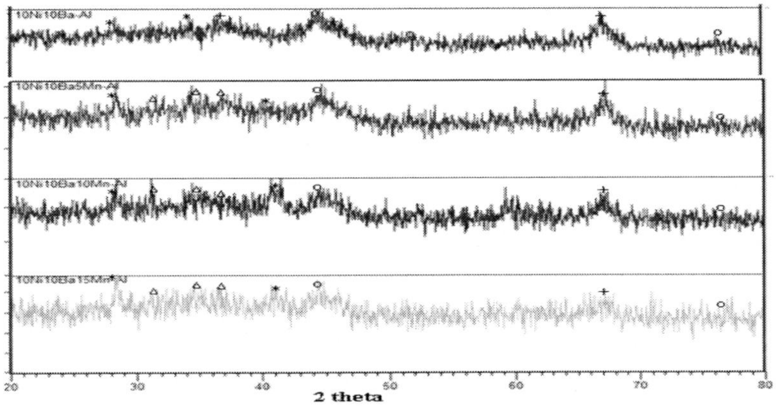

Figure 3. X-ray diffraction pattern of the 10Ni10BazMn/Al catalyst - (°) Ni, (+) γ-Al_2O_3, (*) $BaAl_2O_4$, (Δ) MnO_2

The Mn component expectedly increases anion oxide defected sites and improves heat resistance for catalyst [14]. As expected, the modification with Mn enhances the catalytic activity significantly in most of values. By comparison with non-Mn modified samples, the catalytic performance starts earlier (83°C earlier), reaches T_{60} at a lower temperature (36°C lower), and better still complete the conversion at only 350°C (98%, 47°C sooner). This results are well suited to reports by Iwamoto [14], which will help dissociated adsorption of NO_x takes place better on active Ni sites as well as supporting Mn sites. Compare to a related study of Nguyen et al with the Pt/20BaMn/Al catalyst at an approximate temperature, the NO_x conversion performance of this study is considerably elevated (98% vs 74-84%) [5]. Although the Mn addition makes a considerable decrease in the surface area (17,1 m^2g^{-1}), it only has a slight decline in the pore volume (0,02 cm^3g^{-1}, Table **1**), so as the structure of Mn modified catalysts are more favorable than the previous.

Table 1. Physical properties and characteristics of the NO_x conversion process of tested catalysts

	Begin at (°C)	**T_{60} (°C)**	**At 350°C (%)**	**Max. (%) at (°C)**	**$S_{B.E.T}$ (m^2g^{-1})**	**Pore vol. (cm^3g^{-1})**
Al	/	/	/	/	130	0,25
5Ni/Al	260	494	16	68, 500	122	0,24
10Ni/Al	203	**332**	**63**	**98, 471**	116	0,21
20Ni/Al	**183**	354	59	95, 500	107	0,20
30Ni/Al	188	398	43	94, 500	91	0,16
10Ni5Ba/Al	220	362	54	96, 428	109	0,21
10Ni10Ba/Al	198	**343**	**63**	**98, 407**	86,9	0,17
10Ni15Ba/Al	**196**	362	53	90, 445	75,4	0,14
10Ni20Ba/Al	211	379	43	95, 485	59,2	0,12
10Ni10Ba5Mn/Al	209	383	44	62, 385	73,5	0,15
10Ni10Ba10Mn/Al	**115**	**307**	**98**	**98, 350**	69,8	0,15
10Ni10Ba15Mn/Al	193	381	41	60, 381	65,7	0,13
10Ni10Ba20Mn/Al	209	/	41	55, 376	/	/
10Ni10Ba5Ce/Al	**146**	321	72	85, 387	77,4	0,14
10Ni10Ba10Ce/Al	158	**295**	**79**	**99, 416**	77,0	0,13
10Ni10Ba15Ce/Al	162	335	64	84, 365	73,5	0,13
10Ni10Ba20Ce/Al	179	396	49	80, 411	/	/

3.2.3. Effect of Ce content to the catalytic activity

Beside Mn component which creates a favorable defection on catalyst surface, another component is popularly used in raising NO_x conversion is Ce. According to involved reports, the CeO_2 addition could help other active and supporting sites to be dispersed better on support, creates oxygen trap sites so as increases the product selectivity for N_2 by shifting the balance of $NO \leftrightarrow O_2 + N_2$ reaction to the forward direction [6, 15, 23]. To compare separately the effectiveness of Ce and Mn in changing the physical structure of the catalyst, CeO_2 is added to the sample 10Ni10Ba/Al. With the content of 5%, 10%, 15%, 20% by CeO_2 weight, the samples are denoted as 10Ni10BatCe/Al (with t = 5, 10, 15, 20 respectively).

The modification with Ce alters catalytic activity and the best improvement is performed by sample 10Ni10Ba10Ce/Al with (Table **1**) with 79% NO_x concentration is converted. These results was also more favourable than the drawing of Ha et al that their 15Ni10CeBa/Al catalyst shows only 38% NO_x conversion at the same temperature (350°C) [4].

By comparison with Ba-modified catalyst, the NO_x conversion starts earlier (40°C), reaches T_{60} at a lower temperature (48°C), and gives a higher 350°C conversion efficiency (16%). However, the conversion completes later than the non-Ce sample (9°C, at 99%). The physical structure analysis results (Table **1**) show that the Ce addition reduces both surface area and pore volume of the catalyst (respectively 9,9 m^2g^{-1} less and 0,04 cm^3g^{-1} slightly more than the Mn modified samples). The catalytic improvement (through the better metals dispersion on support) of Ce modified samples taken at the same range of temperature is also agreed by Zhu et al [15].

Nevertheless, the improvement is not more impressive when brought into comparison with 10Ni10Ba10Mn/Al. Although the 10Ni10Ba10Ce/Al reached T_{60} value earlier (12°C earlier), the efficiency at 350°C is only 79% which is 98% by the Mn modified catalyst, worse still the conversion complete much later than both Ba-Mn modified and Ba modified samples.

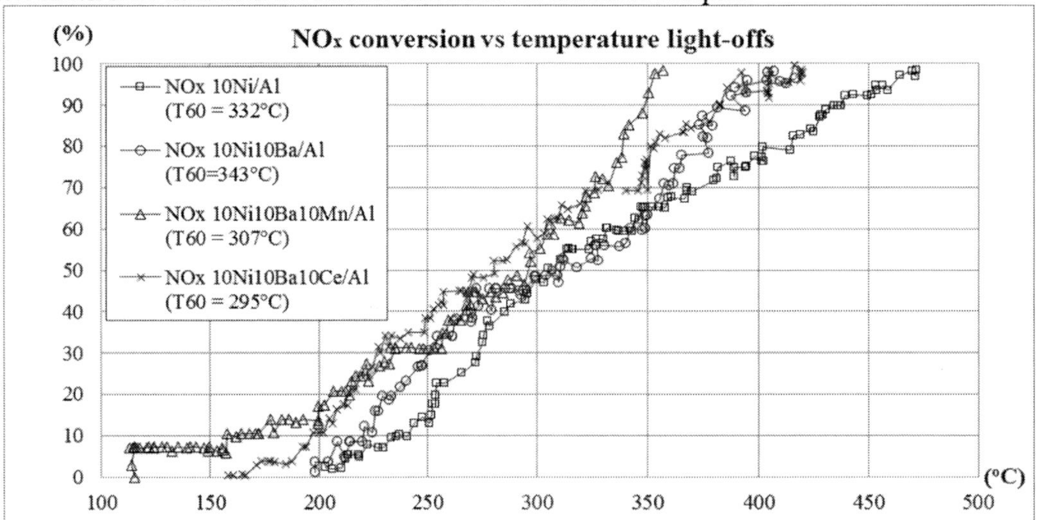

Figure 4. NO_x conversion vs temperature light- offs over modified catalysts

3.3. The correlation of NO_x conversion capabilities of recovered Ni and trade Ni catalysts

Some catalysts are prepared from commercial $Ni(NO_3)_2$ instead of recovered Ni solution with the similar component ratio as the 10Ni10Ba10Mn/Al sample (denoted as $10Ni_{nc}10Ba10Mn/Al$). The NO_x light-off comparative results of these catalysts is shown in **Figure 5**. It can be seen that there is a slightly difference between these catalyst versions' performance. By comparison with the recovered Ni catalyst (°), the commercial Ni catalyst (□) takes a conversion which starts earlier (7°C), reaches T_{60} at a lower temperature (2°C), gets a higher efficiency at 350°C (2%). Due to these extent equivalence, it is capable of putting recovered Ni from fertilizer plants to good use for NO_x conversion.

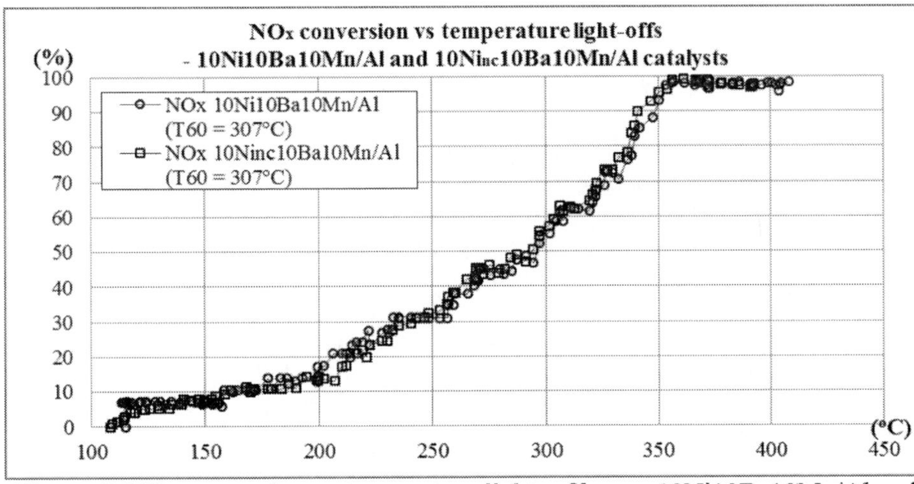

Figure 5. NO_x conversion vs temperature light- offs over 10Ni10Ba10Mn/Al and $10Ni_{nc}10Ba10Mn/Al$ catalysts

4. Conclusion

Among the prepared and tested catalysts, the catalyst 10Ni10Ba10Mn/Al shows the earliest conversion beginning (115°C), reaches T_{60} value promptly (307°C) and achieves the highest performance of 98% at 350°C. This sample has the surface area of 69,8 m^2g^{-1} and 0,15 cm^3g^{-1}.

In the performance of the catalyst 10Ni10Ba10Mn/Al, the NO_x conversion mechanism is very complex, including the reduction, trap and direct thermal-decomposition. The main components of catalyst include active Ni sites which promote the NO_x direct thermal-decomposition process at low temperature due to enhancing the dissociative adsorption of NO molecules on the surface of the catalyst; Ba sites are capable in trapping NO_x molecules, support for the NO_x reduction by CO; Mn sites also support for the NO_x absorption on the catalyst surface through oxygen defects and are potential to improve heat resistance for catalyst.

The use of recovered Ni derived from waste catalyst of fertilizer plants instead of commercial Ni to prepare catalyst for NO_x treatment is highly promising. The comparative results in NO_x conversion of those kind of Ni catalysts show a very little different (93% and 95% at 350°C, respectively).

Acknowledgement

The authors acknowledge with pleasure the supports of this work by the Catalyst Research and Development Department, PetroVietnam Research & Development Centre for Petroleum Processing, Vietnam Petroleum Institute; the Department of Catalytic Materials, Institute of Applied Materials, Vietnam Academy of Science and Technology and the Department of Environmental Geography, University of Social Science and Humanity, Vietnam National University in Ho Chi Minh City.

References

[1] Yang Q Z, et al. 2011 Sustainable recovery of nickel from spent hydrogenation catalyst: economics, emissions and wastes assessment *Journal of Cleaner Production* **19**(4) 365-375
[2] Zhu J, et al. 2005 Effect of Ce on NO direct decomposition in the absence/presence of O2 over La1−x CexSrNiO4 (0≤x≤0.3) *Journal of Molecular Catalysis A: Chemical* **234** 99-105
[3] Huong P T T 2012 Preparation and investigation the catalytic activity of Ni/Diatomite for NO_x treatment in *Inorganic Chemistry* (University of Science : Ho Chi Minh City) 98

[4] Ha N T T and Tung N S 2011 Research on the catalyst containing Ni, CeO_2 on the Al_2O_3 support to apply for the NO_x direct thermal decomposition in *Inorganic Chemistry* (University of Science : Ho Chi Minh City)
[5] Nguyen L P, et al. 2011 NO_x removal efficiency and ammonia selectivity during the NO_x storage-reduction process over Pt/BaO(Fe, Mn, Ce)/Al_2O_3 model catalysts. Part I: Influence of Fe and Mn addition *Applied Catalysis B : Environmental* **102**(3-4) 353-361
[6] Nguyen L P, et al. 2011 NO_x removal efficiency and ammonia selectivity during the NO_x storage-reduction process over Pt/BaO(Fe, Mn, Ce)/Al_2O_3 model catalysts. Part II: Influence of Ce and Mn–Ce addition *Applied Catalysis B : Environmental* **102**(3-4) 362-371
[7] Nguyen L P, et al. 2011 A study of the ammonia selectivity on Pt/BaO/Al_2O_3 model catalyst during the NO_x storage and reduction process *Catalysis Today* **176**(1) 424-428
[8] Choi B and Lee K-S 2013 LNT/CDPF catalysts for simultaneous removal of NO_x and PM from diesel vehicle exhaust *Chemical Engineering Journal* 2
[9] Wang Y, et al. 2009 Synthesis and properties of thermostable γ-alumina prepared by hydrolysis of phosphide aluminum *Journal of Alloys and Compounds* **467**(1-2) 405-412
[10] Tuong B V, et al. 2013 Synthesis of g-Al_2O_3 from Binh Tan $Al(OH)_3$ precursor for petrochemical catalyst preparations *Vietnam Petroleum* 28-35
[11] Masaaki H, et al. 2003 Alkali metal-doped cobalt oxide catalysts for NO decomposition *Applied Catalysis B : Environmental* **46**(3) 473-482
[12] Hieu N V and Huy N M 2011Research on catalysts for atmospheric NO_x treatment via the direct thermal decomposition pathway *Inorganic Chemistry* (University of Science : Ho Chi Minh City)
[13] Masaaki H, Yasuyuki D and Masakuni O 2011 Direct decomposition of NO over Ba–Y_2O_3 catalyst prepared by coprecipitation *Chemical Society of Japan* **84**(12) 1383-1389
[14] Iwamoto S, Takahashi R and Inoue M 2007 Direct decomposition of nitric oxide over Ba catalysts supported on CeO_2-based mixed oxides *Applied Catalysis B : Environmental* **70** 146-150
[15] Junjiang Zu D X, Jing L, Yang X, Wua Y 2005 Effect of Ce on NO direct decomposition in the absence/presence of O_2 over $La_{1-x}Ce_xSrNiO_4$ (x=0-0.3) *Journal of Molecular Catalysis A : Chemical* **234** 99-105
[16] Kirchnerova J, Alifanti M and Delmon B 2002 Evidence of phase cooperation in the $LaCoO_3$–CeO_2–Co_3O_4 catalytic system in relation to activity in methane combustion *Applied Catalysis A: General* **231** 65-80
[17] Belessi V C, Ladavos A K and Pomonis P J 2001 Methane combustion on La–Sr–Ce–Fe–O mixed oxides: bifunctional synergistic action of $SrFeO_{3-x}$ and CeO_x phases **31**(3) 183–194
[18] Belessi V C, et al. 2000 Synergistic effects of crystal phases and mixed valences in La–Sr–Ce–Fe–O mixed oxidic/perovskitic solids on their catalytic activity for the NO+CO reaction *Applied Catalysis B : Environmental* **28**(1) 13-28
[19] Nitadori T and Misono M 1985 Catalytic Properties of $La_{1-x}A'_xFeO_3$ (A' = Sr,Ce) and $La_{1-x}Ce_xCoO_3$ *Journal of Catalysis* **93** 459-466
[20] Nitadori T, Kurihara S and Misono M 1986 Catalytic Properties of La1-xA'xMnO$_3$ (A' = Sr, Ce, Hf) *Journal of Catalysis* **98** 221-228
[21] Thu N T A 2014 Research on Nickel recovery from waste catalyst of Phu My fertilizer plant to synthesize new catalyst to apply for industrial emission treatment *Institute for environment and resources, Vietnam National University - Ho Chi Minh City*
[22] Brunelle J P and Poncelet G 1979 *Studies in Surface Science and Catalysis* 211-232
[23] Hong W-J, Iwamoto S and Inouea M 2011 Direct NO decomposition over a Ce–Mn mixed oxide modified with alkali and alkaline earth species and CO_2-TPD behavior of the catalysts *Catalysis Today* **164** 489-494

UV/Fenton photo-oxidation of Drimarene Dark Red (DDR) containing textile-dye wastewater

T Hudaya*, **J Anthonios and E Septianto**
Chemical Engineering Department, Parahyangan Catholic University, Ciumbuleuit 94 Bandung, Indonesia

E-mail : t.hudaya@unpar.ac.id / t_hudaya@yahoo.com.au

Abstract. Textile dye wastewater contains organic pollutants which are non-biodegradable, characterized by low BOD/COD ratio of typically < 0.01. Therefore to treat such wastewater, the destructive wastewater treatment methods which can be utilized are the Advanced Oxidation Processes (AOPs). One of the AOPs method which is the $UV/H_2O_2/Fe^{2+}$ (or UV/Fenton) offers not only relatively low cost but also quite effective (in terms of color removal and reaction time) treatment. This particular research aimed to optimize the conditions of UV/Fenton photo-oxidation process for Drimarene Dark Red containing textile-dye wastewater. The two main operating conditions to be optimized were the initial concentration of H_2O_2 ranged between 0.022–0.078 %-w and the mol ratio of Fe^{2+} : H_2O_2 was varied from 1 : 13 up to 1 : 45, using the Central Composite Design experimental matrix. The photo-oxidation was carried out at the optimum pH of 3 from some previous experiments. The best processing conditions of the photo-oxidation of Drimarene Dark Red (DDR) were found at the initial concentration of H_2O_2 at 0.050%-w and the mole ratio Fe^{2+}: H_2O_2 of 1 : 22. Under these conditions, the measured 2nd order pseudo-rate constantwas 0.021 $M^{-1}.min^{-1}$. The DDR color removal of 90% was surprisingly achievable within only 10 minutes reaction time.

1. Introduction

Several type of wastewaters come from industrial activity often cause negative as well as serious impact not only to the natural environment, but also to community due to its non-biodegradable and sometimes toxic in nature. Such wastewater can easily be found in Indonesia, especially in heavily populated West Java region. Common problems related to rivers pollution in this area are mostly originated from textile-dye water discharge, if improperly treated to meet the regulation.

Due to its toxic and non-biodegradable characteristics [1], dye containing wastewater can not be easily dealt with commonly used biological treatment. Therefore, as one of the tertiary treatments, Advanced Oxidation Processes (AOPs) offer an alternative yet excellent method of treating such wastewater by effectively oxidizing the dye pollutants. This method can easily and rapidly destroy the non-biodegradable content (with BOD/COD ratio of typically < 0.01) in the wastewater into simpler compounds or intermediates which later on become biodegradable (BOD/COD ratio > 0.1), and thus can be further treated by cheaper conventional biological methods [2], such as using aeration or activated sludge pond.

Various AOPs have something in common, which is the generation of •OH (hydroxyl) radicals whose standard reduction potential is only less positive than Fluorine [3]. Therefore, these radicals will act as very potent oxidizing agents capable of degrading almost any organic pollutants. Hydroxyl radicals may be produced via several ways, but normally by combining UV radiation with common oxidizing agents (such as H_2O_2 or O_3). However, without UV present those radicals may also be generated if certain oxidizing agents meet a catalyst. For example, O_3 (ozone) in alkaline condition or the combination of H_2O_2 with Fe^{2+}, well known as Fenton reagent, may also yield •OH radicals [4].

This research explored the use of UV with Fenton reagent altogether for treating textile-dye (Drimarene Dark Red / DDR) containing wastewater. The combination of UV with Fenton reagent might offers rapid and effective degradation of the dye pollutant due to the synergetic effect of both elements which produced more •OH radicals than separate UV/ H_2O_2 or Fenton reagent as normally used. In addition, the use of Fe^{2+} catalyst to significantly promote the degradation capability of UV/ H_2O_2 system brought attractive economical advantage in comparison to similar process using ozone as additional oxidizing agent, which required expensive capital related cost for the ozone generator.

2. Materials and Methods

2.1. Materials

The DDR containing wastewater was prepared by dissolving 2,5 g dye in 50 L of water resulting in 50 ppm concentration, as normally found in similar (real) textile-dye wastewater. The wastewater was initially mixed with a certain amount of H_2O_2 (50%-w) solution and the pH of the mixture was adjusted to 3 (with 2N HNO_3) as the optimum pH for treating similar textile-dye using UV/ H_2O_2 method [5].

2.2. Photo-oxidation

The mixture was then fed into a 50 L column photo-reactor equipped with 5 low pressure UV lamp (75 Watt each) and Fe^{2+} (FeSO4) solution was subsequently added to the reactor before the UV irradiation began. The mixture was processed batch wisely for 90 minutes, and samples (10 mL each) were withdrawn during the photo-oxidation process within the reactor. Those samples were treated with excess thiosulfate solution to deactivate the remaining H_2O_2 and then centrifuged prior to spectrophotometer analyses (at $\lambda = 525$ nm).

2.3. Experimental design

The effect of the two main variables, namely H_2O_2 initial concentration and the Fe^{2+} : H_2O_2 mole ratio, were investigated using Central Composite Design [6] to determine the optimum conditions, under which the highest reaction rate constant was achievable. Within the experimental design matrix, the center point was at 0.05 %-w H_2O_2 and ratio of 1 : 20, with 4 replications.

3. Results and Discussions

3.1. Response variable and the optimum conditions

The response variable selected as the measure of DDR removal effectiveness was the reaction rate constant, which appeared to follow the pseudo 2nd order kinetic rather than the pseudo 1st order commonly applicable for UV/ H_2O_2 system [5]. The DDR concentration profiles against reaction time showed a 'long tail' or a typical 2nd order reaction curve. The results of the calculated 2nd order reaction constants at various experimental conditions were summarized in Table 1.

Those results were then fitted to the 1st order as well as 2nd order response surface model. The multi linear regression for both models revealed that the response surface followed 2nd order model with R^2 of 0.986 compared to that of 1st order of 0.899. Furthermore, using the 2nd order model the partial derivatization with respect to both x_1 and x_2 resulted in the optimum conditions of 0.0504 %-w H_2O_2 and Fe^{2+} :H_2O_2 ratio of 0.0446 (or 1 : 22.4) under which the reaction rate constant was predicted as 0.021 $M^{-1}.min^{-1}$. The 3-D plot of the results presented in Table 1 was shown in Figure 1, and a validation run was also carried out to confirm the model prediction on the optimum conditions above. The validation result showed an excellent agreement with the 2nd order response surface model prediction.

Table 1. The 2nd order reaction constant at various conditions

RUN	Coded Variable		Rate Constant $[M^{-1}.min^{-1}]$
	x_1 (% H_2O_2)	x_2 (Fe^{2+} : H_2O_2)	k
1	-1	-1	0,0179
2	-1	1	0,0073
3	1	-1	0,0164
4	1	1	0,0195
5	0	0	0,0207
6	0	0	0,0205
7	0	0	0,0183
8	0	0	0,0208
9	1,414	0	0,0115
10	-1,414	0	0,0154
11	0	1,414	0,0178
12	0	-1,414	0,0189

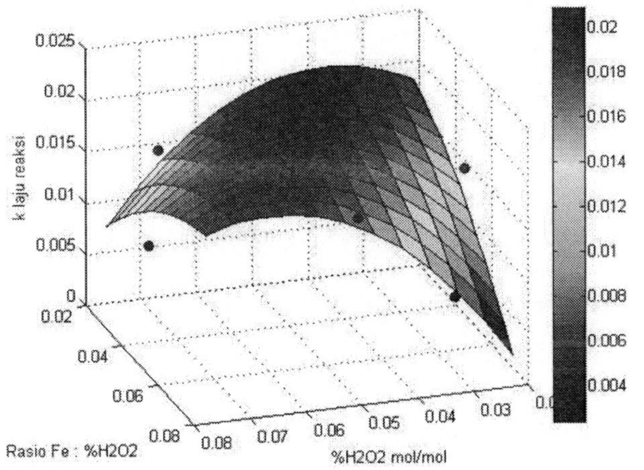

Figure 1. The 3-D plot of the experimental results

Under those optimum conditions, the DDR removal in the wastewater was found to be very rapid as shown in Figure 2. Within only 10 minutes reaction time, the DDR content removal reached 90% and after 20 minutes (97% removal) the DDR concentration showed a plateau which was typical for reactions with 2nd order kinetics.

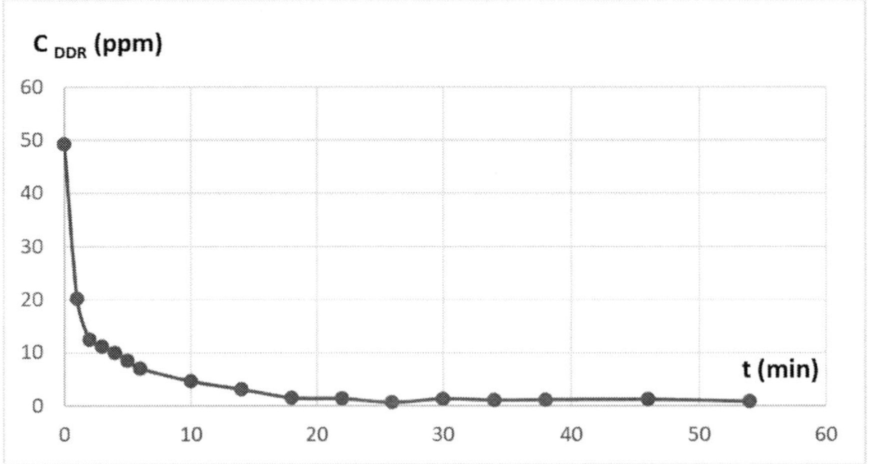

Figure 2. The DDR concentration profile under optimum conditions

3.2. The effect of H_2O_2 concentation
The 3-D plot revealed that the initial H_2O_2 concentration had a significant role, especially at relatively high Fe^{2+} : H_2O_2 ratio, to the DDR removal effectiveness. At lower H_2O_2 concentration, the increasing H_2O_2 concentration brought dramatic advantage due to the high concentration of •OH radicals formed due to the interaction of H_2O_2 with both UV and especially more abundant Fe^{2+}. This fact would finally led to much more quicker DDR degradation which translated to significantly higher k was observed, for example run 4 had k value 2.7 times larger than run 2 due to the increasing %H_2O_2 at relatively high Fe^{2+} : H_2O_2 ratio. However, at higher H_2O_2 concentration, the increasing H_2O_2 concentration led to a detrimental DDR removal performance, i.e. run 9 showed much lower k value (around ½) than that of run 8. Excessive H_2O_2 concentration resulted in a drawback to the DDR degradation / removal because the H_2O_2 itself was reactive to the •OH radicals producing a much less potent HO2• radicals (reduction potential of 1.7 V compared to 2.8 V of •OH radicals) to destroy the pollutant [7]. The reaction between H_2O_2 with the •OH radicals was as follow:

$$H_2O_2 + \bullet OH \rightarrow H_2O + HO_2\bullet$$

3.3. The effect of Fe^{2+} : H_2O_2 ratio
In general, the effect of the Fe^{2+} : H_2O_2 ratio had similar trend with that of H_2O_2 concentration to the DDR degradation effectiveness. An optimum value of around 1 : 22 was ideal ratio for the treatment. Increasing Fe^{2+} concentration was initially beneficial since it was needed for •OH radical generation essential for the DDR degradation. The reaction between Fe^{2+} and H_2O_2 was outlined by Tang [4] as follow:

$$Fe^{2+} + H_2O_2 \rightarrow Fe^{3+} + OH^- + \bullet OH$$

However, at excessive level of Fe^{2+} concentration the UV light available for •OH radical production was partially consumed / absorbed by the Fe3+ (as the product of above reaction) and eventually lower the degradation rate constant. Therefore, it was imperative to operate the degradation under those optimum conditions, so that the treatment could be carried out as effective as possible, in terms of both time and cost.

3.4. DDR photodegradation mechanism
The DDR structure had many aryl (benzene) rings as well as several – N = N – bonds (azobridges) which were known as chromophores [8] responsible for the color of the dye since they contain delocalized electrons able to absorbed some spectrum of the visible light. During the photodegradation, the •OH radicals would initiate the attack to such groups by electrophilic addition

to the Π bonding [4] yielded intermediate radicals which would experience further degradation into simpler species.

The mechanism of the degradation of the aryl ring by •OH radical had been published in several literatures. For example, Christina [9] explained a series of reaction led to the ring opening and oxalic acid as the final product, as shown in Figure 3 below.

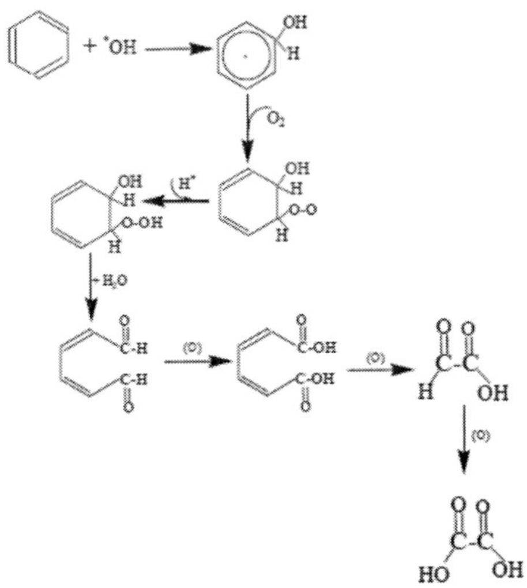

Figure 3. The mechanisms of aryl ring degradation by •OH radical

The aromatic ring would initially turned into hydroxycyclohexadienyl as the result of electrophilic addition. This radical could further react with dissolved O_2 producing unstable hydroxyhydroperoxide which later yield an aliphatic dialdehyde after water molecule elimination step. Subsequently, the dialdehyde species would be further oxidized into glyoxal and finally oxalic acid.

4. Conclusions

The UV/Fenton photo-oxidation process was proven to be effective for treating Drimarene Dark Red (DDR) containing textile-dye wastewater. An optimization using Central Composite Design experimental matrix revealed that the photo-oxidation at pH 3 was best carried out at the initial concentration of H_2O_2 at 0.050%-w and the mole ratio Fe^{2+}: H_2O_2 of 1 : 22. Under these conditions, the measured 2nd order pseudo-rate constant was 0.021 $M^{-1}.min^{-1}$ and the DDR color removal of 90% was surprisingly achievable within only 10 minutes reaction time. The experimental results also showed that it was imperative to operate the degradation under those optimum conditions, so that the treatment could be carried out as effective as possible, in terms of both time and cost.

References

[1] Darsono V 2007 Pengolahan limbah cair tahu secara anaerob dan aerob *Jurnal Teknologi Industri* **XI** 9-20
[2] Green D W and Perry R H 2008 *Perry's Chemical Engineer's Handbook, 8th Ed.*(New York : McGraw-Hill International)
[3] Al-Kdasi A, Idris A, Saed K and Guan C T 2004 Treatment of textile wastewater by advanced oxidation processes - A review *Global Nest : the Int. J.* **6** 222-230
[4] Tang W Z 2004 *Physicochemical treatment of hazardous wastes* (Florida : Lewis Publishers Boca Raton)

[5] Hudaya T, Stefanus M and Agustina M 2011 H_2O_2/UV Photo-oxidation of non-biodegradable DYA textile-dye wastewater in a multi-lamp bubble column photoreactor *Seminar Nasional Teknik Kimia "Kejuangan" 22 Februari 2011, UPN Veteran Yogyakarta*
[6] Montgomery D C 2001 *Design and Analysis of Experiment* 5th Ed. (New York : John Wiley & Sons Inc.)
[7] Legrini O, Oliveros E and Braun A M 1993 Photochemical processes for water treatment *Chem. Rev.* **93** 671-698
[8] Information on http://www.depts.washington.edu
[9] Christina M, Mu'nisatun P, Saptaaji S R and Marjanto D 2007 Warna azo (metil orange) dalam pelarut air menggunakan mesin berkas elektron 350 keV/10 mA *JFN* **1** 31 – 44

IOP Conf. Series: Materials Science and Engineering **162** (2017) 012023 doi:10.1088/1757-899X/162/1/012023

Effect of initiator concentration to low-density polyethylene production in a tubular reactor

A Azmi and N Aziz*
School of Chemical Engineering, Engineering Campus, Universiti Sains Malaysia, Seri Ampangan, 14300 Nibong Tebal, Seberang Perai Selatan, Penang, Malaysia

E-mail : *chnaziz@usm.my

Abstract. Low-density polyethylene (LDPE) is one of the most widely used polymers in the world, which is produced in high-capacity tubular and autoclave reactors. As the LDPE industry turn into more competitive and its market profit margins become tighter, manufacturers have to develop solutions to debottleneck the reactor output while abiding to the stringent product specification. A single polyolefin plant producing ten to forty grades of LDPE with various melt flow index (MFI), therefore understanding the reaction mechanism, the operating conditions as well as the dynamic behavior of tubular reactor is essential before any improvement can take place. In the present work, a steady state mathematical model representing a tubular reactor for the production of LDPE is simulated using MATLAB R2015a®. The model developed is a function of feed inlet, reactor jacket, single initiator injector and outlet stream. Analysis on the effect of initiator concentration (CI) shows sudden declining trend of initiator's concentration which indicates that all of the initiators are exhausted after polymerization reaction and no further reaction occur from this point onwards. Furthermore, the results demonstrate that the concentration of initiator gives significant impact on reactor temperature's profile and monomer conversion rate, since higher initiator concentration promotes greater polymerization rate, and therefore leads to higher monomer conversion throughput.

1. Introduction

LDPE is one of the most important commodity polymer with great industrial significance. The LDPE is known to be possessing extraordinary electrical and impact properties, and excellent resistance to chemical and moisture. Globally, LDPE is widely used for manufacturing various plastic containers, plastic packaging for computer hardware, dispensing bottles, tubing, and various molded laboratory equipment. Nevertheless, its most common use is still in plastic bags. Due to its exclusive properties, LPDE worldwide consumption has been reported to increase. The global LDPE production reached over 35 million metric tons in 1994, and has been reported to growth rapidly at the proportion of 7% per year over the past 15 years [1-2].

The high-pressure polymerization of ethylene in tubular reactors is broadly employed to produce LDPE. The free radical polymerization in a tubular reactor is carried out under extreme conditions. The reactor is operated at very high pressure, typically between 150 – 250 MPa and temperatures of 325 – 625 K [3]. A typical high-pressure LDPE tubular reactor's length is ranging from 500 to 1500 m and characterized by very large ratios of length to diameter (L/D) from 1 – 6 x 10^4. Schematic diagram of a typical industrial LDPE tubular reactor is shown in Figure 1. Ethylene (monomer), transfer agents along with initiators are used as the reactor feed. Ethylene polymerization is a highly exothermic reaction and it takes place in short reaction zones resulting from peroxides injections. The remainder of the reactor is mainly used as heat exchanger, where the reaction mixture is heated up or cooled down to an appropriate temperature for subsequent peroxide addition or downstream process.

Considering the extreme operating conditions and the complexity of ethylene polymerization process, it is apparently known that there are difficulties to obtain good model formulation of the process. An accomplished model should be able to maximize the monomer conversion, minimizing the unwanted product (ethyl, butyl, vinyl and vinylidene groups); while abiding to the stringent polymer product quality with regards to its molecular weight distribution (MWD). However, to achieve all objective functions at one time will be very exhausting, therefore a divide and conquer strategy is adopted in the present work. Here, the effect of initiator concentration variations on reactor temperature profile and ethylene conversion across the tubular reactor is evaluated with the improvement in kinetic scheme used by Yao et al. [1] and Gupta et al. [4] models. It is believed that the results obtained will allow significant improvements in conversion of ethylene.

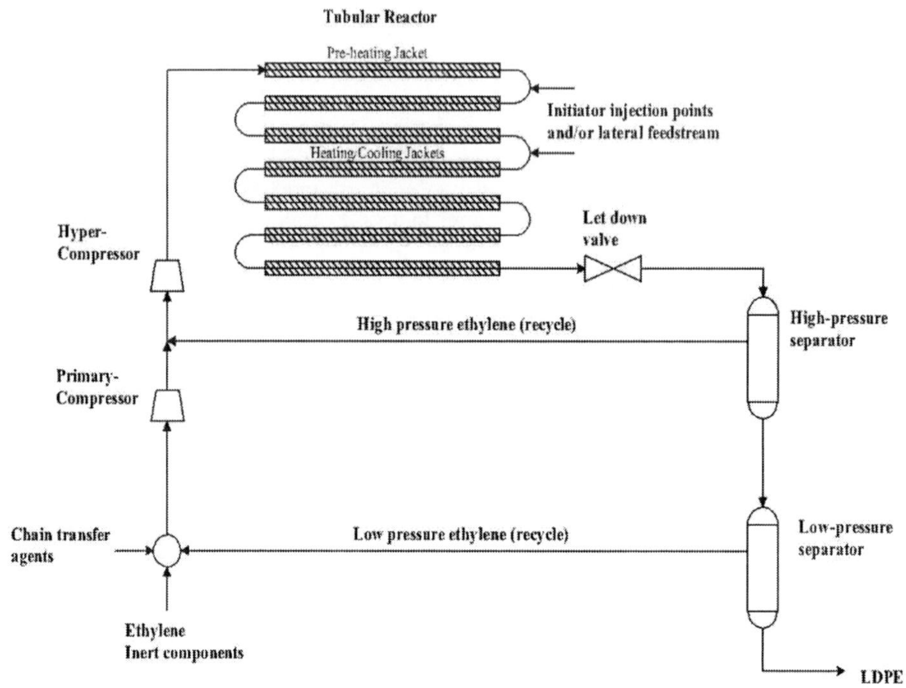

Figure 1. Simplified flowsheet of LDPE plant

Table 1. The kinetic mechanism used in this study

Reaction	Kinetic Mechanism	Equation
Peroxide initiation	$I \xrightarrow{K_d} 2R_{in}$	(1)
Monomer thermal initiation	$3M \xrightarrow{K_{th}} 2R_1$	(2)
Propagation	$R_1 + M \xrightarrow{K_p} R_{l+1}$	(3)
Termination by combination	$R_l + R_k \xrightarrow{K_{tc}} P_{l+k}$	(4)
Thermal degradation	$R_{l+1} \xrightarrow{K_{thd}} P_l + R_1$	(5)
Chain transfer to monomer	$R_l + M \xrightarrow{K_{trm}} P_l + R_1$	(6)
Chain transfer to polymer	$R_l + P_k \xrightarrow{kK_{trp}} P_l + R_k$	(7)
Chain transfer to transfer agent	$R_l + S \xrightarrow{K_{trs}} P_l + R_1$	(8)
β-Scission of secondary radical	$R_{l+1} \xrightarrow{K_\beta} P_l + R_1$	(9)
β-Scission of tertiary radical	$R_{l+1} \xrightarrow{K_{\beta 1}} P_l + R_1$	(10)

Table 2. Model Equations for the LDPE Tubular Reactor

Descriptions	Equation	Equation
Overall mass balance	$\frac{dv}{dz} = -\frac{v}{\rho}\left(\frac{d\rho}{dz}\right)$	(11)
Initiator balance	$v\frac{dC_{I_m}}{dz} = \left(-2fK_{d_m}C_{I_m} - C_{I_m}\frac{dv}{dz}\right) \therefore for\ m = 1,2$	(12)
Monomer balance	$v\frac{dC_M}{dz} = \left(2K_{th}C_M{}^3 - K_{trm}C_M\lambda_0 - C_M\frac{dv}{dz}\right)$	(13)
Solvent balance	$v\frac{dC_S}{dz} = \left(-K_{trs}C_S\lambda_0 - C_S\frac{dv}{dz}\right)$	(14)
Reactor temperature	$\frac{dT}{dz} = \frac{1}{\rho C_P v}\left[-\Delta H K_p C_M\lambda_{00} - \frac{4U(T - T_J)}{D}\right]$	(15)
Reaction mixture density variation	$\frac{d\rho}{dz} = \frac{\partial\rho}{\partial T}\frac{dT}{dz} + \frac{\partial\rho}{\partial C_M}\frac{\partial C_M}{dz}$	(16)
where $\frac{\partial\rho}{\partial T} = \frac{1.436\times10^{-6}C_M - 2.5245\times10^{-4} + (31.6967 - 1.25C_M)\times10^{-8}T - 9.98877\times10^{-11}T^2}{(0.71 - 4.475\times10^{-4}T)^2(0.949 + 4.988\times10^{-4}T)^2}$		(17)
and $\frac{\partial\rho}{\partial C_M} = 0.028\left(\frac{1 - 0.71 - 4.475\times10^{-4}T}{0.949 + 4.988\times10^{-4}T}\right)$		(18)
Overall heat transfer coefficient	$U^{-1} = h_i^{-1} + h_w^{-1}$	(19)
Reaction mixture density (g/cm^3)	$\rho = \frac{1 + 0.028C_M(V_P - V_M)}{V_P}$	(20)
Specific heat of reactant mixture (cal/g.K)	$C_p = 0.518w_m + (1.041 + 8.3\times10^{-4})w_p$	(21)
Heat of reaction (cal/mol)	$-\Delta H = 23000$	(22)
Moment of live polymer radical	$\lambda_0 = \sqrt{\frac{2fK_{d_m}C_{I_m} + 2K_{th}C_M{}^3}{K_{tc}}}$	(23)
Reactor side heat transfer coefficient (cal/cm^2.s.K)	$h_i = \frac{K(Nu)}{D_i}$	(24)
Thermal conductivity (cal/cm.s.K)	$K = (5w_m + 3.5w_p)\times10^{-4}$	(25)
Nusselt Number	$Nu = 0.026Re^{0.8}Pr^{0.33}$	(26)
Reynolds Number	$Re = \frac{\rho D_i v}{A_c\eta_s}$	(27)
Viscosity of reactant mixture, Poise	$\eta_s = \eta_r\eta_o$	(28)
Relative viscosity of monomer	$\eta_r = 10^{0.0313\sqrt{\mu_{00}^{-1}\mu_{00}^3}}$	(29)
Viscosity of monomer, Poise	$\eta_o = 1.98\times10^{-4} + 1.15\times10^2T^{-2}$	(30)
Prandtl Number	$Pr = \eta_S C_P/K$	(31)
Heat transfer coefficient for metal wall (cal/cm^2.s.K)	$h_w = 0.025$	(32)

2. Model formulation

2.1. Assumption

The reactor model is based on the following assumptions:

- Ideal plug flow assumption with no axial mixing [1,4-5]

- The pressure inside reactor is kept constant throughout the reactor since the pressure drop of each zones are less than 10%, which is insignificant [3]
- There is no pulse valve effect in the model as inferred from pilot plant studies by Donati et al. [7]
- Homogenous phase of reaction mixture is assumed as suggested by Ehrlich and Mortimer [11]
- The heat capacity of reaction mixture is denoted as total heat capacities of pure components as prescribed by Chen et al. [5]
- The model utilized the axial length (z) of tubular reactor as the independent variable, and the ordinary differential equations (ODE) are integrated along the reactor length.
- Volumetric flow rate and temperature of jacket fluid is kept constant [1,5]
- Ten reaction mechanism is chosen and listed in Table 1
- Single feed and initiator injection point where the axial location of feed and initiator, $z = 0.0$ mm

2.2. *Simulation*

The developed model consist of ODEs, are solved using MATLAB R2015a®. The differential equations have been solved using ode23solver due to its suitability for stiff differential equations, furthermore it is a variable order solver based on the numerical differentiation formulas (NDFs). The reaction mechanisms, reaction rate constants, design features and operating conditions for based case, are given in Tables 1 - 3. Equations 1 – 10 list all the reaction mechanisms chosen in this study. Equations 11 – 14 constitute the mass balance on each components in the reactor, whereas Equation 15 represents the heat transfer from the reaction mixture to the jacket coolants. The density variation was present in Equation 16 – 18 followed by the physical properties of the reaction mixture and tubular reactor which are described in the Equations 19 – 32.

Table 3. Rate constants and parameters used in this study

Rate of reaction constant (1/s)		Source
Monomer thermal initiation, K_{th}	$6.04E3 \exp[-36.707E3(T+273.15)^{-1}]$	[8]
Propagation, K_p	$9.9E5 \exp[-1.982E3(T+273.15)^{-1}]$	[8]
Termination by combination, K_{tc}	$4.4E4 \exp[-3.772E2(T+273.15)^{-1}]$	[7]
Chain transfer to monomer, K_{trm}	$1.2E5 \exp[-6.758E3(T+273.15)^{-1}]$	[8]
Chain transfer to polymer, K_{trp}	$1.8E8 \exp[-4.7303E3(T+273.15)^{-1}]$	[8]
β-scission to sec. radical, K_β	$1.4E9 \exp[-9.371E3(T+273.15)^{-1}]$	[8]
β-scission to tertiary radical, $K_{\beta 1}$	$4.4E9 \exp[-9.371E3(T+273.15)^{-1}]$	[8]
Thermal degradation, K_{td}	$7.7E4 \exp[-9.618E3(T+273.15)^{-1}]$	[9]
Chain transfer to solvent, K_{trs}	$5.6E7 \exp[-5.29E3(T+273.15)^{-1}]$	[8]
Peroxide initiation, K_d	$6.5E11 \exp[-1.28E4(T+273.15)^{-1}]$	[10]
Parameters		
Tubular reactor length, L (m)	1200	
Reactor diameter, D (m)	0.05	
Pressure, P (atm)	2000	
Initial concentration of monomer,	19.54	
Initial concentration of initiator,	1.5E-3, 3.0E-3, 4.5E-3	
Initial reaction mixture velocity, v_o (m/s)	18.8	
Feed Temperature, T_F (°C)	140	
Jacket Temperature, T_J (°C)	180	

3. Results and Discussions

The simulation of the model are solved in MATLAB R2015a® and the results are presented in the Figures. 2 – 5. The initiator concentration (CI) of 1.5 x 10^{-6}mol/cm^3, ethylene concentration (CM) of 19.54 x 10^{-3}mol/cm^3, reaction mixture velocity of 18.8 m/s, tube length of 1200 m, reactor inner diameter of 0.05 m, feed temperature of 140°C and constant reactor pressure of 2000 atm are chosen as the reference condition in this simulation. Table 2 list out almost all parameters which are applied in this work. Temperature profiles and monomer conversion across the tubular reactor are selected as dependent variables for this particular analysis.

Figure 2 shows the temperature profiles within the reactor. It can be observed that reactor temperature rises sharply to a peak value in the 0 to 15000 mm axial range. For CI of reference condition, the peak value is observed at z = 12030 mm, whereas for CI of 3.0 x 10^{-6}mol/cm^3 and 4.50 x 10^{-6}mol/cm^3, the ethylene conversions are 27.05% and 29.34%, respectively. Judging from Figure 2 and Figure 3, both reactor temperature and ethylene conversion profiles are observed to be in agreement to one another. This is true since higher initiator concentration promotes greater ethylene conversion and therefore increasing the peak temperature. This is due to exothermic nature of ethylene polymerization.

As the reaction temperature achieve its highest peak, conversion of ethylene become static indicating that no further reaction take place from this point forwards. This is due to the initiator depletion rapidly to zero after its introduction in the tubular reactor as seen in Figure 4. As a result, concentration of monomer become constant from this point until the reactor outlet. From Figure 4, it is noticeable that 100% of initiator had been consumed during the reaction, whereas for monomer there is some remaining amount left even after the process is completed as seen in Figure 5. The final concentration for CI of 1.5 x 10^{-6} (reference condition), 3.0 x 10-6 mol/cm^3 and 4.50 x 10^{-6} mol/cm^3 are 15 x 10^{-3} mol/cm^3, 14.25 x 10^{-3} mol/cm^3 and 13.81 x 10^{-3} mol/cm^3 respectively.

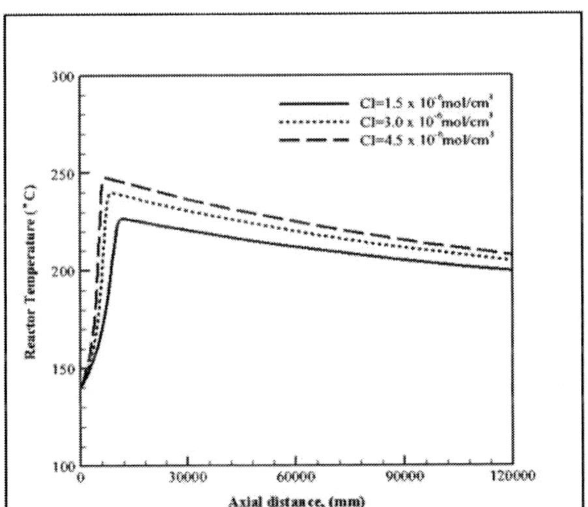

Figure 2. Reactor temperature profiles along reactor length

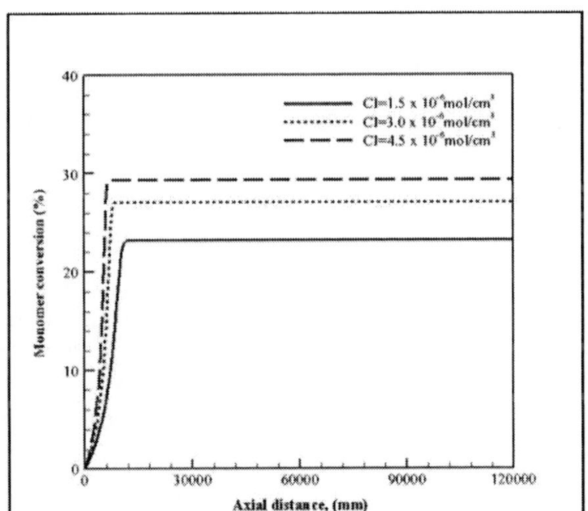

Figure 3. Monomer conversions along reactor length

IOP Conf. Series: Materials Science and Engineering **162** (2017) 012023 doi:10.1088/1757-899X/162/1/012023

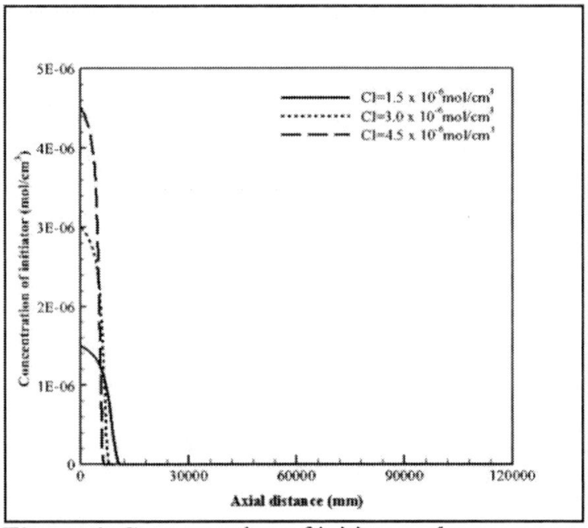

Figure 4. Concentration of initiators along reactor length

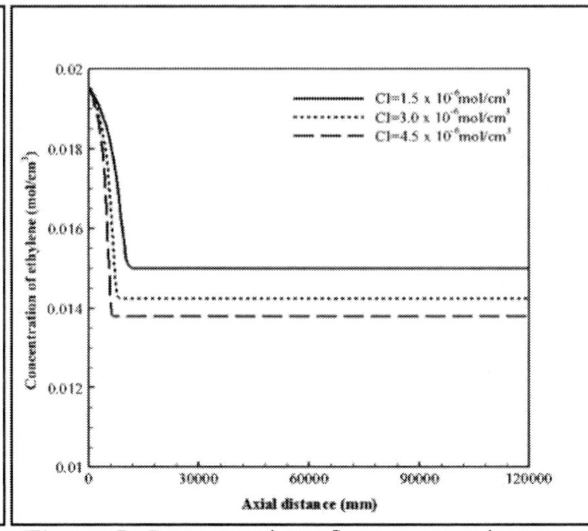

Figure 5. Concentration of monomers along reactor length

4. Conclusion

The simulation of high pressure LDPE production has been carried out using selected reaction mechanism and kinetic constant from various literature sources. The present study reveals that initiator concentration amount plays significant role to the ethylene conversion rate. Higher initiator concentration leading to a higher concentrations of free radical and thus increasing the ethylene conversion rate. In the future, it will be very beneficial to conduct a comparison study of varying the amount of initiator versus adding extra injector to a LDPE tubular reactor.

Acknowledgements

The authors gratefully acknowledge the financial support given by Universiti Sains Malaysia through Research University Grant No. 1001/PJKIMIA/814237 and Malaysian government through MyBrain 15 funding.

Nomenclature

Ac	Cross-sectional area of tubular, m^2
C_{Im}	concentration of initiator, mol/l
C_M	monomer concentration, mol/l
C_S	solvent concentration, mol/l
C_p	specific heat of reactant mixture, cal/g·K
D	inside diameter of reactor, cm
I	initiator
M	monomer
P_l	dead polymer with chain length l
P_k	dead polymer with chain length k
R_{in}	primary initiator radical
R_l	radical of chain length l
R_k	radical of chain length k
Re	Reynolds number
S	solvent

References

[1] Yao F Z, Upreti S R, Lohi A and R. Dhib 2004 Modeling, simulation and optimal control of ethylene polymerization in non-isothermal, high-pressure tubular reactors *International Journal of Chemical Reactor Engineering* **2** 1542 – 6580

[2] Kim D M and Iedema P D 2008 Modeling of branching density and branching distribution in low-density polyethylene polymerization *Chemical Engineering Science* **63** 2035 – 2046

[3] Azmi A and Aziz N 2016 Simulation studies of low-density polyethylene production in a tubular reactor *Procedia Engineering* **148** 1170-1176

[4] Gupta K, Kumar A and Krishnamurthy M V G 1985 Simulation of tubular low-density polyethylene *Polym. Eng. Sci.* **25** 37-47

[5] Chen C H, Vermeychuk J G, Howell J A and Ehrlich P 1976 Computer model for tubular high-pressure polyethylene reactors *AIChE J.* **22** 463 – 471

[6] Azmi A and Aziz N 2015 Low density polyethylene tubular reactor modeling : Overview of the Model developments and future directions *International Journal of Applied Engineering Research* **11** 9906 - 9913

[7] Donati G, Marini M, Marziano L, Mazzaferri C, Spampinato M and Langianni E 1982 Mathematical-model of low-density polyethylene tubular reactor *Chemical Reaction Engineering* **196** 579–590

[8] Yao F Z 2003 Modeling, simulation and optimal control of ethylene polymerization in a high-pressure tubular reactor *Master thesis* Ryerson University

[9] Agrawal N, Rangaiah G P, Ray A K and Gupta S K 2006 Multi-objective optimization of the operation of an industrial low-density polyethylene tubular reactor using genetic algorithm and its jumping gene adaptations *Ind. Eng. Chem. Res.* **45** 3182 – 3199

[10] Agrawal S and Han C D 1975 Analysis of high-pressure polyethylene tubular reactor with axial mixing *AIChEJ.* **21** 449 – 465

[11] Ehrlich P and Mortimer G A 1970 Fundamentals of the free radical polymerization of ethylene *Adv. Polym. Sci.* **7** 386 – 448

Transesterification of palm oil using sodium silicate base catalyst from geothermal sludge

I Perdana*, N Nugrahanti, Sofiyah and I M Bendiyasa
Department of Chemical Engineering, Universitas Gadjah Mada
Jl. Grafika 2, Bulaksumur, Yogyakarta 55281, INDONESIA

E-mail : iperdana@ugm.ac.id

Abstract. The use of solid base catalysts in biodiesel synthesis is becoming more preferable because of their superiority over homogeneous catalysts. In the present work, a strong base catalyst of sodium silicate synthesized from silica-rich geothermal sludge was used in a transesterification of palm oil with methanol. The catalyst was calcined at 400°C for three hours with a temperature ramp of 20°C/min. The transesterification was carried out at varying temperature in the range of 50 - 70°C for 60 minutes with a methanol-palm oil molar ratio of 8.8:1. The catalyst-palm oil ratio was varied in the range of 1 - 5% (w/w). In order to investigate kinetics of reaction, at a certain interval of time samples were taken consecutively during the reaction. Experimental results showed that the sodium silicate was very active in the transesterification of palm oil with methanol. Reaction temperature at 60°C was sufficient to reach a conversion level as high as 93% in a relatively short reaction period. Meanwhile, the high conversion was still achievable with the use of 1 % (w/w) catalyst. In addition, a lumped model of reaction kinetics was adequate to approach the experimental data with a calculated activation energy of 15.73 kcal/mole. Results of the present work suggested that sodium silicate synthesized from local resources of geothermal sludge would become potential solid base catalyst in biodiesel synthesis.

1. Introduction

In the last few years, a lot of efforts have been done to search potential renewable energy in order to find solution for the depletion of crude oil which is at the moment still the main source of energy. Among various types of renewable energy sources, biodiesel is one of the interests due to the availability of its raw material i.e. vegetable oil, animal fat, algae, etc. The use of homogeneous base catalysts of NaOH and KOH is very common to produce biodiesel through transesterification of triglycerides with alcohol. The process is already well developed and commercially implemented [1-3]. Even though, high conversion is achievable at a moderate condition, the process still faces problematic separation and purification and it requires intensive energy that causes high production cost [4]. In addition, the need to use inexpensive feed-stocks that might contain high concentration of free-fatty acids and water can lead to saponification that further restraints purification process [5].

As an alternative to homogeneous base catalyst, solid base catalysts for transesterification have drawn a lot of interest. The use of solid catalysts can ease separation/purification of the product, avoid formation of soap and therefore, reduce separation cost that eventually leads to a more competitive biodiesel prices [6]. Several solid base catalysts for transesterification have been investigated including earth alkaline oxides such as calcium oxide [7, 8] and calcium methoxides [9], Mg-Al hydrotalsite [10, 11], $K_2CO_3/\gamma\text{-}Al_2O_3$[12], TiO_2-MgO [13], Zr-La oksid [14], Eu_2O_3/Al_2O_3 [15], La/β zeolite [16], $Na/NaOH/\gamma\text{-}Al_2O_3$ [17], etc. Even though these catalysts are found active in the transesterification of different kinds of triglycerides, the activity is generally less than that of homogeneous catalysts so that a longer reaction time is needed to reach high reaction conversion.

Alkali type catalysts of sodium silicates have also been studied and used as solid base catalysts in biodiesel synthesis. Commercial sodium silicates can be thermally treated and used as catalysts in transesterification of soybean and rapeseed oils both in sub-critical and atmospheric conditions [18-21]. Calcination temperature during pretreatment was found affecting the activity of the catalyst. It was found that the optimum activity was reached from calcination at 400°C for 2 hours [20]. In transesterification of soybean oil with methanol, the catalysts could serve a maximum yield of nearly 100% after 60 minute reaction at 60°C with the use of 3% (w/w) catalyst, 7.5:1 methanol-oil molar ratio and at constant agitation of 250 rpm. Meanwhile, similar catalyst used in transesterification of rapeseed oil with methanol resulted in a maximum yield of 99.6% under optimized conditions of 3% (w/w) sodium silicate, 9:1 methanol-oil molar ratio, 60 min reaction time, 60°C reaction temperature, and 250 rpm stirring rate [19]. The reaction mechanism of soybean-methanol transesterification was studied by Guo, et al.[21]. They concluded that calcined sodium silicate was more active as proton donor in the reaction. The activity of sodium silicates synthesized from commercial amorphous silica and sodium hydroxide in transesterification of corn oil with methanol was investigated [22]. It was found that temperature rise during pretreatment had significant effect on the catalyst activity. The catalyst calcined at 400°C with a temperature ramp of 20°C/min could lead to the maximum conversion of corn oil by about 96%. In our other works, we synthesized sodium silicates from silica rich geothermal sludge and used it as a catalyst in transesterification of corn oil with methanol [23].

The use of widely available feedstocks as raw material in biodiesel synthesis in crucial. In the present work, sodium silicates synthesized from silica-rich geothermal sludges was further studied as a solid base catalyst in transesterification of palm oil with methanol. The effect of temperature and amount of catalyst on the reaction conversion were investigated in a batch reactor. In addition, in the present work a simplified model of over-all reaction kinetics was proposed.

$$\mathrm{TG} + 3\mathrm{M} \leftrightarrows 3\mathrm{ME} + \mathrm{G} \quad (1)$$

Assuming that the catalyst is always saturated with methanol and the reaction is reversible, the rate of lumped reaction can be expressed as in eq. (2)

$$-\frac{d[\mathrm{TG}]}{dt} = k_f([\mathrm{TG}] - 1/K[\mathrm{ME}]^3[\mathrm{G}]) \quad (2)$$

where k_f is a rate constant of forward reaction and K is an equilibrium constant. Triglyceride is denoted as TG, methyl ester ME and glycerol G. The rate of reaction and equilibrium constants are temperature (T) dependent as expressed in eq. (3) and (4), respectively.

$$k_f = k_o e^{-\frac{E}{R}\left(\frac{1}{T}-\frac{1}{T_0}\right)} \quad (3)$$

$$K = \alpha e^{\beta/T} \quad (4)$$

where E is activation energy, R universal gas constant, T_0 reference temperature, whereas K_0, α and β are constants.

2. Methodology

Geothermal sludge from PT Geo Dipa Energi in Dieng, Centre Java was initially washed with distilled water, filtered and then dried in an oven at 150°C for 16 hours. The remaining dried material was pulverized to obtain -60 mesh powder fractions so called dry sludge. The dry sludge was underwent silica extraction following procedure developed in the previous works [24]. 10 g silica resulted from the extraction step was slowly added to a solution that was made by dissolving 11.7 g NaOH (99.9%, Merck) in 12 mL distilled water. The solution was continuously stirred at a constant temperature of 60°C while evaporation was taking place to finally form a gel phase. The gel-phase was dried in an oven at 120°C for two hours. The dried substance was then calcined at 400°C with a ramp rate of 20°C/min while the temperature was kept constant for three hours. The resulting dried material was then pulverized for use as a catalyst. The resulting catalyst was used in transesterification of palm oil with methanol. As much as 120 g commercial palm oil in a beaker glass was heated to 60°C. Meanwhile, 158 mL methanol (Merck) (8.8 methanol-palm oil molar ratio) and 6 g sodium silicate (5% w/w) were placed into a three-neck flask equipped with a mercury sealed stirrer, a thermometer and a condenser. The mixture in the three-neck flask was heated to 60°C at a constant stirring of 410 rpm and was kept at the same temperature for one hour for methanol activation. After that, the preheated palm oil was added to the flask and transesterification started. Samples as much as 5 mL each were taken during the reaction. The samples were centrifuged and the residual oil in the heavy phase was analyzed using standard iodometric method to determine the amount of glycerol produced and total conversion of the triglycerides. The experiments were repeated at varying temperatures of 50, 60 and 70°C; and varying catalyst-palm oil ratios of 1%, 3% and 5% (w/w). The experimental data was then used to verify the proposed kinetics model.

3. Results and Discussion

In the present work, the activity of sodium silicate catalysts in transesterification of palm oil with methanol was studied. The ratio of catalyst-palm oil was varied at 1%, 3% and 5% (w/w) while other parameters i.e. temperature, agitation speed and palm oil-methanol molar ratio were kept constant at 60°C, 410 rpm and 8.8:1, respectively. Figure 1 shows conversion of reaction during the time at various catalyst-palm oil ratios. The conversion increases with time and it approaches about the same high conversion level at prolong reaction time. For all catalyst-palm oil ratios, the conversion can reach as high as ~ 93% after about 20 minute-reaction. This result is much better compared to that of other sodium containing catalysts such as $Na/NaOH/Al_2O_3$ as reporter by Lee, et.al.[25].The experimental results show that the activity of the catalyst is even comparable with NaOH of conventional homogeneous catalyst. In sodium silicates, the sodium is anchored in the silicate frameworks. Transesterification reaction is activated by formation of surface CH_3O- species due to interaction between methanol and sodium on the catalyst surface [21].

Figure 1 also shows that catalyst-palm oil ratio influences the early state of reaction conversion. It is evidence that the reaction rate is lower with the use of a lower amount of catalyst. The rate is apparently faster with the use of higher amount of catalyst. However, the use of catalyst-palm oil ratio above 3% (w/w) results in insignificant difference of reaction rate. As seen in the Figure 1, the maximum conversion is already achieved in 10 minutes of reaction. Similar work had been conducted by Guo, et al. (2010) which used commercial sodium silicate treated at 400°C for 2 hours for transesterification of soybean oil at 60°C [20]. They found that the maximum conversion was reached with the use of 3% (w/w) catalyst at 7.5:1 methanol-oil molar ratio and at a constant agitation of 250 rpm.

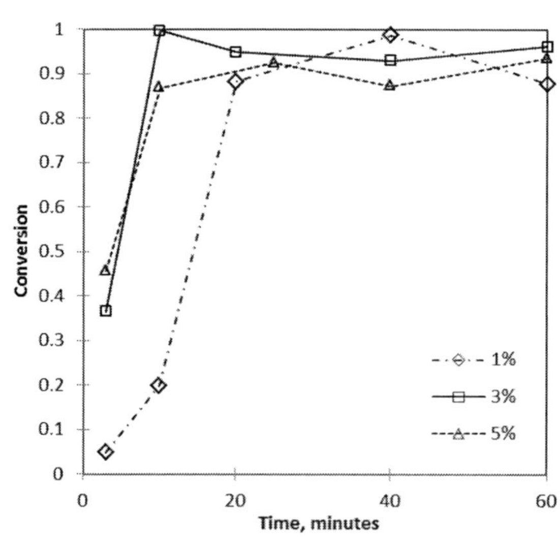

Figure 1. Reaction conversion at various catalyst-palm oil ratios
(reaction at 60°C, 410 rpm agitation and 8.8:1 methanol-palm oil molar ratio)

The activity of sodium silicates as strong base catalyst in transesterification of palm oil with methanol is activated by temperature. Figure 2 shows the reaction conversion profile at different reaction temperatures. Even though the conversion always increases with reaction time, the increase has different rate. It is obvious that the rate is faster at higher temperatures to reach the maximum conversion. At high temperatures, interaction intensity among reactant molecules is expected to escalate that increases the reaction rate. However, as seen in Figure 2, different level of maximum conversion is apparently reached at different reaction temperatures. At lower temperatures, the reaction resulted in a maximum conversion at a lower level. This indicates that at the operating conditions used in the current study, transesterification of palm oil with methanol is limited by an equilibrium condition. Since the reaction is endothermic, the level of maximum conversion is expected to be higher at elevated temperatures. As depicted in Figure 2, the average maximum conversions are 0.85, 0.92 and 0.97 at 50°C, 60°C and 70°C, respectively. It is evidence that even though the amount of methanol with a methanol-palm oil molar ratio of 8.8:1 is much larger than its stoichiometric requirement, the reaction is determined not merely by the amount of methanol in the solution. Since the number of active sites on the catalyst surface is limited, only a certain amount of methanol molecules can be adsorbed and the catalyst surface is always saturated by the adsorbed methanol.

Based on the aforementioned maximum conversion levels, reaction equilibrium constant at each temperature can be determined. The relation between the equilibrium constants and temperatures is shown in Figure 3. As expected, the equilibrium constant increases with temperature. This means that the reaction conversion can be maximized at elevated temperatures. This also indicates that the reaction is endothermic so that it requires heating to increase the equilibrium conversion. The relation between equilibrium constant (K) and temperature (T in Kelvin) can be written as in eq. (5).

$$\ln K = 37.804 - \frac{11824}{T} \tag{5}$$

Since the reaction is thermodynamically preferable at high temperatures, the reaction condition is also desirable to increase the rate of reaction. As depicted in Figure 2, at elevated temperatures the maximum reaction conversion is reachable in a relatively shorter reaction time. At high reaction temperatures, a steep conversion change is observable which indicates a fast reaction rate. A certain activation energy level is needed to overcome for the reaction to proceed. As the results, at high temperatures not only the reaction conversion can be enhanced, but duration of the reaction can also be shortened to reach an acceptable reaction conversion.

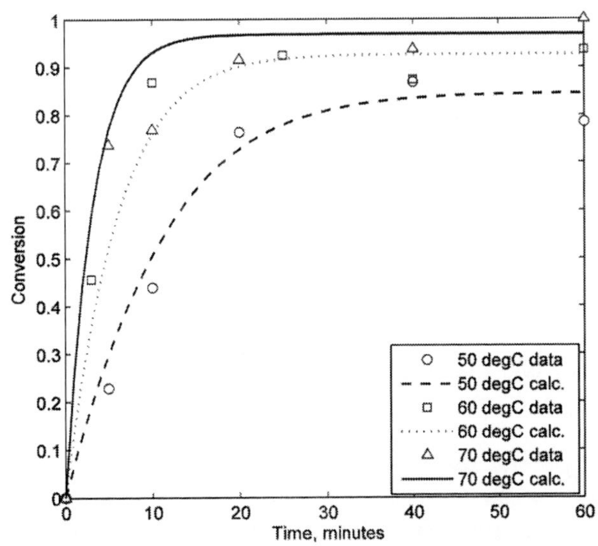

Figure 2. Reaction conversion at various reaction temperatures
(5% w/w catalyst, 410 rpm agitation and 8.8:1 methanol - palm oil molar ratio)

Figure 2 also shows results from simulation based on reaction kinetics model proposed in the present work. The model which considers a reversible reaction scheme can apparently simulate the reaction conversion in a good agreement with the experimental. It is clearly indicated that the reaction is activated by temperature. The reaction is faster at higher temperatures with an activation energy of 15.73 kcal/mole. The reaction rate constant (k_f, minutes^{-1}) as a function of temperature (T, Kelvin) can then be expressed as in eq. 6.

$$k_f = 0.1484e^{-7914(1/T-1/333)} \quad (6)$$

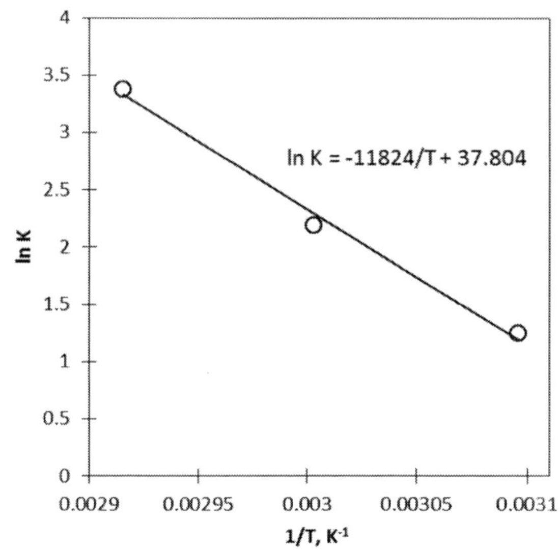

Figure 3. Equilibrium constant (K) at various temperatures (T)

As suggested in the model, even though the amount of methanol is much larger than stoichiometric requirement (8.8 times larger), reaction conversion is still limited by equilibrium condition. Although, the catalyst is always saturated with methanol, it seems that the amount of methanol adsorbed on the surface of catalyst is limited to react with palm oil. This eventually leads to a certain equilibrium reaction conversion which level is determined by reaction temperature.

3. Conclusions

Sodium silicates synthesized from silica rich geothermal sludge used as a solid base catalysts was found very active in transesterification of palm oil with methanol. Even though the amount of methanol in the reaction was much higher than that of its stoichiometric requirement, the reaction was limited by equilibrium which maximum conversion was temperature dependent. The highest conversion level of about 93% could be achieved by the use of as low as 1% catalyst (w/w) in a relatively short period of reaction time. A lumped reversible reaction kinetics model assuming that the catalyst is always saturated with methanol is sufficient to approach reaction conversion profile.

Acknowledgment

The authors are very grateful for supports from PT Geo Dipa Energi (Persero) for providing Dieng's geothermal sludge. The authors also thank the Ministry of Research, Technology and Higher Education for financial support through PUPT research grant 2015-2016. A lot of thanks also for NZAID-UGM CaRED Programme and Geothermal Research Center, UGM. that have been facilitating the researchers for conducting geothermal related research activities.

References

[1] Furuta S, Matsuhashi H and Arata K 2006 Biodiesel fuel production with solid amorphous-zirconia catalysis in fixed bed reactor *Biomass and Bioenergy* **30**(10) 870-873

[2] Ma F and Hanna M A 1999 Biodiesel production : A review *Bioresource Technology* **70**(1) 1-15

[3] Shibasaki-Kitakawa N, Honda H, Kuribayashi H, Toda T, Fukumura T and Yonemoto T 2007 Biodiesel production using anionic ion-exchange resin as heterogeneous catalyst *Bioresource Technology* **98**(2) 416-421

[4] Lee A F and Wilson K 2015 Recent developments in heterogeneous catalysis for the sustainable production of biodiesel *Catalysis Today* **242, Part A** 3-18

[5] Ganesan D, Rajendran A and Thangavelu V 2009 An overview on the recent advances in the transesterification of vegetable oils for biodiesel production using chemical and biocatalysts *Reviews in Environmental Science and Biotechnology* **8**(4) 367-394

[6] Di Serio M, Tesser R, Pengmei L and Santacesaria E 2008 Heterogeneous Catalysts for Biodiesel Production *Energy & Fuels* **22**(1) 207-217

[7] Zhu H, Wu Z, Chen Y, Zhang P, Duan S, Liu X and Mao Z 2006 Preparation of biodiesel catalyzed by solid super base of calcium oxide and its refining process *Chinese Journal of Catalysis* **27**(5) 391-396

[8] Liu X, He H, Wang Y, Zhu S and Piao X 2008 Transesterification of soybean oil to biodiesel using CaO as a solid base catalyst *Fuel* **87**(2) 216-221

[9] Liu X, Piao X, Wang Y, Zhu S and He H 2008 Calcium methoxide as a solid base catalyst for the transesterification of soybean oil to biodiesel with methanol *Fuel* **87**(7) 1076-1082

[10] Xie W, Peng H and Chen L 2006 Calcined Mg–Al hydrotalcites as solid base catalysts for methanolysis of soybean oil *Journal of Molecular Catalysis A: Chemical* **246**(1–2) 24-32

[11] Deng X, Fang Z, Liu Y-h and Yu C-L 2011 Production of biodiesel from Jatropha oil catalyzed by nanosized solid basic catalyst *Energy* **36**(2) 777-784

[12] Liu X, Piao X, Wang Y and Zhu S 2010 Model Study on Transesterification of soybean oil to biodiesel with methanol using solid base catalyst *The Journal of Physical Chemistry A* **114**(11) 3750-3755

[13] Wen Z, Yu X, Tu S-T, Yan J and Dahlquist E 2010 Biodiesel production from waste cooking oil catalyzed by TiO_2–MgO mixed oxides *Bioresource Technology* **101**(24) 9570-9576

[14] Sun H, Ding Y, Duan J, Zhang Q, Wang Z, Lou H and Zheng X 2010 Transesterification of sunflower oil to biodiesel on ZrO_2 supported La_2O_3 catalyst *Bioresource Technology* **101**(3) 953-958

[15] Li X, Lu G, Guo Y, Guo Y, Wang Y, Zhang Z, Liu X and Wang Y 2007 A novel solid superbase of Eu_2O_3/Al_2O_3 and its catalytic performance for the transesterification of soybean oil to biodiesel *Catalysis Communications* **8**(12) 1969-1972

[16] Shu Q, Yang B, Yuan H, Qing S and Zhu G 2007 Synthesis of biodiesel from soybean oil and methanol catalyzed by zeolite beta modified with La3+ *Catalysis Communications* **8**(12) 2159-2165

[17] Kim H-J, Kang B-S, Kim M-J, Park Y M, Kim D-K, Lee J-S and Lee K-Y 2004 Transesterification of vegetable oil to biodiesel using heterogeneous base catalyst *Catalysis Today* **93–95** 315-320

[18] Yin J-Z, Ma Z, Hu D-P, Xiu Z-L and Wang T-H 2010 Biodiesel production from subcritical methanol transesterification of soybean oil with sodium silicate *Energy & Fuels* **24**(5) 3179-3182

[19] Long Y-D, Guo F, Fang Z, Tian X-F, Jiang L-Q and Zhang F 2011 Production of biodiesel and lactic acid from rapeseed oil using sodium silicate as catalyst *Bioresource Technology* **102**(13) 6884-6886

[20] Guo F, Peng Z-G, Dai J-Y and Xiu Z-L 2010 Calcined sodium silicate as solid base catalyst for biodiesel production *Fuel Processing Technology* **91**(3) 322-328

[21] Guo F, Wei N-N, Xiu Z-L and Fang Z 2012 Transesterification mechanism of soybean oil to biodiesel catalyzed by calcined sodium silicate *Fuel* **93**(0) 468-472

[22] Chendrika L 2013 Pengaruh suhu dan laju kenaikan suhu kalsinasi terhadap karakteristik natrium silikat sebagai katalisator basa padat, in *Postgradutae Programe, Department of Chemical Enginerring* Universitas Gadjah Mada : Yogyakarta

[23] Perdana I, Nugrahanti N, Sofiah and and Bendiyasa I M 2015 Sodium silicate as a solid base catalyst from Dieng's geothermal sludge. in *Seminar Nasional Teknik Kimia Indonesia 2015, Sustainable Energy and Mineral Processing for National Competitiveness* Yogyakarta, Indonesia

[24] Reningtyas R, Perdana I and Bendiyasa I M 2012 purification of silica recovered from Dieng's geothermal sludge in *the 19th Regional Seminar on Chemical Engineering (RSCE) 2012* Denpasar, Indonesia

[25] Lee D-W, Park Y-M and Lee K-Y 2009 Heterogeneous base catalysts for transesterification in biodiesel synthesis *Catal Surv Asia* **13** 63 - 77

Indonesian low rank coal oxidation : The effect of H_2O_2 concentration and oxidation temperature

S S Rahayu[1], F Findiati[2] and F Aprilia[2]
[1]Lecturer and [2]Undergraduate students in Chemical Engineering Department, Universitas Gadjah Mada, Indonesia

Email : ssrahayu@chemeng.ugm.ac.id

Abstract. Extraction of Indonesian low rank coals by alkaline solution has been performed to isolate the humic substances. Pretreatments of the coals by oxidation using H2O2 prior to extraction are required to have higher yield of humic substances. In the previous research, only the extraction process was considered. Therefore, the effects of reaction temperature and residence time on coal oxidation and composition of extract residues are also investigated in this research. The oxidation temperatures studied were 40°C, 50°C, and 70°C and the H_2O_2 concentrations studied were 5%, 15%, 20 %, and 30 %. All the oxidation variables were studied for 90 minutes. The results show that the higher the concentration of H_2O_2 used, the less oxidized coal produced. The same trend was obtained by using higher oxidation temperature. The effect of H_2O_2 concentration, oxidation temperature and reaction time to the yield of humic substances extraction have positive trends.

1. Introduction

Coal is an important fossil energy source for electricity. This material consists of a wide range of organic structures in its complex macromolecular, primarily composed by carbon, hydrogen, oxygen, nitrogen, and sulfur. The molecular structure depends on its rank and the environmental conditions during coalification of plant matter. The rank classification is usually based on the contents of volatile matters, moisture, calorific value, and ash. Indonesian low rank coals, have calorific value: 4300 – 5000 kcal/kg are abundant resources and commonly have been utilized for electric power generation [1]. Others application of this low rank coal is limited due to the difficulty (auto ignition) for handling or transportation and storage.

Coal also regarded as the source of humic acids, organic acids are commonly used as a soil supplement in agriculture, and sometimes as a human nutritional supplement [2]. The non energy utilization of the coal, especially for humic acid, is still unpopular. Hence, it is important to study the methods for utilizing it efficiently.

From the published structural model, humic acid is a complex of many different acids, mainly composed of hydrophilic functional groups, such as carboxyl –COOH and hydroxyl – OH groups [3]. Humic acid is a principal component (the most chemically active component) of humic substances (humus), the major organic constituents of soil, which usually also present in sediments and coals. Humic substances can be divided into three main fractions: humic acids, fulvic acids, and humin. Humic and fulvic acids are highly soluble in alkaline, while humin is insoluble in alkaline [2]. Hence, humic and fulvic acids extracted from the sources into an alkaline solution (NaOH or KOH). Humic acids precipitated from the solution by adjusting the pH to 1 with HCl, leaving the fulvic acids in solution.

Humic substances are obtained from decomposition or biodegradation of plant or dead organic matter which happen during coalification process. Progressively the humic substances are converted to peat, lignite, subbituminous coal, bituminous coal and finally into anthracite. Consequently, the amounts of extractable humic acids depend on the degree of coalification. Peat and lignite are coal with lower degree of coalification having higher humic substances contents than bituminous. So, sufficient extraction of humic acids depends on the chemical composition of the source, and the method of extraction.

To increase the extraction yield, pretreatment process prior to extraction the low rank coals are performed. Oxidizing is pretreatment for the low rank coal which result higher content of extractable humic substances [4]. Oxidation with nitric acid indicate a substantial improvement in the yield of the coal to humic acids from 20% to 85%. In addition to nitric acid, the oxidators might be used are $KMnO_4$ and H_2O_2. By oxidating the coal with hydrogen peroxide in liquid phase at 60 °C for 2 h, some weak covalent bonds are broken and oxygen functional groups are introduced [3]. A comparative evaluation of the oxidators indicated that weight losses upon HNO_3 oxidation is relatively lower than H_2O_2 oxidation, while $KMnO_4$ was not recommended for coal oxidation, as a substantial amount of inorganic compounds was introduced to the products and the essential yield generated was low [4]. Even though, the weight losses involved in nitric acid oxidation was acceptable, but this reaction generates nitro humic acid and produces a large amount of waste water which causing serious environmental problems. Therefore H_2O_2 is chosen as oxidator for extraction of humic acids. In addition, H_2O_2 is an effective oxidator to convert Low Rank Coal into useful product[5]. The objective of this research is to study effect of oxidation parameters (H_2O_2 concentrations, temperatures, and reaction times) on the yield of humic acids.

2. Materials and Experimental Procedure

2.1. Materials

Low rank coal used for this study came from East Kalimantan. It has calorific value 4828.51 cal/g. Before use, the coal was ground into smaller size, screened into -70 +80 mesh, and dried at 60°C for 48 hours. The proximates analysis (% adb): moisture 12.73, ash content 14.53, volatile matter 45.75, and fixed carbon 26.99.

2.2. Experimental Procedure

2.2.1. Oxidation of coal

Oxidation of the coal was performed as follows: 6 times 2.5 g of dried coal were weighed accurately. Each portion (= m_c) was for oxidation of 15, 30, 45, 60, 75, and 90 minutes, respectively. Each portion of dried coal was mixed individually into 25 mL of certain concentration (varied from 5%, to 20%) of H_2O_2 in a 250 mL beaker glass. Mixing speed was 80 shakes/ min in a water bath (Figure 1).

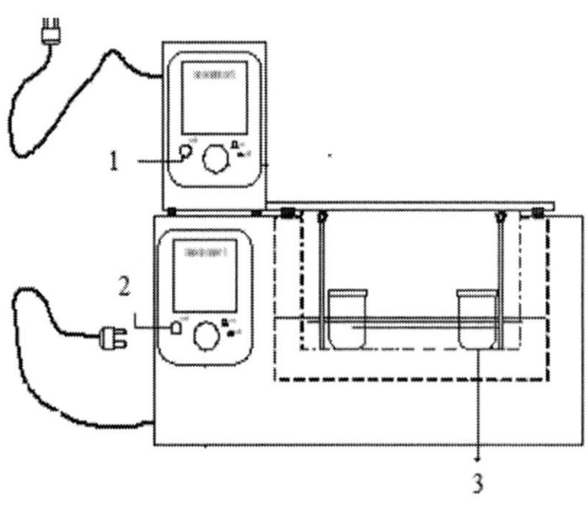

Figure 1. Experimental set up

The water bath was operated at a constant temperature (e.g. 60°C) during the time of oxidation (15, 30, 45, 60, 75, and 90 minutes). When the certain time of extraction has completed, the beaker glass was taken off individually from the bath, then filtered immediately through 42 Whatman's paper (the weight of the paper is known before used), then washed with distilled water to remove the water soluble matters. The filtrate was collected, while the solid residue (oxidized coal) was dried at 105°C until constant weight. The nett weight of oxidized coal was noted (= m_{OC}). To examine the effect of the temperature, the oxidation was also performed at various temperature (40°C, 50°C, and 70°C). The weight ratio of the coal after and before oxidation is:

$$Y_o = \frac{m_{OC}}{m_C} \quad (1)$$

2.2.2. Oxidized coal extraction

In order to analyze the effect of oxidation process parameter on the yield of extractable humic acid, every sample of oxidized coal was extracted at certain condition, i.e : 0.1 g oxidized coal sample, 50 mL 0.1 N standard solution of KOH as solvent. Extractions were performed for 7 samples, in 7 beaker glasses which placed in a water bath operated at 60°C. Five milliliter of extract sample was taken off after 60 minutes extraction, diluted to 100 mL in a volumetric flask. The diluted samples were titrated using 0.001 N HCl to find the remaining KOH. By assuming that the humic acid completely reacted, then its amount was equivalent to the amount of KOH reacted (=C_{HA}). If all the humic acid in the oxidized coal was assumed to be completely extracted, then the yield

$$Y_E = \frac{m_{OC}\, c_{HA}}{m_C} \, x \, 100\% \quad (2)$$

3. Results and Discussions

3.1. Oxidation

In a solid-liquid system, H_2O_2 will be adsorbed into the coal surface and oxidation happened, which resulted, long chains of humic substances cracked into short. Table I summarized the weight ratio of coal after and before oxidation (Y_O) at various process conditions.

Tabel I. Weigh Ratio of Coal, After to Before Oxidation

Oxidation time, min	Yield of oxidation					
	40°C & 10% H_2O_2	50°C & 10% H_2O_2	70°C & 10% H_2O_2	60°C & 5%H_2O_2	60°C & 15%H_2O_2	60°C at 20%H_2O_2
0	1.0000	1.0000	1.0000	1.0000	1.0000	1.0000
15	0.9078	0.8725	0.6761	0.8561	0.7199	0.6770
30	0.9044	0.8644	0.6599	0.8532	0.7185	0.6561
45	0.8884	0.8558	0.6356	0.8532	0.7189	0.6432
60	0.8169	0.8066	0.5815	0.8501	0.7141	0.6136
75	0.8041	0.7842	0.5652	0.8491	0.7072	0.6463
90	0.8029	0.7912	0.5495	0.8558	0.7110	0.6233

Table 1 shows that there are weight losses of the coal during oxidation using H2O2. The weight lost due to solubilization of various products of oxidation. Due to H_2O_2 is acidic, so the humic acids are not dissolved out of coal while fulvic acid and other water-soluble organics dissolved and diffuses into the solvent of H_2O_2. The lost is quite significant (45.05%) when oxidation conducted for 90 min using 10 % H_2O_2 at 70°C. Eventhough, this value is comparable to the oxidation of low rank

IOP Conf. Series: Materials Science and Engineering **162** (2017) 012025 doi:10.1088/1757-899X/162/1/012025

coal of Mukah. At the same H2O2 concentration and oxidation temperature, Fong et.al [4] reported that for 120 min oxidation the weight lost was 50.33 %. At another condition (5 % H_2O_2, 70°C, and 120 min) Fong, et.al [4] reported the loss was 32.23 %, while this research (5 % H_2O_2, 60°C, 90 min) only 14.42%. The higher the temperature of oxidation or the H_2O_2 concentration, the more the weight lost.

3.2. Extraction

The humic acid contained in oxidized coal which prepared under various conditions extracted using 0.1 N KOH solution at 60°C. The extractions were conducted for 60 minutes. The yields of humic acid on the extraction of oxidized coal summarized at Figure 3 for yield of humic acid based on oxidized coal weight (Y'_E) while Figure 4. Yield of humic acid based on original coal weight (Y_E). Compared to the original coal (non - oxidized), the yield of humic acids prepared with oxidation using 20% H_2O_2 at 60°C for 90 min, increased from 8.37 mgeq KOH to 19.65 mgeq KOH/gram of coal oxidized for coal or from 46.95% to 68.57 %. This value was comparable with from 46 to 63 % [5]. Although appeared to be promising, the considerable weight losses of coal upon oxidation as much as 45.05% render it incomparable. The losses were also observed to correspond to the concentration possibly due to prominent occurrence of desulfurization and demineralization. These processes were commonly reported for oxidation with H_2O_2 [4].

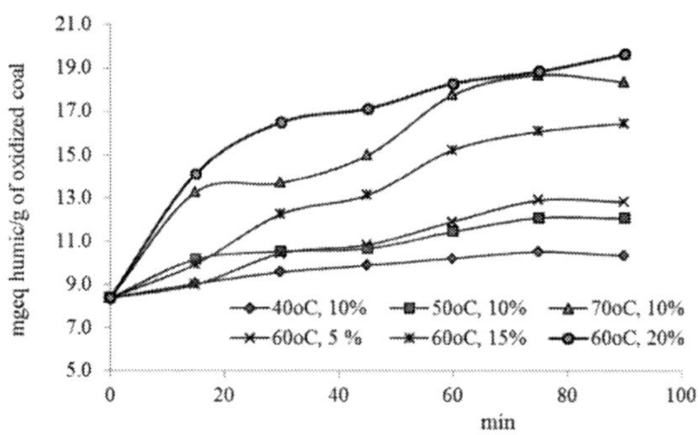

Figure 2. Yield of oxidized coal to humic acid

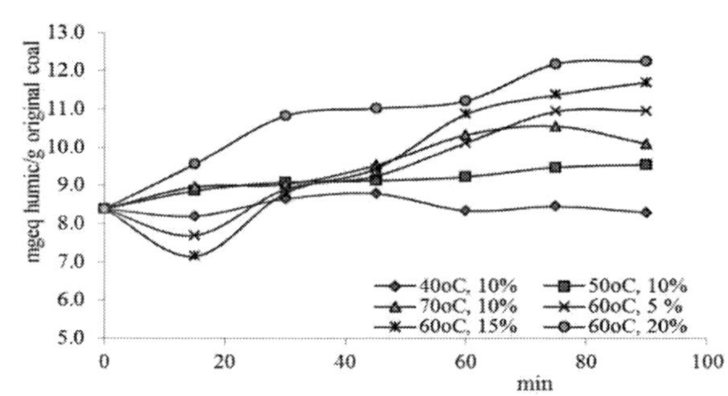

Figure 3. Yield of original coal to humic acid

The increasing yield of humic acid proved that the low rank coal have lots of hydrophilic functional groups such as -COOH and –OH. The increasing yield as the oxidation resulted more acidic groups in the coal, and as the coal is extracted with a base, the more acidic groups are reacted and dissolved in the solvent. More carboxylic functional groups, more water solubles usually obtained, so that the humic acid produced will increase. Figure 4 -7 present the FTIR for analyze the absorbance of the carboxylate (1700 1/cm) and phenolate (3300 1/cm).

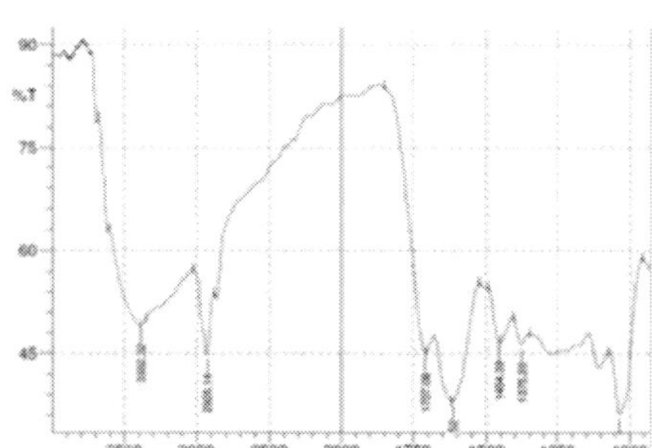

Figure 4. FTIR spectra for non oxidized coal

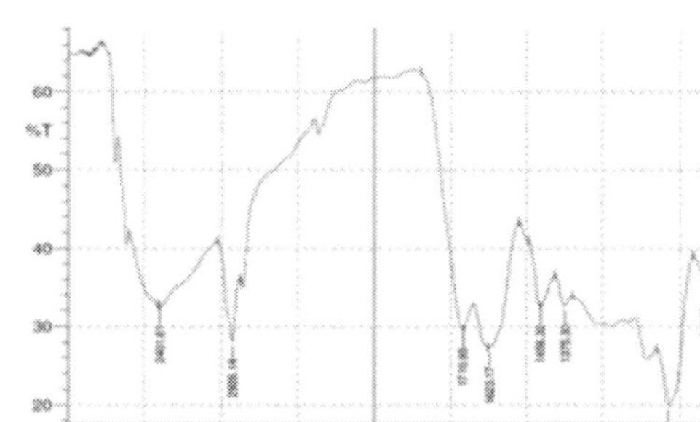

Figure 5. FTIR spectra oxidized coal (5%, 60°C, 60 min)

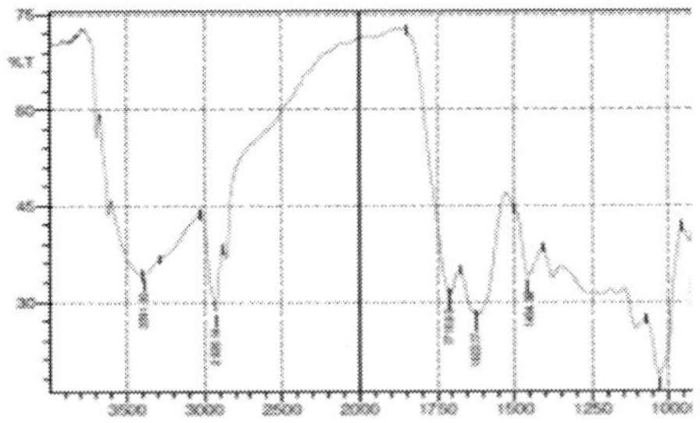

Figure 6. FTIR spectra oxidized coal (10%, 50°C, 60min)

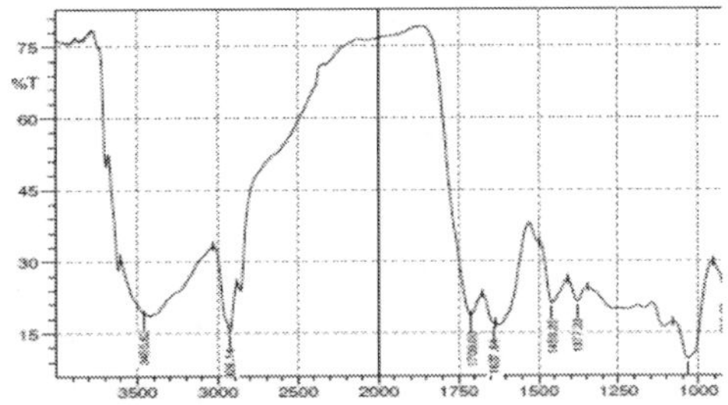

Figure 7. FTIR spectra oxidized coal (10 %,60°C, 90 min)

4. Conclusions

The effect of variation of H_2O_2 concentration, oxidation temperature, and time of oxidation to the weight loss of original coal have positive trends. The same trends to the yield of humic acids. Eventhough the trends are not too significant especially for lower temperature or lower concentration. Liquid phase oxidation of Indonesian low rank coal using H_2O_2 increases the yield of humic acid from 46.95% to 68.57 % of the original coal. Oxidation using H_2O_2 resulted more water soluble materials, but less overall yield of humic acids.

References

[1] Retrieved August 20, 2014 from http://www.esdm.go.id/index-en.html
[2] Retrieved August 20, 2014 from http://en.wikipedia.org/wiki/Humic_acid
[3] Mae K, Maki T, Araki J and Miura K 1997 Extraction of Low-Rank Coals Oxidized with Hydrogen Peroxide in Conventionally Used Solvents at Room Temperature *Energy & Fuels* **11** 825-831
[4] Fong S S, Seng L, Majri N and Mat H 2007 A Comparative Evaluation on the Oxidative Approaches for Extraction of Humic Acids from Low Rank Coal of Mukah, Sarawak *J. Braz. Chem. Soc.* **18**(1) 34-40
[5] Yu J, Jiang Y, Tahmasebi A, Han Y, Li X, Lucas J, Wall T 2014 Coal Oxidation under Mild Condition Current Status and Application *Chem Eng. Technol.* **37** 10
[6] Rahayu S S, Sulistyorini A and Perdana B A 2014 The Effect of H_2O_2 in the extraction of Indonesia Low Rank Coal *Proceeding 27th Symposium Of Malaysian Chemical Engineers (Somche 2014) In Conjunction With 21st Regional Symposium On Chemical Engineering (RSCE 2014) 29 October – 30 October 2014 Taylor's University Lakeside Campus, Subang Jaya*

IOP Conf. Series: Materials Science and Engineering **162** (2017) 012026 doi:10.1088/1757-899X/162/1/012026

Application of a hybrid Electrocoagulation-Fenton process in yarn dye wastewater : Kinetic study

L.Riadi[1,2], L.Sapei[1], T.Lidiawati[1,2] and Y.E.Agustin[1]
[1]Chemical Engineering Department, University of Surabaya
[2]Center for Environmental Studies,University of Surabaya
Jalan Raya Kalirungkut , Surabaya, Indonesia

E-mail : lieke@staff.ubaya.ac.id

Abstract. Reactive dyes contain a significant portion of colorants used in yarn dying process and also in textile industry. Since the COD content is usually high in such wastewater,we conducted a hybrid electrocoagulation-fenton method to treat the wastewater. This work describes the application of the hybrid system to the removal of chemical oxygen demand and color from the wastewater in a batch reactor. Having worked with initial pH of 3,0; temperature at 30°C, molar ratio of Fe^{2+}/H_2O_2 =1/10 and the mol ratio H_2O_2/COD = 4, we got 88.3% COD conversion and 88.5% color removal. The COD degradation process can be explained in two phases, the first phase is instantaneous reaction and the second phase is first order reaction. The kinetic constant was 0.0053 minute^{-1} and the rate of COD degradation was 0.0053[COD] mg/L minute.

1. Introduction

The disposal of yarn dye wastewater is a problematic issue in the area of high density population in Surabaya city. The wastewater is dark black to red colored, since it contains a lot of dye substances. Since the wastewater can not be treated biologically, we proposed a hybrid method to remove color, Chemical Oxygen Demand (COD) and Total Suspended Solid (TSS) from the wastewater. Advanced Oxidation Processes (AOPs) has been known for its capacity in reducing COD from recalcitrant wastewaters [1-5] . A hybrid system consists of electrocoagulation-fenton process will give a good combination to solve the wastewater. Electrocoagulation will help fenton oxidation process since high TSS will impair the oxidation. Hydrogen peroxide has been widely used for oxidizing agent, it is not effective to oxidize the organic compounds if it is used alone. The presence of metal salts especially ferrous ions can enhance hydrogen peroxide to form hydroxyl radicals. The standard redox potential of hydroxyl radical is 2.80 V and hydrogen peroxide is 1.77 V [6], so the radicals are more powerful oxidizing agent. Ferrous ions and hydrogen peroxide are known as Fenton's reagent and used in ambient temperature..The Fenton used in this study is a homogenous catalytic oxidation process involving the reaction of hydrogen peroxide with ferrous ions. The hydroxyl radicals are generated from this reaction :

$$Fe^{2+} + H_2O_2 \rightarrow Fe^{3+} + HO\bullet + OH^- \quad (1)$$

In the presence of organic contaminants (RH), the contaminants will be oxidized by the radicals as following

$$RH + HO\bullet \rightarrow R\bullet + H_2O \quad (2)$$

The hydrogen peroxide also can regenerate Fe^{3+} to form Fe^{2+} ,

$$Fe^{3+} + H_2O_2 \rightarrow Fe^{2+} + HO_2\bullet + H^+ \quad (3)$$

$$Fe^{3+} + HO_2\bullet \rightarrow Fe^{2+} + O_2 + H^+ \quad (4)$$

IOP Conf. Series: Materials Science and Engineering **162** (2017) 012026 doi:10.1088/1757-899X/162/1/012026

The new radical (R•) and also the hydroxyl radical (HO•) can degrade organic compound to carbon dioxide and water. Result from our previous study showed that the best performance of COD removal was at molar ratio of Fe^{2+}/H_2O_2 =1/10 and the mol ratio of H_2O_2/COD = 4. The aim of this work is to investigate the application of hybrid method Electrocoagulation-Fenton in the removal of color and COD from yarn dye wastewater at various pH. The kinetic of the Fenton process in degrading the COD was also studied.

2. Materials and Methods

2.1. Yarn dye wastewater

The wastewater in this study was collected from a yarn dye industry utilizing variety of textile color with the COD content of 872 ppm, BOD (Biological Oxygen Demand) = 108 ppm, TSS 6220 ppm, color (Optical Density) =1.65 and pH 4.8.

2.2. Methods

2.2.1. Experiments

A flexi glass reactor with 1500 cm^3 volume was used. Anode and Cathode Al/Al was used with a distance between two electrodes was 2 cm. The size of Alumunium plate was 8cm x 8 cm, with a thickness of 2 mm.The electrocoagulation process used a method as described by Riadi *et.al* [7]. The effluent from electrocoagulation process was then used as influent for Fenton oxidation process. A 1.5 L reactor equipped with stirrer and cooling jacket was used for the experiment. The cooling jacket was used to maintain the experiment at room temperature. The molar ratio of Fe^{2+}/H_2O_2 =1/10 and molar ratio of H_2O_2 to COD = 4. The initial pH in these experiments were varied. Samples were withdrawn periodically up to 90 minutes of experiment.

2.2.2 Analytical methods

Chemical Oxygen Demand was measured using closed reflux, colorimetric method [8]. Total Suspended solid was measured using dried method based on a method from APHA, color measurement was conducted using UV-Vis spectrophotometer from APHA [8]. The BOD (Biological Oxygen Demand) was measured by respirometer using BOD system BD600,Lovibond.

3.Results and Discussion

Effluent from electrocoagulation process has characteristic as presented in Table 1. Electrocoagulation will help fenton oxidation process since high TSS will hinder fenton oxidation. Suspended solid in the wastewater will interfere fenton process, hence electrocoagulation was introduced prior fenton process. Hence, we proposed hybrid system electrocoagulation-fenton processes in this study. Before performing a batch process for Fenton oxidation, the effluent from electrocoagulation process was examined.

Table 1. Characteristic of yarn dye wastewater after electrocoagulation

Parameters	Value	% removal
pH	4.8	n/a
COD	515 ppm	41
TSS	5810 ppm	6.6
Color (Optical Density)	0.192	88.4

3.1.Effect of intial pH

An important variable process in Fenton oxidation is pH, the best operation pH in the experiment was pH=3.0 (Table 2) which gave the best of COD removal. Result from the experiment is in line with a review reported by Wang.*et.al* [9]. The process is called homogenous Fenton process. The production of hydroxyl radicals is influenced by pH [10]. Since it is a homogenous Fenton process, the optimum of initial pH range is 2.5-3.0 for degradation of the most organic compounds. In higher pH (pH>3.0) the hydrolysis and precipitation of Fe^{3+} in the liquid phase can reduce the capacity of the the catalyst itself [11]. However, if the pH is too low (< 2.50) , there will be a strong scavenging effect of hydroxyl radical by H^+ as reaction below :

$$HO\bullet + H^+ + e^- \rightarrow H_2O \quad (5)$$

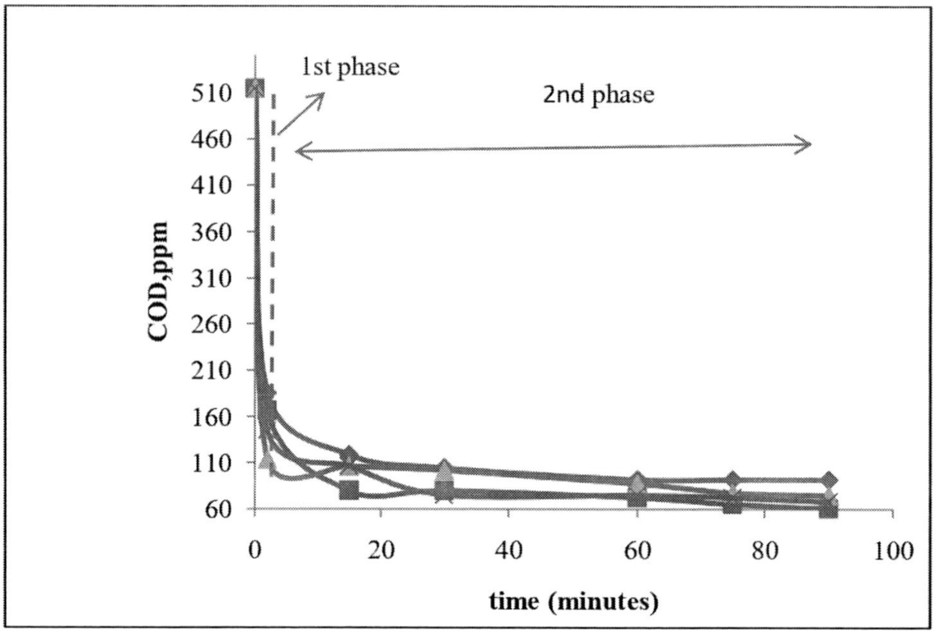

Figure 1. COD profile in a fenton oxidation batch system at molar ratio of Fe^{2+}/H_2O_2 =1/10; molar ratio of H_2O_2 to COD =4.0, x= pH 5.0, ▲=pH4.0,■ =pH=3.0, ♦ pH=2.0

Figure 1 shows the COD profile at different initial pH of wastewater. The first two minutes was a instantaneous reaction which showed a sharply decrease of COD which we mentioned as first phase. After 2 minutes of reaction the reduction of COD was continue up to 90 minutes of experiment,but the degradation was not as fast as that at the first two minutes. The poor performance for COD removal was at initial pH of 2.0, since at lower pH the reaction of hydrogen peroxide with Fe^{2+} is affected because of complex species formation ($Fe(H_2O)_6)^{2+}$ [3].

Table 2. Percentage removal of COD at different initial pH of wastewater

Initial pH	% removal of COD for the first two minutes	% total removal of COD
2.0	63.9	82.1
3.0	67.8	88.3
4.0	78.1	85.4
5.0	71.8	86.0

Fenton process can improve the treatment of COD removal for additional 47.3 % removal since the hydroxyl radicals which is produced from Fenton process is a strong oxidation agent. The color removal almost didn't occur in the Fenton process since all color has been removed in electrocoagulation process.

3.2..Kinetic study

The kinetic removal of COD during Fenton oxidation process was studied based on the assumption that hydrogen peroxide was in excess as the molar ratio of H_2O_2 to COD was 4.0.Hence, the kinetic removal of COD in batch system will follow the pseudo first order and can be written as follows :

$$\frac{-dCOD}{dt} = k\,[COD] \qquad (6)$$

The equation can be integrated between time of experiment, t=0 and t = t_{final}, which giving equation 7:

$$ln\frac{COD_0}{COD} = k\,t \qquad (7)$$

The first two minutes of the reaction is an instantaneous reaction, which is predicted as written in equation (1) and (2). The reaction after two minutes was slower, which is predicted that the hydrogen peroxide reacted with Fe^{3+} to form Fe^{2+} and hydroperoxy radical $HO_2\bullet$. The standard redox potential of hydroperoxy radical is lower than that of hydroxyl radical. Hence, the oxidation of organic compound is slower compare to that in the first two minutes. The values of k(kinetic constant) and R^2(R-squared) are presented in Table 3.

Table 3. The parameter kinetic constant obtained from Fenton oxidation at various intial pH.

Initial pH	k (kinetic constant, min^{-1})	R-squared value
2.0	0.0033	0.8348
3.0	0.0053	0.9718
4.0	0.004	0.983
5.0	0.0046	0.6716

The value of k decreases with the pH decrease below 3.0.The best performance of the Fenton oxidation is for pH=3.0. The degradation of COD can be written as $\frac{-d\,COD}{dt} = 0.0053\,(COD)$ mg/L min. The R-squared value is lower in Initial pH 5.0 since the precipitation of Fe^{3+} in higher pH can hinder the oxidation process and impair the COD analysis which gave a poor data of COD and lead to the unfit model of pseudo first order.

4.Conclusion

The overall results of this study show that the application of hybrid Electrocoagulation-Fenton oxidation is a promising method to treat yarn dye wastewater. The treatment achieved a significant decrease of COD which has been shown as 88.3 % removal. The color was also removed by 88.5 %. The best performance for the Fenton oxidation in batch system was at initial pH 3.0 and molar ratio of Fe^{2+}/H_2O_2 =1/10 ; molar ratio of H_2O_2 to COD = 4. The kinetic study based on COD degradation showed that the reaction is first order with kinetic constant (k) 0.0053 min^{-1}.

Acknowledgement
The authors greatly appreciate financial support from Indonesia Directorate General of Research and Higher Education Research Grant 2016 under a PUPT scheme. The valuable contribution of R.Wisanto and A.R.Herlambang are also acknowledged.

References

[1] Barbusinski K and Fajkis S 2011 *Environmental Progress and Sustainable Energy* **30** 4 620-631
[2] Guclu D, Sirin N, Sahinkaya S, Sevimli MF 2013 Environmental Progress and Sustainable Energy **32** 2 176-180
[3] Lucas M S and Peres J A 2009 *Journal. of Hazardous Materials* **168** 1253-1259
[4] Patil N N and Shukla S R 2015 *Environmental Progress and Sustainable Energy* **34** 6 1652-1661
[5] Wang Z, Chen K, Li J, Mo L and Wang Q 2011 *Environmental Progress and Sustainable Energy* **30** 2 168-176
[6] Tarr M A 2013 Fenton and Modified Fenton Methods for pollutant degradation, in *Chemical Degradation Methods for wastes and Pollutants* ed.MA Tarr (Marcell Dekker Inc,New York) pp 165-178
[7] Riadi L, Ferydhiawati W, Loeman L D S 2014 *Reaktor* **15** 2 73-78
[8] APHA, 1998. *Standard methods for the examination of water and wastewater*,.20th ed (APHA)
[9] Wang N, Xheng T, Zhang G, Wang P 2016 *Journal of Environmental Chemical Engineering* **4** 762-787
[10] Qiao R P, Li N, Qi X H, Wang Q S and Zhuang Y Y 2005 *Toxicon* **45** 745-752
[11] Fassi S, Djebbar K, Bousnoubra I, Chenini H and Sehili T 2014 *Desalination and Water Treatment* **52** 4982-4989

IOP Conf. Series: Materials Science and Engineering **162** (2017) 012027 doi:10.1088/1757-899X/162/1/012027

Pulverization of coffee silverskin extract as a source of antioxidant

S Tan[1,a], S P Kusumocahyo[2,b], and D I Widiputri[3,c]
[1,2,3] Department of Chemical Engineering, Faculty of Life Sciences and Technology, Swiss German University, EduTown BSD City, Tangerang 15339, Indonesia

E-mail : [a]suwi_chen@yahoo.co.id, [b]samuel.kusumocahyo@sgu.ac.id, [c]diah.widiputri@sgu.ac.id

Abstract. Coffee silverskin (CS) is waste from coffee roasting process that has a value as source of antioxidant. In this research, two types of variant coffee Robusta and Arabica CS were extracted for their phenolic content, flavonoid content, and antioxidant activity. The extraction was done at 40°C for 60 minutes using hydroalcoholic solvent. The phenolic, flavonoid content, and antioxidant activity of Robusta CS extract were 816.75 ± 63.24 mg GAE/L and 32.82 ± 2.47 mg QE/L, and 54.80% inhibition respectively, while for Arabica CS extract were 473.51 ± 56.70 mg GAE/L, 18.58 ± 2.47 mg QE/L, and 26.30% inhibition respectively. Thus, the Robusta coffee silverskin extract has higher value of total phenolic content, flavonoid content, and antioxidant activity than Arabica coffee silverskin extract. To produce high antioxidant powder of CS extract, the effect of drying method (freeze drying and spray drying) affecting the phenolic content, flavonoid content, and antioxidant activity was evaluated. The effect of evaporation prior to both drying processes was also evaluated. Evaporation caused up to 23% of total phenolic content degradation. Spray drying resulted in dried CS extract with degradation of total phenolic content up to 17%. On the other hand, freeze drying resulted no major degradation of total phenolic content. However, the coffee silverskin extract can be directly spray dried without evaporation resulting in higher amount of phenolic content in the powder than the one which was evaporated first.

1. Introduction

Recently, coffee has become one of the most popular beverages in the world. Indonesia itself is the third biggest coffee producer in the world, after Brazil and Vietnam [1], but only two species are commercially traded worldwide. There are *Coffeaarabica* (Arabica) and *Coffeacanephora* (Robusta) [1]. Robusta is considered to have a higher yield because it is easier to tend on the farm and less sensitive to insects. Also from the price perspective, Robusta's greenbean is about half the price of Arabica's green bean on the commodity market. However, Arabica has more preferable taste and aroma than Robusta. Robusta is often used in instant coffee while Arabica is used in coffee shops [15].

Commonly, coffee beans must go through several steps of processing until it is served. First, coffee cherries are harvested from the coffee plant, they are each composed of two coffee beans covered by a thin parchment like hull and further surrounded by pulp [16]. When the pulps are removed through pulping process, the parchment is removed through either wet or dry process which produces the dried beans, commonly known as the green bean. Then, the green bean enters another process called roasting that affects the overall color, aroma, and flavor of the final product [12].

The roasting process of green beans into brown is a crucial step of coffee processing. It is a process where a high temperature (around 210°C-240°C) is applied to the coffee beans. The water contained is evaporated and chemical reactions like Maillard reactions [17], oxidation, thermal degradation and hydrolysis of the coffee's chemical components takes place. There are four main degrees of coffee roasting that will affect the organoleptic properties of the final product, which are light roast, medium roast, medium dark roast and dark roast [8].

Furthermore, the process of obtaining coffee bean from the whole fruit leaves the silverskin as waste. Silverskin, or commonly known as chaff in the industries is usually produced during the roasting process. It is actually the outer layer of the green coffee beans that is usually still intact during the pulp and hull removing process. When the coffee beans come into contact with high temperature, the coffee beans crack as the result of swelling due to loss of moisture content, and also make the silverskin detach from the beans. Due to its lightness, silverskin usually escapes from the roasting container through the roaster exhaust or cyclone [11]. The silverskin, or the chaff is usually thrown away by the roasters due to some problems related to roaster cleanliness and air pollution. On the other hand, the silverskin is mostly used only as combustible or fertilizer [17]. However, coffee silverskin actually has a potential of antioxidant activity because it is highly rich in phenols [18]. Phenolic compound from coffee silverskin can be obtained by extraction and the extract has many application possibility for food and pharmaceutical based on previous research by Costa el al. [6] in which the total phenolic content and total flavonoid content of the coffee silverskin extract was 302.5 ± 7.1 mg GAE/L and 83.0 ± 1.4 mg ECE/L. Therefore, the extract of coffee silverskin has potential as antioxidant food or cosmetic products. Hence, the extract of coffee silverskin could be more applicable for commercial uses while in powder form instead of liquid form in terms of distribution and storage. The desired powder should be high in phenolic content, flavonoid content, and antioxidant activity. However, drying process affects the phenolic content based on previous research by Orphanides et al. [14]. The characteristics of the final powder form of coffee silverskin extract is then identified.

2. Methodology

The raw material was homogenized using blender. Then, the homogenized coffee silver skin was extracted. The extraction was done on a hotplate inside a beaker and stirred with a magnetic stirrer. The extraction condition was set at 40°C for 60 minutes using hydroalcoholic solvent with 1:1 ratio of alcohol and water. There were two ratio of solid to solvent which are 1:50 and 1:10. The extract with 1:50 solid to solvent ratio was analyzed for comparison of Robusta and Arabica variant, while the extract with 1:10 solid to solvent ratio was used for the drying step. The extract was filtered using filter paper and stored in a freezer prior to analysis. The two drying processes were done separately. Spray drying was done directly to the extract at inlet temperature 170°C, outlet temperature 125°C with feed flow 20ml/min. In the case of freeze drying, the alcoholic content of the extract was first removed using vacuum rotary evaporator at 50°C for 3 hours. Then, the extract was freeze dried at -50°C for 24 hours. As a comparison, spray drying was also done to the extract whose alcoholic content had been removed using vacuum rotary evaporator at the same condition. The resulting powder was immediately stored in desiccator until analysis.

3. Results and Discussion

The extract of coffee silver skin from both Arabica and Robusta were compared in terms of total phenolic content, total flavonoid content, and antioxidant activity. As shown in Figure 1, total phenolic content of Robusta coffee silver skin extract was 816.76 ± 63.24 mg GAE/ L while Arabica coffee silver skin was 473.51 ± 56.70 mg GAE/ L. Thus, Robusta coffee silver skin has the value of phenolic content almost twice than Arabica coffee silver skin. This finding is similar to the study conducted by Farah (2012) which reported higher amount of chlorogenic acid is originally present in the green bean of Robusta rather than in Arabica. Chlorogenic acid belongs to the phenolic group, therefore contributing to the absorbance reading of phenolic.

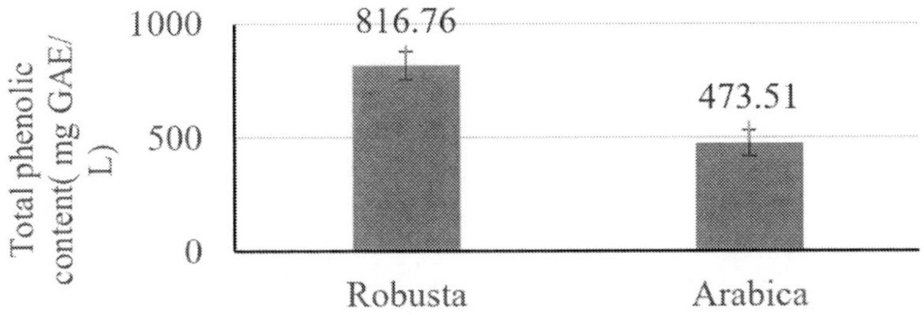

Figure 1. Total phenolic content of Arabica and Robusta coffee silver skin extract.

As shown in Figure 2, total flavonoid content of Robusta coffee silver skin extract was 32.82 ± 2.47 mg QE/ L while Arabica coffee silver skin extract was 18.58 ± 2.47 mg QE/ L. This result is also in agreement with the total phenolic content because flavonoid is one of the phenolic group. Robusta coffee silver skin extract has twice total flavonoid content than Arabica coffee silver skin extract.

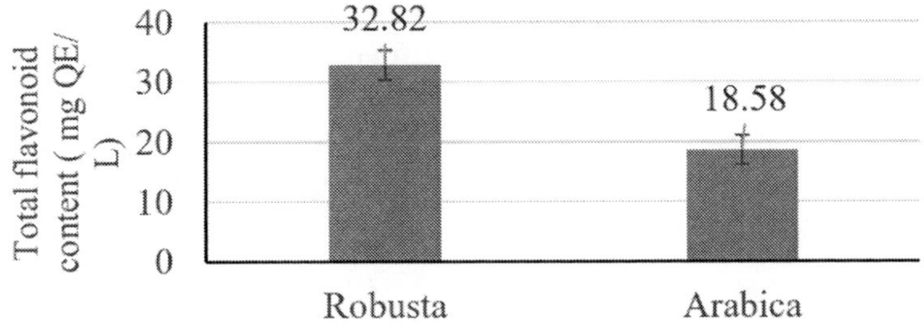

Figure 2. Total flavonoid content of Arabica and Robusta coffee silver skin extract

The antioxidant activity of the Arabica and Robusta coffee silver skin extract can be seen in Figure 3. The inhibition to DPPH activity of Robusta coffee silver skin extract was 54.80 % and Arabica coffee silver skin extract was 26.30 %. As the total phenolic content gets higher, it also increases the antioxidant activity. The antioxidant activity of Robusta coffee silver skin extract is higher than of Arabica coffee silver skin extract. This indicates that Robusta coffee silver skin contain high amount of antioxidant activity.

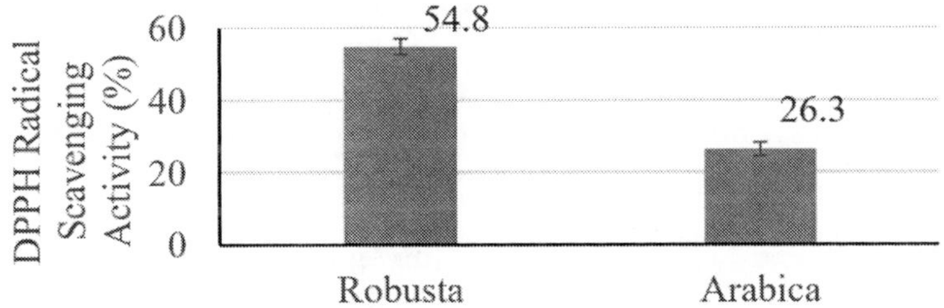

Figure 3. Antioxidant activity between Robusta and Arabica coffee silver skin extract

Total phenolic content and total flavonoid content were analysed in this research as an approach to identify the amount of chlorogenic acid in coffee silver skin. Both chlorogenic acid and flavonoid belong to the phenolic group. Most of phenolic group has antioxidative property, so do chlorogenic acid and flavonoid. Based on the results from this research, the total flavonoid content is only around 4% from the total phenolic content found in Robusta and Arabica coffee silver skin extract. Therefore, it can be concluded that flavonoid does not have much contribution to the high antioxidant activity which means the antioxidative property comes from other compounds from the phenolic group. This other compounds are assumed to be dominated mostly by chlorogenic acid, corresponding to 74% of chlorogenic acid detected in coffee silverskin based on Bresciani et al. [4].

The products from both drying method were compared. The total phenolic content results show in Table 1. First, the total phenolic content of the Robusta coffee silver skin after extraction with solid to solvent ratio 1:10 was 210.27 mg GAE/ g solid extract. The extract was evaporated using rotary vacuum evaporator. The total phenolic content was degraded by 22.61% to 162.72 mg GAE/g solid extract. After evaporation, the Robusta coffee silver extract was dried with freeze drying and spray drying. Total phenolic content of freeze dried extract was 162.91 mg/ g powder while spray dried extract was 134.76 mg/ g powder. Another spray dried product using extract without evaporation was also analysed and the result was 188.83 mg / g powder in total phenolic content. From the results, the total phenolic content from freeze drying process does not give major degradation while spray drying process degrade 17%. The result shows, phenolic compound affected by high temperature and in accordance to study by Orphanides et al. [14] on effect of drying method on the phenolic content and antioxidant capacity of Spearmint.

Table 1. Total phenolic content of Robusta coffee silverskin extract from different drying method

Total Phenolic Content of Robusta coffee silverskin extract			
Drying Method	**After Extraction** (mg/ g solid extract)	**After Evaporation** (mg/ g solid extract)	**After Drying** (mg/ g powder)
Freeze drying	210.27	162.72	162.91
Spray drying	210.27	162.72	134.76
	210.27	Without evaporation	188.83

The total flavonoid content of Robusta coffee silver skin extract results show in Table 2. Total flavonoid content after extraction was 8.09 mg QE/ g solid extract. Thus, total flavonoid content of the extract was degraded 23.85% after evaporation to 6.16 mg QE/ g solid extract. After that, the extract was went to drying processes which are freeze drying and spray drying. Then, total flavonoid content of freeze dried product was 6.31 mg QE/ g powder while spray dried extract was 4.03 mg QE/ g powder. Another spray dried product using extract without evaporation was also analysed and the result was 7.22 mg/ g powder.

Table 2. Total flavonoid content of Robusta coffee silverskin extract from different drying method

Total Flavonoid Content of Robusta coffee silverskin extract			
Drying Method	**After Extraction** (mg/ g solid extract)	**After Evaporation** (mg/ g solid extract)	**After Drying** (mg/ g powder)
Freeze drying	8.09	6.16	6.31
Spray drying	8.09	6.16	4.03
	8.09	Without evaporation	7.22

The antioxidant activity of the product was determined using IC_{50}. The result of IC_{50} each product can be seen in Figure 4. IC_{50} is the amount of sample to inhibit 50% of Free Radical or in this case is DPPH. The IC50 of Robusta CSE freeze dried was 2139.80 ppm, Robusta CSE (with evaporation) spray dried was 2596.73 ppm, Robusta CSE (without evaporation) spray dried was 1721.32 ppm, Arabica CSE (without evaporation) spray dried was 2236.77 ppm, Arabica CSE (with evaporation) spray dried was 2923.88 ppm, and Arabica CSE freeze dried was 2391.50 ppm. The smallest value means, the better the antioxidant. Robusta CSE (without Evaporation) Spray dried gives smallest value between the other Robusta CSE. This value correlated with the amount of phenolic content in the sample. The higher phenolic content means the better the antioxidant activity. Other study also revealed that the antioxidant activity was directly proportional to the content of phenolic compounds [3].

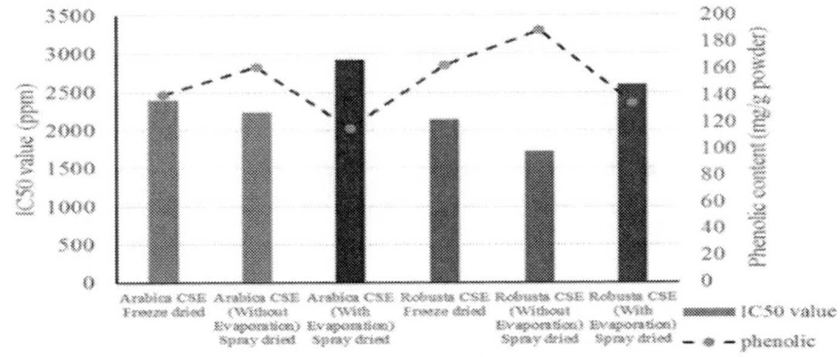

Figure 4. IC_{50} and the phenolic content of the spray dried and freeze dried of the Arabica and Robusta coffee silverskin extract

The ascorbic acid was compared with the phenolic content in the spray dried and freeze dried of Robusta and Arabica coffee silverskin extract in terms of their antioxidant activity. As the result shown in Figure 5, ascorbic acid has greater antioxidant activity than the spray dried and freeze dried of the Robusta and Arabica coffee silverskin extract with the IC_{50} value of ascorbic acid at 141.33 ppm and IC50 value of the phenolic content in the spray dried and freeze dried of Arabica and Robusta coffee silverskin extract at around 350 ppm. These results show that the coffee silver skin extract can substitute ascorbic acid in terms of antioxidant activity with twice more amount.

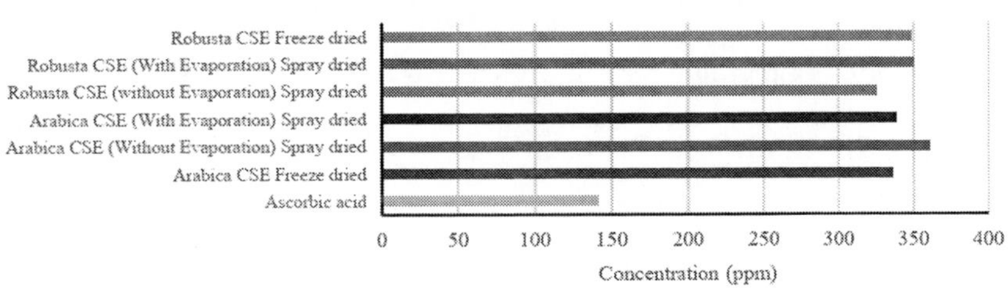

Figure 5. The IC_{50} of the phenolic content in the spray dried and freeze dried of the Arabica and Robusta coffee silverskin extract and Ascorbic acid

The extract of coffee silverskin extract which was pulverized using spray dried gives moisture content, solubility value, and mass mean diameter were 5.30%, 8.47 mg/ml, and 240.53µm respectively while pulverized using freeze dried were 3.86%, 9.28 mg/ml, and 1343.33µm respectively. The physical properties of coffee silverskin extract which was pulverized using spray dried was light brown fine powder with rancid and medicinal aroma while using freeze dried was dark brown crystalline powder with mild rancid and medicinal aroma.

4. Summary

In summary, the total phenolic content, the total flavonoid content, and antioxidant activity of Robusta coffee silverskin extract were 816.76 ± 63.24 mg GAE/ L, 32.82 ± 2.47 mg QE/ L, and 54.80 % inhibition of DPPH respectively while Arabica coffee silverskin extract were 473.51 ± 56.70 mg GAE/ L, 18.58 ± 2.47 mg QE/ L, and 26.30 % inhibition of DPPH. Thus, the Robusta coffee silverskin extract has higher value of total phenolic content, the flavonoid content, and antioxidant activity than Arabica coffee silverskin extract.

Prior to drying, the extract was evaporated first with total phenolic content 162.72 mg GAE/g solid extract, and total flavonoid content 6.16 mg QE/ g solid extract. Decrease of total phenolic content was 22.61% through evaporation. The total phenolic content, total flavonoid content, and IC50 value of spray dried of Robusta coffee silverskin extract were 134.76 mg GAE/ g powder, 4.03 mg QE/ g powder, and 2596.73 ppm respectively, while for freeze dried Robusta CS extract were 162.91 mg GAE/ g powder, 6.31 mg QE/ g powder, and 2139.80 ppm respectively. Moreover, freeze drying does not give major degradation to the total phenolic content and the total flavonoid content in the coffee silverskin extract while spray drying gives 17% of degradation to the total phenolic content and flavonoid content. However, the coffee silverskin extract can be directly spray dried. The resulting powder has higher amount of phenolic content than the one which was evaporated first.

References

[1]. Anwar I 2014 Indonesia Coffee Market
[2]. Augustin M A, Hemar Y 2009 Nano- and micro-structured assemblies for encapsulation of food ingredients *Chem. Soc. Rev.* **38(4)** 902–912
[3]. Ballesteros L F, Teixeira J A and Mussatto S I 2014 Selection of the Solvent and Extraction Conditions for Maximum Recovery of Antioxidant Phenolic Compounds from Coffee Silverskin *Food Bioprocess Tech.* **7(5)** 1322–1332
[4]. Bresciani L, et al. 2014 Phenolic composition, caffeine content and antioxidant capacity of coffee silverskin *Food. Res. Int.* **61** 196–201
[5]. Clifford M N 1999 Chlorogenic acids and other cinnamates–nature, occurrence and dietary burden *J. Sci. Food Agr.* **79(3)** 362–372
[6]. Costa A S G, et al. 2014 Optimization of antioxidants extraction from coffee silverskin, a roasting by-product, having in view a sustainable process *Ind. Crops Prod.* **53** 350–357
[7]. Farah A, et.al. 2008 Chlorogenic acids from green coffee extract are highly bioavailable in humans **1, 2** 2309–2315
[8]. Farah A 2012 Coffee : Emerging Health Effects and Disease Prevention *Coffee Constituents* 21–58
[9]. Manach C, et al. 2004 Polyphenols : Food sources and bioavailability *Am. J. Clin. Nutr.* **79(5)** 727–747
[10]. Christ M Smart freeze drying with system
[11]. Moore K 2008 Instant Coffee *Chem. Eng. News*
[12]. Mussatto S I et al. 2011 Production, composition, and application of coffee and its industrial residues *Food Bioprocess Tech.* **4(5)** 661–672
[13]. Narita Y, Inouye K 2014 Review on utilization and composition of coffee silverskin *Food Res. Int.* **61** 16–22
[14]. Orphanides A, Goulas V and GekasV 2013 Effect of drying method on the phenolic content and antioxidant capacity of spearmint *Czech J. Food Sci.* **31(5)** 509–513
[15]. Pabari S 2015 10 differences between robusta & arabica coffee
[16]. Pandey A et al 2000 Biotechnological potential of coffee pulp and coffee husk for bioprocesses *Biochem. Engi. J.* **6(2)** 153–162
[17]. Sacchetti G, et al. 2009 Effect of roasting degree, equivalent thermal effect and coffee type on the radical scavenging activity of coffee brews and their phenolic fraction *J. of Food Eng.* **90(1)** 74–80
[18]. Saenger M, et al. 2001 Combustion of coffee husks *Renew. Energ* **23(1)** 103–121
[19]. dos Santos M D, et al. 2006 Evaluation of the anti-inflammatory, analgesic and antipyretic activities of the natural polyphenol chlorogenic acid *Biol Pharm. Bull* **29(11)**2236–2240

IOP Conf. Series: Materials Science and Engineering **162** (2017) 012028 doi:10.1088/1757-899X/162/1/012028

Extraction and identification of bioactive compounds from agarwood leaves

N Y Lee[1, a], M A C Yunus[1, 2, b *], Z Idham[1], M S H Ruslan[1], A H A Aziz[1] and N Irwansyah[2]

[1] Centre of Lipids Engineering & Applied Research (CLEAR), IbnuSina Institute for Scientific and Industrial research, UniversitiTeknologi Malaysia, 81310 UTM Johor Bahru, Johor

[2] Faculty of Chemical and Energy Engineering, UniversitiTeknologi Malaysia, 81310 UTM Johor Bahru, Johor

E-mail :[a]nianyian913@gmail.com, [b]azizi@cheme.utm.my

Abstract. Agarwood commonly known as gaharu, aloeswood or eaglewood have been used as traditional medicine for centuries and its essential oil also being used as perfumery ingredients and aroma enhancers in food products. However, there is least study on the agarwood leaves though it contains large number of biomolecules component that show diverse pharmacological activity. Previous study showed that the extracted compounds from the leaves possess activities like anti-mutagenic, anti-tumor and anti-helminthic. The main objectives of this research were to determine bioactive compounds in agarwood leaves; leaves extract and oil yield obtained from maceration and soxhlet extraction methods respectively. The maceration process was performed at different operating temperature of 25°C, 50°C and 75°C and different retention time at 30, 60, 90 and 120 minutes. Meanwhile, various solvents were used to extract the oil from agarwood leaves using soxhlet method which are hexane, water, isopropanol and ethanol. The extracted oil from agarwood leaves by soxhlet extraction was analyzed using gas chromatography mass spectrometry. The results showed that the highest extract of 1.53% was obtained when increase the temperature to 75 °C and longest retention time of 120 minutes gave the highest oil yield of 2.10 % by using maceration. This is because at higher temperature enhances the solubility solute and diffusivity coefficient, thus increase the extract yield while longer retention time allow the reaction between solvent and solute occurred more rapidly giving higher extract. Furthermore, the soxhlet extraction using n-hexane as the solvent gave the highest oil yield as compared to other solvent due to the non-polar properties of n-hexane increase the efficiency of oil which is also non-polar to soluble in the solvent. In addition, the results also reported that the oil extracted from agarwood leaves contains bioactive compounds which are phytol, squalene, n-hexadecanoic acid and octadecatrienoic acid. Therefore, oil extracted from agarwood leaves has the potential to be applied in food, pharmaceutical, nutraceutical and cosmetics industries.

1. Introduction

Agarwood (*Aquilariasinensis*, *Aquilariacrasna*) is widely known as incense in Thailand, Cambodia and Taiwan. The agarwood leaves also being processed as health tea product in some countries such as Taiwan and Thailand. The term “agarwood” refers to resin-impregnated pieces of wood that have been at least partially shaved from the non-impregnated woods [1, 2]. According to Nor Azah [3], this resin impregnated woody tissues produced in the heartwood area mainly found in certain species of *Aquilaria*which has been a highly priced commodity for more than 2000 years. Agarwood is an evergreen tree that can grow up to 40 m high and 60 cm in diameter while the leaves are 5 to 9 cm long and oblong lanceoloate [4]. These trees usually found in low-land tropical forests with optimal sunlight, shade and moisture.

It is also known by many names; in Malaysia and Indonesia, it is called as “gaharu”, “chenhsiang” or “chenxiang” in China; “jin-koh” in Japan; “agar” in India; “chim-hyuang” in Korea; “kristsananoi” in Thailand; "tram huong" in Vietnam; "bolsd'agle", or 'calambour" in French and "oud" in the Middle East [5, 6, 7].

The study of bioactive compounds is very important in discover of therapeutic agent in plant [8]. The most significant bioactive compounds of these plants are alkaloids, tannins, flavonoids and phenolic compounds [9]. Agarwood is used in different of communities and majority of its medicinal uses involved in anti-inflammatory, used to treat rheumatism, arthritis, body pain, asthma and gout [2]. There are some studies revealed that agarwood has remarkable anticancer activity [10]. However, there is least study on the identification of bioactive constituents in agarwood leaves. Therefore, this study aimed to extract and identify the presence of bioactive compounds in agarwood leaves by different extraction method along with gas chromatography-mass spectrometry (GC-MS).

2. Methodology

2.1 Materials

The fresh leaves of agarwood were collected from Masjid Tanah Melaka, Malaysia.

2.2 Preparation of plant materials

The leaves were dried in the oven at the temperature of 60 °C and ground into powder form using POLYMIX® PX-s MFC 90 D grinder and sieved into average particle size of 215μm. The powdered sample was kept in freezer to prevent the growth of microorganisms and protect the sample from light prior to extraction.

2.3 Extraction of agarwood leaves by maceration process

5.0 g of dried powder leaves of Agarwood were placed into macerated in 75 mL of distilled water and placed into shaking bath (Clifton Range®) at constant speed of 220 rpm and retention time of 30 minutes. The shaking maceration process was performed at different temperature of 25 °C, 50 °C and 75 °C. Meanwhile, shaking maceration process was also done at different retention time of 30, 60, 90 and 120 minutes at constant temperature of 75 °C with constant speed of 220 rpm. The crude extracts were freeze dried.

2.4 Extraction of agarwood leaves by soxhlet extraction

Soxhlet extraction method was used to extract the agarwood leaves and the extraction process was performed using different solvents included distilled water, hexane, isopropanol and ethanol. All the solvents used were analytical grade. The extraction process was done for 6 hours at constant temperature of 60 °C. The crude extract of agarwood was evaporated using a rotary evaporator (Heidolph-insturments, Rotavapor, Germany). The weight of all extracts were measured after solvent evaporation and kept into air tight container for analysis purpose.

2.5 Gas chromatography-mass spectrometry (GC-MS) analysis

The extracts from soxhlet extraction were analysed using gas chromatography-mass spectrometry of Hewlett Packard 6890 series Gas Chromatography with Auto-sampler and 5973 N Mass Selective Detector. A HP-5 crosslinked 5% phenyl methyl siloxane fused silica capillary column, (30 m long * 0.25 mm internal diameter) with 0.25 μm film thickness was used. The extracts were dissolved in ethanol and filtered using 0.45 μm pore size whatman polyethersulfone membrane filter. Helium gas was used as carrier gas at a flow rate of 0.9 mL/min.The column temperature was programmed to 50 °C and hold for 2.5 min to 150 °C with the increase of 15°C per min. The temperature of injector and detector were maintained at 250 °C. The temperature of mass spectral ionization was set at 230 °C and operated in the electron impact ionization mode at 70 eV while the detector was performed in scan mode from 20-600 atomic mass unit (amu). Hence, the identification of compounds in the extracts were based on the mass spectral matching with the standard compounds in NIST (National Institute of Standards and Technology) and Wiley libraries. The relative amount of individual components was expressed as percentage of peak area relatives to total peak area.

3. Results and Discussion.

3.1 Effect of operating temperature on overall leaves extract by maceration

Operating temperature and time of extraction are important parameters to be optimized even in order to minimise energy cost of the process [11]. The overall extract obtained from agarwood leaves was increased with the increasing of operating temperature from 25 °C to 75 °C. This is because the increase of operating temperature enhances the solubility of solute and diffusivity coefficient of compounds, thus favours the extraction [12, 13]. In addition, the solubility of leaves extract in the solvent was increased and also the mass transfer was enhanced at higher temperature due to the increase of vapour pressure of solutes with increasing of operating temperature [14]. The solvent viscosity was decreased when increasing the temperature and this allows penetration of solvent into the plant matrix, thus accelerating the whole extraction [15]. Moreover, heating is known to affect the morphological changes in the plant sample matrix and thus increase the mass transfer rate of leaves extracts [16]. Consequently, the effects of higher temperature and mass transfer rates synergistically increase the rate of the extraction from agarwood leaves. The percentage of extracts from Agarwood leaves were shown in Table 1.

Table 1. Percentage of leaves extract obtained at different operating temperature by maceration process

Operating Temperature, °C	Extract Yield, g	Percentage of leaves extract, %
25	0.0190	0.38
50	0.0282	0.56
75	0.0765	1.53

3.2 Effect of retention time on overall leaves extract by maceration

As shown in the Table 2, the percentage of leaves extract increased with the increasing of retention time. By increasing the retention time, prolong the reaction between solvent and matrix interaction, hence favour the extraction process. The swelling of plant cell wall also increase the penetration of the solvent into the plant material, so increase the leaves extract. The percentage of leaves extract at 90 minutes and 120 minutes do not show significant difference due to the extraction process reaching the equilibrium state when all the solute had completely extracted, hence extend the extraction time would not affect the rate of extraction because the extraction process has been completed [17].

Table 2. Percentage of leaves extract obtained at different retention time by maceration process

Retention time, min	Extract Yield, g	Percentage of leaves extract, %
30	0.0889	1.78
60	0.0917	1.83
90	0.1030	2.06
120	0.1050	2.10

3.3 Effect of various solvents used on overall oil yield by soxhlet extraction

Soxhlet extraction traditionally was used in determining the total oil [18]. The extraction yield and extraction activity are strongly dependent on the solvent, thus the selection of extraction solvents is critical for the complex plant samples [15]. According to Wang *et al.* [19], an extraction solvent system is generally selected based on the purpose of extraction, polarity of the interested compounds, polarity of undesirable components, overall cost, safety and environmental concern. As shown in the Fig.1, the use of hexane as the solvent gives the highest percentage of oil yield in comparison to the other solvents in this order: hexane > isopropanol > ethanol > water. This result is in accordance with polarity of the solvents used in the extraction and solubility of oil from agarwood leaves in solvents [15]. The polarity index of hexane, isopropanol, ethanol and water are 0.1, 3.9, 5.2 and 10.2 respectively. A solid solute tends to dissolve in a liquid solvent when the molecules of both are similar enough in polarity. Therefore, oil from agarwood leaves is non-polar tend to dissolves in non-polar solvent according to "like dissolves like" adage, thus highest oil yield obtained by using hexane as the solvent.

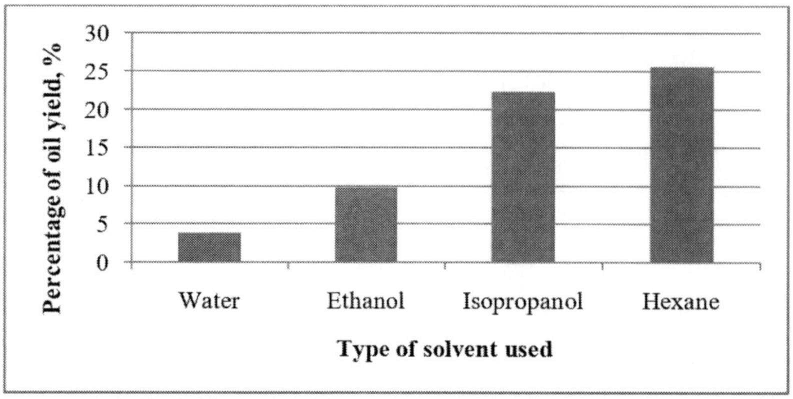

Figure 1. Percentage of oil yield obtained by soxhlet extraction with various solvents

3.4 Gas chromatography-mass spectrometry (GC-MS) analysis

The oil yield obtained from soxhlet extraction were analysed using GC-MS. Table 3 shows the results of the GC-MS analysis of the oil yield extracted from Agarwood leaves using isopropanol as the solvent. The analysed results showed that the extracted oil contained squalene, n-hexadecanoic acid, tetramethyl-2-hexadecen-1-1 and octadecatrienoic acid. Squalene also known as 2, 6, 10, 15, 19, 23-hexamethyltetracosa-2, 6, 10, 14, 18, 22-hexaene is a highly unsaturated hydrocarbon and a chain-like triterpene [20]. Squalene has many physiologic functions, such as promotion of superoxide dismutase activity in vivo, enhancement of the immune responses, anti-aging, anti-fatigue and anti-tumour activities [20]. Besides, n-hexadecanoic is a saturated fatty acid and also known as palmitic acid. Meanwhile, octadecatrienoic acid is C18 polyunsaturated fatty acid. Most of the fatty acids are known to have antibacterial, antifungal properties and also exhibited anti-inflammatory effect [21]. Furthermore, phytol is an acyclic diterpene alcohol and it had been used as the precursor for the production of synthetic form of Vitamin E [22] and Vitamin K1 [23].

Table 3. Results of the GC-MS analysis of the oil yield extracted from agarwood leaves using isopropanol as the solvent

No	Retention time (min)	Percent of total (%)	Bioactive Compound	Structure
1	12.07	28	Tetramethyl-2-Hexadecen-1-ol (Phytol)	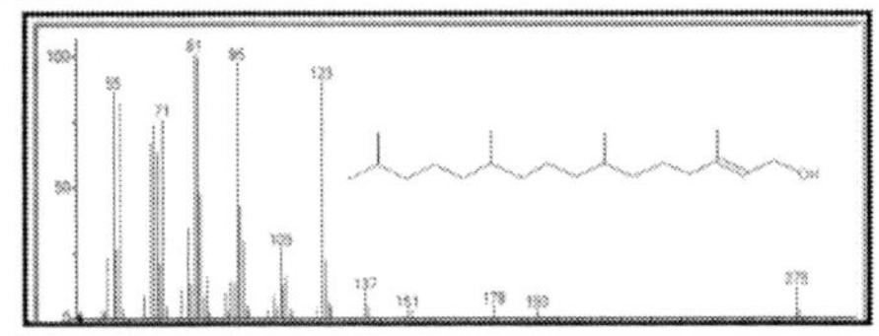
2	12.32	76.3	n-hexadecanoic acid , methyl ester	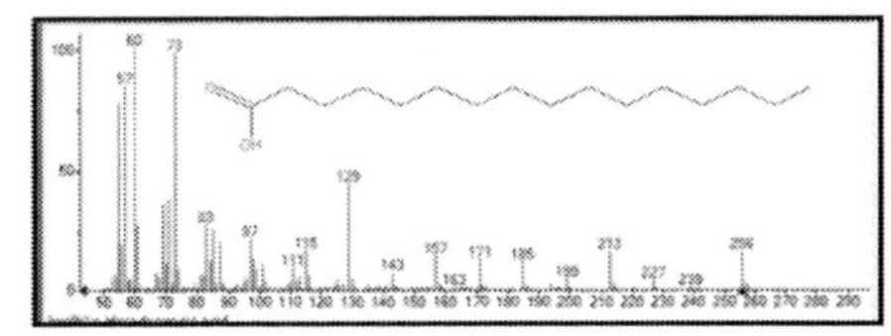
3	12.75	30.0	Octadecatrienoic acid	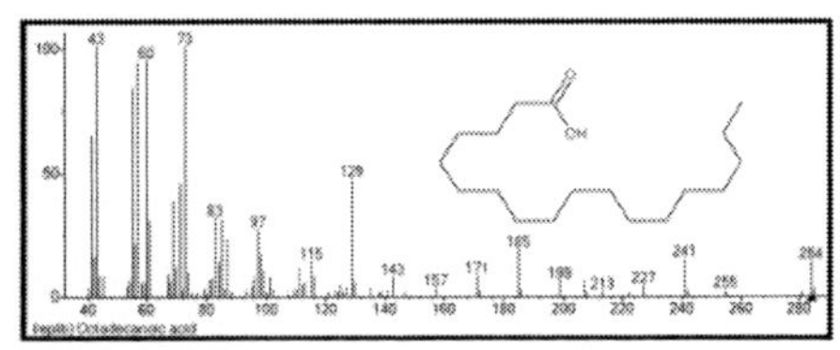
4	14.79	32.8	Squalene	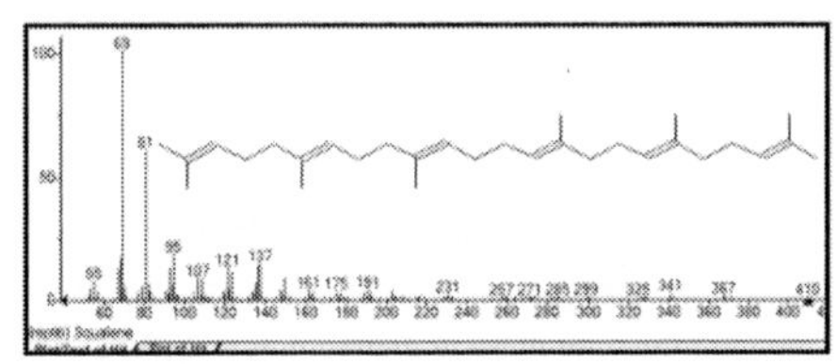

4. Conclusion

In this study, maceration process was applied to obtain the leaves extracts from Agarwood leaves and the effects of operating temperature and retention time on the overall leaves extracts were investigated. The results show that leaves extracts increased with increasing of operating temperature and retention time. Meanwhile, soxhlet extraction method with various solvents was used to study the effect of different solvents used on oil yield extracted. Extraction using hexane as solvent gave highest percentage of oil yield. The GC-MS analysis showed that oil of agarwood leaves contains important bioactive compounds of squalene, n-hexadecanoic acid, tetramethyl-2-hexadecen-1-1 and octadecatrienoic acid which very valuable to be applied in pharmaceutical, nutraceutical and cosmetics products.

Acknowledgement
We would like to acknowledge the financial support from the Ministry of Higher Education (MOHE) for fundamental research grant scheme (R.J130000.7809.4F766) and UniversitiTeknologi Malaysia for professional development research university grant (Q.J130000.21A2.02E85).

References

[1] Swee L L C 2008 Agarwood (*Aquilariamalaccensis*) in Malaysia *International Expert Workshop on CITES Non-Detriment Findings, Cancun, Mexico*

[2] Hashim Y Z H, Kerr P G, Abbas P 2016 *Aquilaria* spp. (Agarwood) as Source of Health Beneficial Compound: A Review of Traditional Use, Phytochemistry and Pharmacology, *J Ethnopharmacol* **189** 331-360

[3] Azah N, Chan Y S, Mailina J, Abu S A, Majid J A, Saidatul H S, Nor HH, Nik Y Y 2008 Comparison of chemical profiles of selected gaharu oils from Peninsular Malaysia *The Malaysiana J. of Anal. Sci.* **12 (2)** 338-340

[4] Subasinghe S M C U P, Hettiarachchi D S 2013 Agarwood resin production and resin quality of *Gyrinopswalla*Gaertn*Int J AgrSci* **3(1)** 356-362

[5] Burkil I H 1935 *A Dictionary of the Economic Products of the Malay Peninsula* Crown Agents for the Colonies London

[6] Ng L, Chang Y, Kadir A 1997 A review of agar (gaharu) producing *Aquilariaspps J. Trop. Prod.* **2** 272-285

[7] Sidiyasa K 1986 Exploration and study of regeneration of gaharu producing substances in Kintal Forest Region, South Kalimantan, *Bul. Penelit. Hutan* 59-66

[8] Kahlil A.S, Rahim A A, Taha K K, Abdallah K B 2013 Characterization of methanolic extracts of agarwood leaves *J. Appl. Ind. Sci.***1 (3)** 78-88

[9] Satapathy A K, Gunasekaran G, Sahoo S C, Amit K, Rodriques P V 2009 Corrosion inhibition by JusticiaGendarussa plant extract in hydrochloric acid solution *Corros. Sci.* **51** 2848-2856

[10] Gunasekera S P, Kinghorn A D, Cordell G A, Farnsworth N R 1981 Plant cancer agents, XIX. Constituents of *AquilariaMalaccensis*, *J. Nat. Prod.* **44** 569-572

[11] Spigno G, Tramelli L, Faveri D M D 2007 Effects of extraction time, temperature and solvent on concentration and antioxidant activity of grape marc phenolics*J. Food Eng.* **81** 200-208

[12] Spigno G, Faveri D M D 2007 Antioxidants from Grape Stalks and Marc: Influence of Extraction Procedure on Yield, Purity and Antioxidant Power of the Extracts J. *Food Eng.* **78** 793-801

[13] Al-Farsi M A, Lee C Y 2008 Optimization of phenolics and dietary fibre extraction from date seeds *Food Chem.* **108 (3)** 977-985

[14] Zhao S, Zhang D 2013 A parametric study of supercritical carbon dioxide extraction of oil from *Moringaoleifera*seeds using response surface methodology *Sep. Purif. Technol.* **113** 9-17

[15] Tan M C, Tan C P, Ho C W 2013 Effects of extraction solvent system, time and temperature n total phenolic content of henna (*Lawsoniainermis*) Stems *Int. Food Res. J.* **20 (6)** 3117-3123

[16] Hemwimon S, Pavasant P, Shotipruk A 2007 Microwave-assisted extraction of antioxidativeanthraquinon from roots of *Morindacitrifolia Sep. Purif. Technol.* **54** 44-50

[17] Geankoplis C J 2003 *Transport Processes and Separation Process Principles (Includes Unit Operations), fourth ed.* (USA : Prentice Hall Press)

[18] Rajaei A, Barzegar M, Yamini Y 2005 Supercritical fluid extraction of tea seed oil and its comparison with solvent extraction *Eur. Food Res. Technol.* **220 (3-4)**401-405

[19] Wang J, Sun B G, Cao Y, Tian Y, Li X H 2008 Optimization of ultrasound-assisted extraction of phenolic compounds from wheat bran *Food Chem.* **106** 804-810

[20] Xiao H, Yao Z, Peng Q, Ni F, Sun Y, Zhang C X, Zhong Z X 2016 Extraction of squalene from camellia oil by silver ion complexation *Sep. Purif. Technol.* **169** 196-201.

[21] Aparna V, Dileep K V, Mandal P K, Karthe P, Sadasivan C, Haridas M 2012 Anti-Inflammatory Property of N-hexadecanoic Acid: Structural Evidence and Kinetic Assessment *Chem Biol Drug Des.* **80** 434-439.

[22] Netscher T 2007 Synthesis of vitamin E*Vitam. Horm.***76** 155-202.

[23] Daines A M, Payne R J, Humphries M E, Abell A D 2003 The synthesis of naturally occuring vitamin K and vitamin K analogues *Curr. Org. Chem.***7 (16)** 1625-1634.

IOP Conf. Series: Materials Science and Engineering **162** (2017) 012029 doi:10.1088/1757-899X/162/1/012029

Natural red dyes extraction on roselle petals

H M Inggrid, Jaka and H Santoso
Chemical Engineering Department, Parahyangan Catholic University, Ciumbeluit 94 Bandung, Indonesia

E-mail : inggrid@unpar.ac.id / jakasanjaya.9@gmail.com

Abstract. Roselle (*Hibiscus sabdariffa L.*) has a high quantity of anthocyanin pigment and is a good colorant. The anthocyanin pigment can be used as a natural colorant and antioxidant. An antioxidant is an organic compound that has the ability to inhibit free radical reactions in the human body. The objective of this research is to study the effect of pH and temperature on total anthocyanin and antioxidant activity in roselle extract, and to evaluate the effect of temperature and sunlight on the stability of the red color from roselle. Dried roselle petals were extracted with solid liquid extraction method using water as solvent. The variables in this study are temperature (5°C, 30°C, and 55°C) and pH (2, 7, and 12). Total anthocyanin was analysed using the pH differential method. The antioxidant activities were determined using the DPPH method. The highest total anthocyanin in the roselle petals was 80.4 mg/L at a temperature of 5°C and pH 2. The highest antioxidant activity and yield content in the roselle were 90.4% and 71.6 % respectively, obtained at 55°C and pH 2.

1. Introduction

Roselle is a plant of the malvaceae family and is native to the tropical areas such as Indonesia. This plant has good potential to be developed in Indonesia and is commonly found in the highlands in West Java. Along with agricultural technologic progress, anthocyanin compound is found in roselle petals and it also can serve as a source of antioxidants. Roselle is good for health because of its high antioxidant, and anthocyanin can be used as natural dye. Using natural dyes in food products is better and more safe compared to synthetic dyes [1]. Synthetic dyes give a negative impact to our body's health, potentially being poisonous to our body's metabolism. The most dominant anthocyanins of roselle petals are delphinidin-3-sambubioside and cyanidin-3-siambubioside [2]. These compounds are the ones that give roselle petals their red colour. Anthocyanin's structure will be presented in Fig. 1. Anthocyanins are derivatives of the basic flavylium cation structure, which are highly reactive. The rate of anthocyanin destruction depends on many factors such as temperature, pH, ultraviolet light, and oxygen [2]. Anthocyanin dyes can be applied to food and drink products, such as jam, soft drinks, milk, and yoghurt [3].

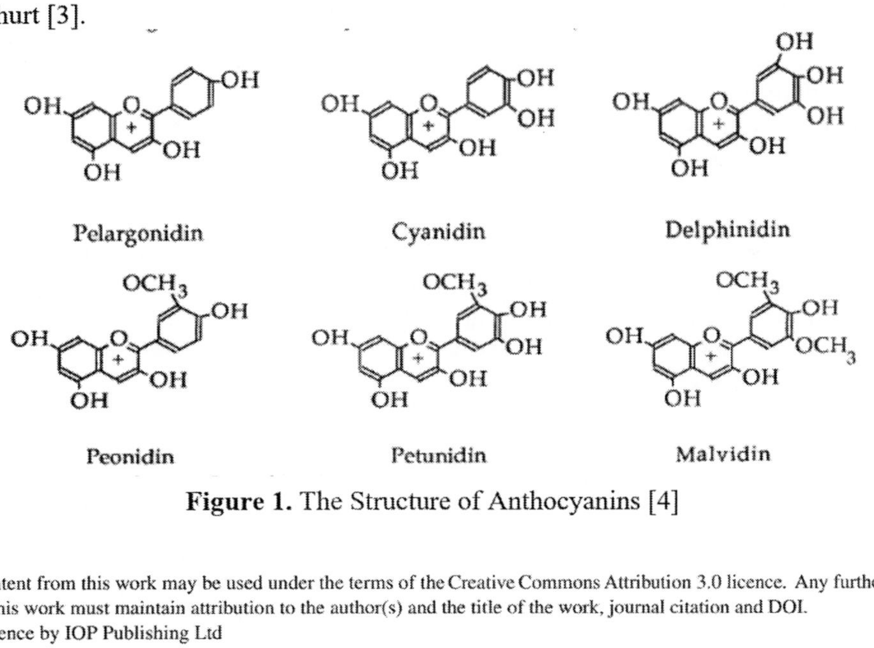

Figure 1. The Structure of Anthocyanins [4]

In this experiment, water was used as a solvent because it produces a brilliant red color extract. Antioxidant activity of roselle extract is also pH dependent, the activity decreases as pH increases [2]. This experiment aimed to study the effect of pH and temperature on total anthocyanin and antioxidant activity of roselle petals, and also to determine the effect of temperature and sunlight on the stability of red color in roselle petals.

2. Materials and methods

Dried roselle petals were extracted using water as a solvent and extracted on F:S (1:15) with variation of temperature and pH conditions. These variations were analyzed to discover the highest total anthocyanin and antioxidant activity. The samples were analysed using spectrophotometry at anthocyanin wavelength (520 nm).

2.1. *Sample preparation*

Samples of roselle petals (*Hibiscus sabdariffa L.*) were taken from North of Bandung, Indonesia. The roselle petals were dried at 45°C for 24 hours to maintain moisture content below 10% before being subjected to anthocyanin extraction.

2.2. *Sample extraction*

The extraction for each sample was performed in duplicate. Briefly, 20 grams of dried roselle petals (-50+60 mesh) was transferred to 1 L batch extractor to which buffer (pH 2, pH 7, and pH 12) was added, and the mixture was mixed at 200 rpm, while temperature condition used is 5°C, 30°C, and 55°C. The mixture was stirred for 5 hours and then filtered through Whatman no. 41 paper under vacuum. After being filtrated, the sample was evaporated at 45°C using rotary evaporator under high vacuum. The resulting extract were used for all the determinations during this study.

2.3. *Anthocyanin content*

A pH differential method was used to determine the anthocyanin content. Solution of KCl buffer pH 1 and Natrium Acetate buffer pH 4,5 were prepared. Then, the sample concentration was adjusted to 5000 ppm by diluting 50 mg extract with the buffer solution until 10 ml (DF = 10). After that, sample was analysed by measuring the absorbance at 520 nm and 700 nm. The reason for measuring the absorbance at 700 nm is to correct for haze. The determination of the absorbance of a sample can be calculated by :

$$\text{Total Anthocyanins (mg/L)} = \frac{(A \times MW \times DF \times 1000)}{(\varepsilon \times l)}$$

Note :

ε	= molar absorptivity of cyanidin-3-glucoside (26,900 L/mol.cm)
l	= width of cuvette (1 cm)
MW	= molecular mass of cyanidin-3-glucoside (449.2 g/mol)
DF	= dilution factor (10)

2.4. *Antioxidant activity*

In determination of antioxidant activity, the DPPH method is used by measuring the sample at wavelenght of 517 nm. The principle of this method is the reaction of the hydrogen which donates hydrogens atom to neutralize free radicals of DPPH and it will turn the color from purple to yellow [5]. These reaction produce non-radical compouds DPPH-hydrazine which is presented in Figure 2.

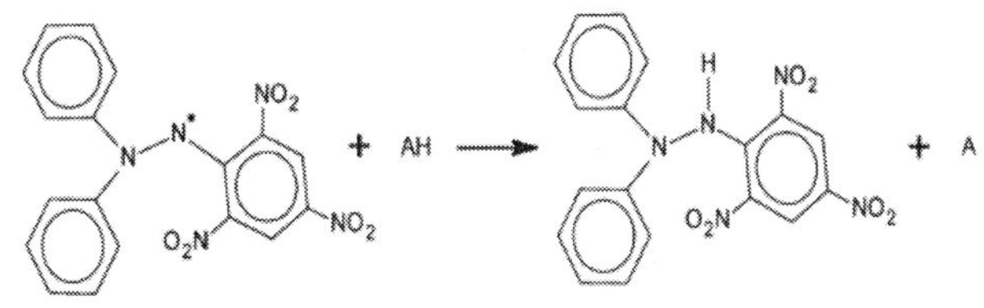

Figure 2. Neutralization Reaction of DPPH [6]

2.5. *Determination of red dye's stability on temperature and ultraviolet rays*

In this study, the extract was diluted with distilled water to 5000 ppm. Then, the treatment of temperature was given at 55°C and 80°C for three hours. While in treatment of ultraviolet rays, the samples were placed in a dark room and a place with ultraviolet rays for a day. After treatments, absorbance of samples were measured to evaluate the changes of the intensity of the color.

3. Results and Discussion

3.1. *Anthocyanin content*

The total pigment in roselle extract was measured by the pH differential method of Giusti and Wrolstad, which used buffer solution of pH 1 and pH 4.5. In strongly acidic solution (below pH 2), the most stable structure for anthocyanin is in the form of red flavylium cation. While in weakly acid (pH 3-5), the red flavylium cation mostly converts to the more stable colorless hemiketal or carbinol (Figure 3) [7].

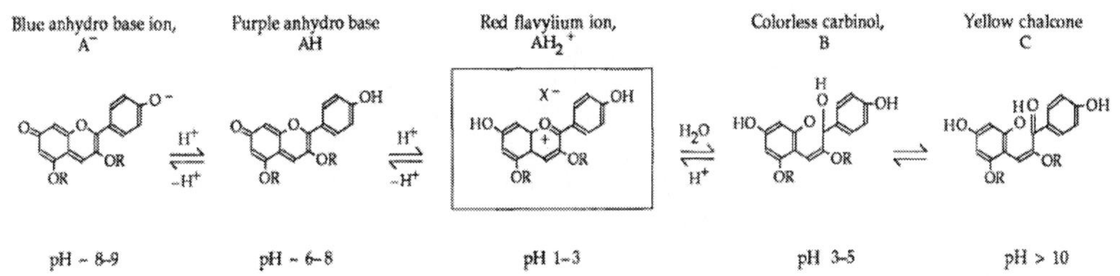

Figure 3. Structural Changes of Anthocyanins with Changes of pH [8]

Based on the results of this study,the highest total anthocyanin was found at pH 2 and temperature of 5°C (80.4 mg/L). In Figure 4, the effect of pH and temperature on anthocyanin contents is presented.

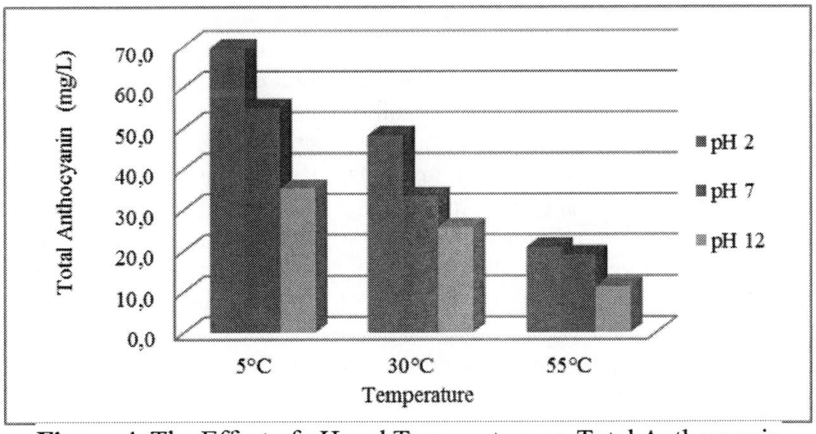

Figure 4. The Effect of pH and Temperature on Total Anthocyanin

IOP Conf. Series: Materials Science and Engineering **162** (2017) 012029 doi:10.1088/1757-899X/162/1/012029

Figure 4 shows that the lower the pH and temperature, the higher the quantity of anthocyanins obtained. At high temperature, it leads to higher anthocyanins degradation because anthocyanins degrade faster at high temperature compared to low temperature [9]. The damages of the heating can occur through two stages. First, hydrolysis happens in anthocyanin glycosidic bonds so the aglycone become unstable, and then the rings of aglycone are opened and become carbinol groups and chalcone [2]. Analysis of variance (ANOVA) was done for every variation to determine the effect of pH and temperature on total anthocyanin. Result of experiments was obtained by using design expert software. It show that the p-value for pH and temperature to total anthocyanins is <0.0001 smaller than 0.05 which means the effect of pH and temperature is significant as an influence variable of total anthocyanins. It also showed that there is an interaction between pH and temperature.

3.2. *Antioxidant Activity*

Determination of antioxidant activity was analysed by the DPPH method. The principle of this method is to donate hydrogen atoms to neutralize purple radicals of DPPH, so when the absorbance is measured, the colour of DPPH will be reduced and turn to pale yellow or colorless. The higher antioxidant activity indicates that the ability of samples to reduce free radicals is also high [10]. From the results of this study, the highest antioxidant activity is found at pH 2 and 55°C (90.4%), while the lowest antioxidant activity is found at pH 12 and 5°C (80.1%). In Figure 5, the effect of pH and temperature on antioxidant activity is presented.

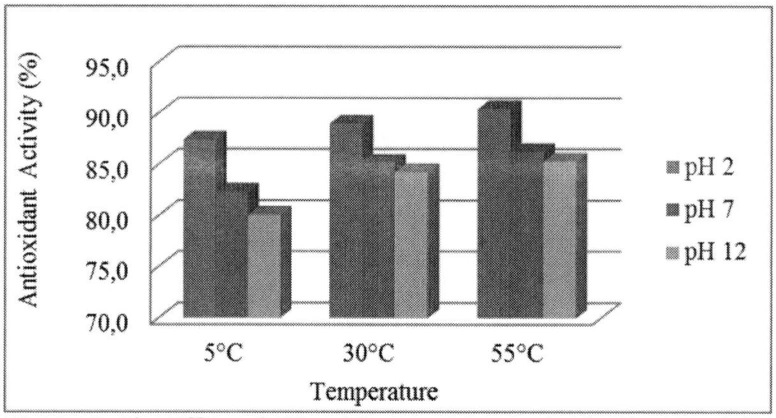

Figure 5. The Effect of pH and Temperature on Antioxidant Activity

Figure 5 shows that a temperature of 55°C produced relatively high antioxidant activity because at high temperatures, it would increase the solubility of solute so many compounds like flavonoids and phenols, which were potentially antioxidant sources, could be extracted. Antioxidant activity depends on total electrons that can be donated. The more antioxidants that can be extracted, it will make the more electrons that can be donated so the antioxidant activity become higher. Lower temperature (5°C) decreases the solubility of active compound, thus decreasing the active compound extracted. Antioxidant activity obtained in acidic conditions (pH 2) was relatively higher than other pH conditions (pH 7 and pH 12). According to Chumsri et al, antioxidant activity of roselle extract is also pH dependent, the activity decreases as pH increases [2]. It happens because in acidic conditions, the increasing number of vacoules cell walls broke out so the anthocyanins and other active compounds like flavonoid and phenols could be well extracted. Result of experiments was obtained by using design expert software. It show that the p-value for pH and temperature to antioxidant activity are <0.0001 and 0.0003, it is smaller than 0.05 which means the effect of pH and temperature is significant as an influence variable of antioxidant activity. It also showed that there is no significant interaction between pH and temperatures, because the p-value for AB is 0.5055, greater than 0.05.

3.3. *Yield Content*

The yield of roselle's extract was calculated as X = (weight of extract after evaporating/weight of dried roselle before extracting) x 100%. Based on the results of this study, the highest yield contents were obtained at pH 2 and a temperature of 55°C (71.6%). In Figure 6, the effect of pH and temperature on yield content is presented.

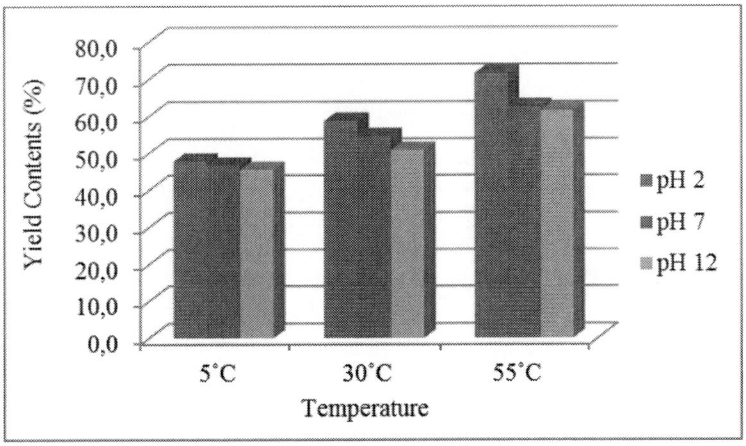

Figure 6. The Effect of pH and Temperatures on Yield Content

Figure 6 shows that the higher the temperature, the higher yield content obtained. It is because high temperature increases the solubility of the solute, so that more solute was extracted. But in terms of quality, not only anthocyanin compounds were extracted, but also other compounds like a mixture of citric acid and malic acid and phenols. It caused the obtained yield contents to outgrow the yield content at temperatures of 5°C and 30°C. Result of experiments was obtained by using design expert software. It show that the p-value for pH and temperature to antioxidant activity are 0.0073 and 0.0001 respectively, it is smaller than 0.05 which means the effect of pH and temperature is significant as an influence variable of antioxidant activity. It also showed that there is no significant interaction between pH and temperatures, because the p-value for AB is 0.3044, greater than 0.05.

3.4. *Determination of Red Dye's Stability on Temperature*

The heat treatment that has been given to this study was at temperatures of 30°C, 55°C and 80°C. The initial transmittance (%T) of roselle extract was 67.5%. The heat treatment on temperatures of 55°C and 80°C made the %T of the sample increase to 70.8% and 78.5%, but %T of roselle extract at a temperature of 30°C did not change which showed anthocyanins are still stable at this temperature. The increasing of %T on roselle extracts at temperatures of 55°C and 80°C happened because anthocyanin compounds were degraded thermally. The degradation mechanism of anthocyanin compounds is presented in Figure 7.

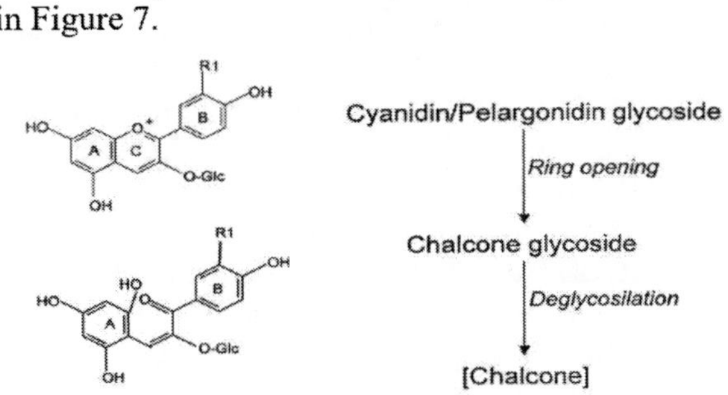

Figure 7. The Degradation Mechanism of Anthocyanin Compounds [10]

Figure 8 shows that the increasing of temperature cause the anthocyanin compounds became unstable. In the initial phase, hydrolysis of the glycosidic bond of anthocyanins will occur to form unstable rings of aglycone. Then, at a later stage, the rings of aglycone will be opened and form carbinol and chalcone groups [12]. From the study, heat treatment at temperature of 55°C did not change the color significantly so it can be concluded that red color from roselle extract is stable at temperature of 55°C. However, at temperature of 80°C, it causes discoloration from red becaming pale red. It shows that thermal degradation on anthocyanin occured at a temperature of 80°C.

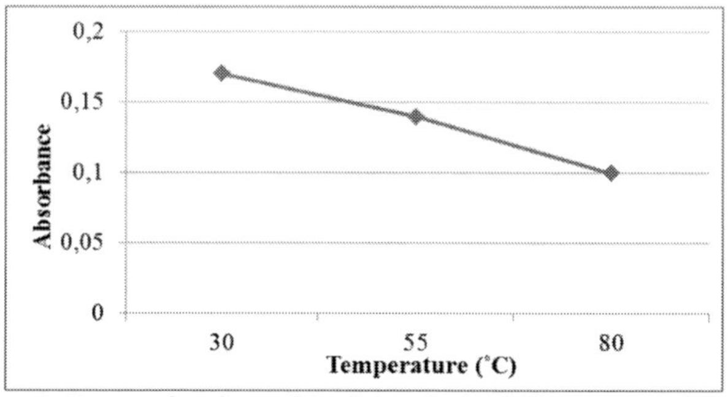

Figure 8. Determination of Red Dye's Stability on Temperature

3.5. *Determination of Red Dye's Stability on Ultraviolet Rays*
This analysis was observed on discoloration of roselle extract that was stored in a dark space and space equipped with ultraviolet rays. From the study, the red color from roselle extract did not change color significantly in dark space treatment, while in the space equipped with ultraviolet rays treatment experienced a significant change in color from red to reddish yellow. It shows that thermal degradation occurs due to the energy of ultraviolet rays that damages the structure of anthocyanins, which the red cation flavium became colorless carbinol [13].

4. Conclusion
The highest total anthocyanin content in the roselle extract is 80.4 mg/L and it was obtained at 5°C and pH 2. Meanwhile, the highest antioxidant activity and yield content in the roselle extract were 90.4% and 71.6% respectively at 55°C and pH 2. From this study, it is known that temperature and pH significantly affect total anthoycanin, antioxidant activity, and yield content. In addition, temperature and ultraviolet rays also affect the stability of red color in roselle petals.

References
[1] Hayati E K, Budi U S and Hermawan R 2012 Konsentrasi Total Senyawa Antosianin Ekstrak Kelopak Bunga Rosella (*Hibiscus Sabdariffa L.*) : Pengaruh Temperatur dan pH *Jurnal Kimia* **6 (2)** 138-147
[2] Chumsri P, Sirichote A and Itharat A 2008 Studies on The Optimum Conditions fot The Extraction and Concentration of Roselle Extract *Songklanakarin J. Sci. Technol* **30 (1)** 133-139
[3] Tsai P J, McIntosh J, Pearce P, Camden B and Jordan B R 2002 Anthocyanin and Antioxidant Capacity in Roselle (*Hibiscus Sabdarifffa L.*)Extract *Food Research International* **35** 351-356
[4] Bhagwat S, Haytowitz D B and Holden J M 2014 USDA Database for the Flavonoid Content of Selected Foods *U.S. Department of Agriculture* 1-173
[5] Molyneux P 2004The Use of The Stable Free Radical Diphenylpicryl-Hydrazyl (DPPH) for Estimating Antioxidant Activity *Songklanakarin Journal of Science Technology* **26 (2)** 211-219
[6] Pyrzynska K and Pekal A 2013 Application of Free Radical Diphenylpicryhydrazyl (DPPH) to Estimate The Antioxidant Capacity of Food Samples *Royal Society of Chemistry Publishing* **5** 4288-4295

[7] Jordheim M 2007 Isolation, Identification and Properties of Pyranaoanthocyanins and Anthocyanins Forms *Department of Chemistry, University of Bergen, Norway*

[8] Ananga A, Georgiev V, Ochieng J, Phills B and Tsolova V 2013 Production of Anthocyanins in Grape Cell Cultures: A Potential Source of Raw Material for Pharmaceutical, Food, and Comestic Industries *The Mediterranean Genetic Code-Grapevine and Olive* 248-287

[9] Moldovan B, David L, Chisbora C, and Cimpoiu C 2012 Degradation Kinetics of Anthocyanins from European Cranberrybush (*Vibrium Opulus L.*) Fruit Extracts. Effect of Temperature, pH, and Storage Solvent *Molecules* **17** 11655-11666

[10] Zhong Y and Shahidi F 2015Methods For The Assessment of Antioxidant Activity in Foods *Handbook of Antioxidants for Food Preservation (USA : Elsevier)* 287-333

[11] Sadilova E, Carle R and Stintzing F C 2007 Thermal Degradation of Anthocyanins and Its Impact on Color and *In Vitro* Antioxidant Capacity *Mol.Nutr.Food Res.* **51** 1461-1471

[12] Suzery M, Lestari S and Cahyono B 2010 Penentual Total Antosianin Dari Kelopak Rosella (*Hibiscus sabdariffa L*) dengan Metode Maserasi dan Sokshletasi *Jurnal Sains dan Matematika* **18 (1)** 1-6

[13] Ni Q, Xu G, Lu G, Gao Q, Zhou C and Zhang Y 2012 Investigation of the Stability and Antioxidant Properties of Anthocyanins-based Purple Potato Colorant After Processing *African Journal of Biotechnology* **11 (14)** 3379-3387

Supercritical carbon dioxide (SC-CO_2) extraction of essential oil from *Swietenia mahagoni* seeds

N S M Norodin[1, a *], L M Salleh[2, b], Hartati[3,c] and N M Mustafa[4,d]

[1] Department of Bioprocess and Polymer Engineering, Faculty of Chemical and Energy Engineering, UniversitiTeknologi Malaysia, 81310 UTM Johor Bahru, Johor, Malaysia
[2] Center of Lipids Engineering and Applied Research (CLEAR), Ibnu Sina Institute Scientific & Industrial Research (Ibnu Sina ISIR), Universiti Teknologi Malaysia, 81300 Johor Bahru, Johor, Malaysia
[3] Department of Biology, Universitas Negeri Makassar, South Sulawesi, Indonesia
[4] Natural Product Division, Forest Research Institute Malaysia (FRIM), 52109 Kepong, Selangor Darul Ehsan, Malaysia

E-mail : [a]nursalsabilamdnorodin@gmail.com, [b]i.liza@cheme.utm.my, [c]tati_biounm@yahoo.co.id, [d]musaadah@frim.gov.my

Abstract. *Swietenia mahagoni* (Mahogany) is a traditional plant that is rich with bioactive compounds. In this study, process parameters such as particle size, extraction time, solvent flowrate, temperature and pressure were studied on the extraction of essential oil from *Swietenia mahagoni* seeds by using supercritical carbon dioxide (SC-CO2) extraction. *Swietenia mahagoni* seeds was extracted at a pressure of 20-30 MPa and a temperature of 40-60°C. The effect of particle size on overall extraction of essential oil was done at 30 MPa and 50°C while the extraction time of essential oil at various temperatures and at a constant pressure of 30 MPa was studied. Meanwhile, the effect of flowrate CO_2 was determined at the flowrate of 2, 3 and 4 ml/min. From the experimental data, the extraction time of 120 minutes, particle size of 0.5 mm, the flowrate of CO_2 of 4 ml/min, at a pressure of 30 MPa and the temperature of 60°C were the best conditions to obtain the highest yield of essential oil.

1. Introduction

Swietenia mahagoni is also known as 'tunjuk langit' in Malaysia is used traditionally to treat various diseases such as diabetes and high blood pressure [1]. *S. mahagoni* tree is 30 meters or taller [2] and the wood, usually being used for making furniture [3]. Meanwhile, the bark can be used for natural colorant [4]. The fruit of *S. mahagoni* is woody and consisting of capsules containing winged seeds [5]. Whereas, the seed of *S. mahagoni* can be obtained by removing the wing. In Malaysia, the raw seeds were used for treating hypertension and diabetes [6]. In addition, *S. mahagoni* seeds have been reported to have various biological activities such as anti-inflammatory activity, anticancer and antitumor activity [7] and also antidiabetic activity [8].

Also, carbon dioxide (CO_2) is the most frequent solvent used that is environmental friendly (fairly non-toxic), low cost and can be easily removed from the extract [9]. The elimination of CO_2 is easily achieved since CO_2 is in a gas state at room temperature. In addition, CO_2 in the supercritical state is in a moderate critical temperature (31.3°C) and pressure (7.38 MPa). Supercritical state is when gas and liquid are indistinguishable where at this state it is compressible but possessing a density of a liquid. In a word, supercritical CO_2 makes a good solvent because of the gas-like state that attributed the low viscosity and high diffusion coefficient and the liquid-like state that gave the solvating power [10].

2. Methodology

2.1. Materials and reagents

Swietenia mahagoni seeds were bought in the local market. The seeds were rinsed with tap water to remove any foreign particles and dirt prior to drying. Then, the cleaned seeds were cut into small pieces and dried by using oven at temperature of 50°C for a week to remove moistures. The seeds were ground by using a blender (merck Panasonic) and sieved to approximate 0.25, 0.50 and 0.75 mm of particle size. Commercial grade liquid carbon dioxide (purity 99.99%) used in supercritical carbon dioxide extraction was purchased from Kras, Instrument and Services, Johor, Malaysia.

2.2. Supercritical Carbon dioxide (SC-CO_2) extraction

CLEAR supercritical fluid extraction (SFE) machine in Center of Lipids Engineering and Applied Research (CLEAR), Universiti Teknologi Malaysia consisted of CO_2 gas cylinder, CO_2 controller pump (Lab Alliance), co-solvent pump (Lab Alliance), oven (Memmert, Germany), 10 ml stainless steel extraction vessel, pressure gauge (Swagelockk, Germany), automatic back pressure regulator (Jasco BP 2080- Plus) and restrictor valve. A schematic diagram of CLEAR SC-CO_2 unit is illustrated in figure 1.

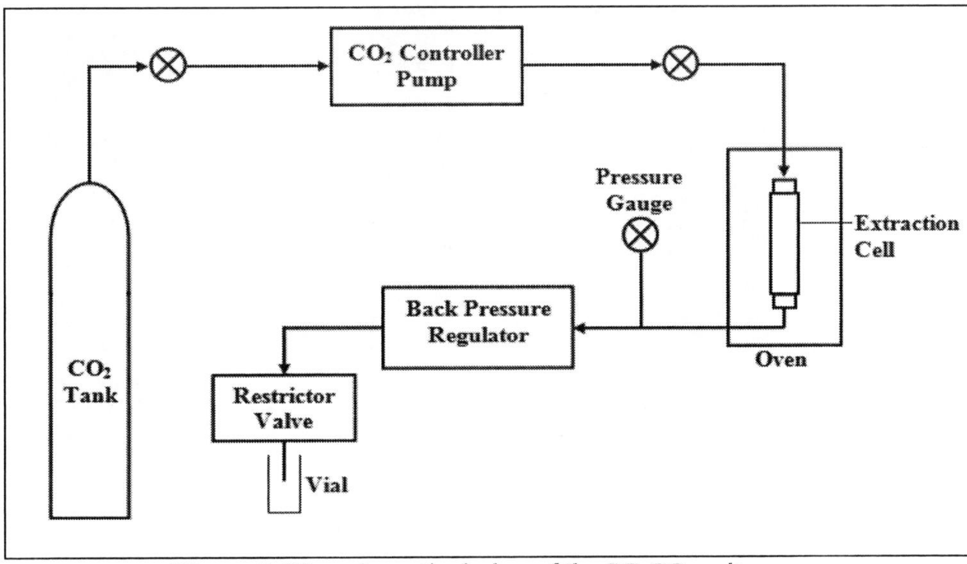

Figure 1. The schematic design of the SC-CO_2 unit

Five gram of sample was placed in 10 ml stainless steel extraction vessel and sealed tightly in the oven. Set all the parameters (temperature, pressure and flowrate of CO_2), the extraction process started after all the parameters were attained. Lastly, depressurized the system and the oil yields were collected. The parameters used in extraction process is presented in table 1.

Table 1. The process parameters for SC-CO_2 extraction

Parameter	Range/ value
Temperature (°C)	40-60
Pressure (MPa)	20-30
Flowrate of CO_2 (ml/min)	2.00-4.00
Particle size (mm)	0.25-0.75
Mass of sample (g)	5.00
Extraction time (min)	0-180

Collected oil yield was calculated as percentage of oil yield by using equation below:

$$\text{Oil yield (\%)} = \frac{\text{Mass of oil extract (g)}}{\text{Mass of sample (g)}} \times 100 \qquad \text{(Eq. 1)}$$

3. Result and Discussion

3.1. Particle size

Particle size is one of the important factors to be considered in the extraction of *S. mahagoni* seeds. In a word, as the surface area increases, the extraction rate also increases. Hence, smaller particle size will increase the extraction yield because the mass transfer will increase with smaller particle size of the sample. But the downside is as the sample is ground to small, the extraction rate may decrease due to channeling effect inside the extraction vessel [11]. Channeling effect in the extraction vessel is due to over fine particle size of the sample that reduce the contact between the sample and solvent. Supported by Salleh (2010), fine powder with smaller particle size can increase the extraction rate but it is also lead to the difficulty in maintaining the flow rate to obtain a good optimization of $SC\text{-}CO_2$ extraction [12].

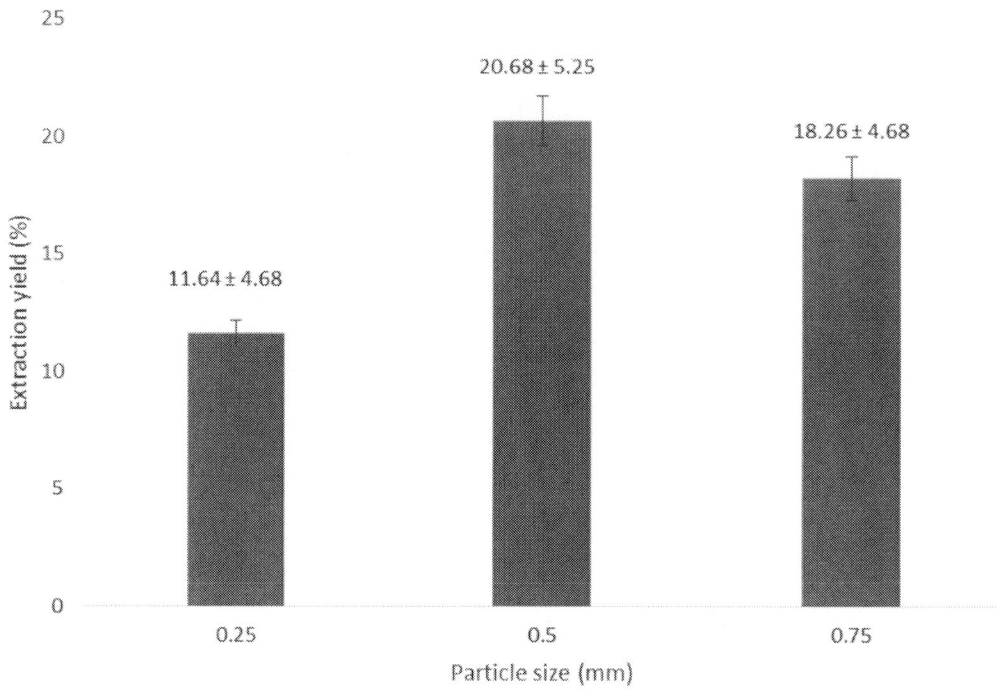

Figure 2. Effect of particle size on extraction yield

Figure 2 shows that particle size of 0.5 mm gave the highest percentage of oil yield which is 20.68% ± 5.25 compared to 0.75 mm and 0.25 mm. From the figure, it shows that when the particle size is too small, the extraction yield is decreased. This is because when the particle size is too fine, the porosity of the sample is reduced and slower CO_2 penetration into the sample, hence lower the extraction yield. In this study, *S. mahagoni* seeds are naturally oily thus it may easily clumped together in the extraction vessel hence reduce the surface area. Similarly with the study in the extraction of rubber seed, smaller particle size (≤0.355 mm) can easily clump together, however larger particle size (0.50mm) do not clump together easily and provide larger surface area [13]. It is resulted that the particle size of 0.5 mm is considered for this SC-CO2 extraction.

3.2. Extraction time

The determination of the extraction time for the extraction of *S. mahagoni* seeds by using $SC\text{-}CO_2$ extraction was performed. The conditions used were pressure of 20, 25 and 30 MPa, with the temperature of 50°C. The flow rate used in this study was 2 ml/min. The yield of extract was collected every 20 minutes over 180 minutes of extraction time. The results of the best extraction time are presented in Figure 3.

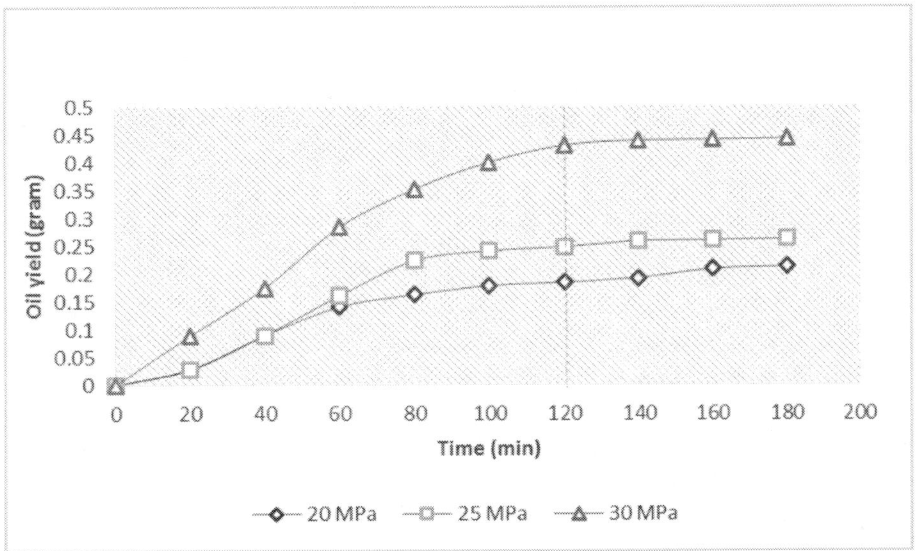

Figure 3. The effect of extraction time on extraction yield of SC-CO_2 at 20, 25 and 30 MPa and the temperature of 50°C

The figure displays the oil yield in percent versus the extraction time in minutes. At pressure of 20 MPa, it took more than 120 minutes to achieve the asymptomatic value. The highest yield (0.4306%) was at the highest pressure of 30 MPa within 120 minutes. Lina *et al.*, (2010) also obtained similar result with this study where the seed oil of *Microula sikkimensis* increased as extraction time increased [14]. The same trend was found in the study of saffron extract by Lozano *et al.* (2000) [15]. In conclusion, the extraction time of *S. mahagoni* seeds was performed at 120 min constantly for each sample throughout this study in order to obtain asymptotic value to ensure maximum extraction yield was extracted.

3.3. Flowrate

Flow rate of solvent (example: CO_2) is another factor affecting the extraction process to produce a high yield of extract by using supercritical carbon dioxide (CO_2) [16]. By increasing the flow rate of solvent, also reduce the extraction time and mass transfer resistance.

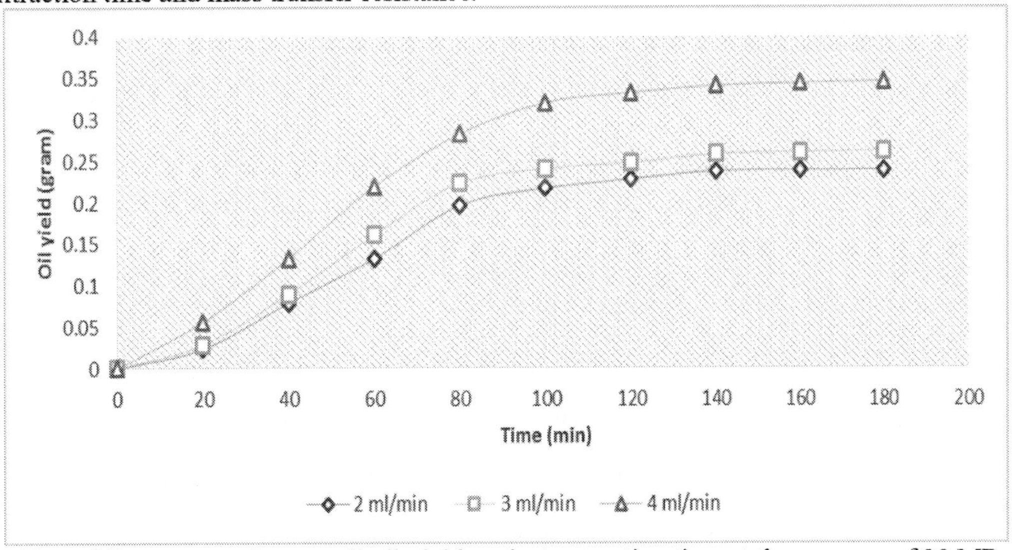

Figure 4. Effect of flow rate on the overall oil yield against extraction time at the pressure of 30 MPa and temperature of 50°C

As shown in figure 4, the overall oil yield increase with the increasing of flow rate at constant pressure of 30 MPa and temperature of 50°C. The extraction of borage oil by SC-CO_2 extraction showed higher flow rate gives higher yield of borage seed oil [17]. Similarly, in the extraction of flaxseed showed that at higher flow rate (3 L/min) recovered higher oil yield [18]. Furthermore, by increasing the SC-CO_2 flowrate, the intermolecular interaction between solvent and the solute increase due to the increase of CO_2 molecules per volume in the extraction vessel [19]. In this condition, the mass transfer increases as the flowrate increases. Hence, 4 ml/min of CO_2 was selected as the best flowrate for the extraction of *S. mahagoni* seeds.

3.4. Temperature and Pressure

The extraction of *S. mahagoni* seeds by using SC-CO_2 extraction shows that the highest yield (14.45 %) obtained was at the maximum level (30MPa, 60°C) meanwhile lowest (1.49%) at (20MPa, 60°C) as shown in Table 2. At higher temperature, the decomposition of cell walls occurred thus extracted oil produced is higher [19]. Operating at higher pressure will influence the solvent density, thus enhanced the solvating power (the interaction of inter-molecules and solutes increase) [11]. Similar result was reported in the extraction of *S. mahagoni* seeds by using SC-CO_2 extraction [20].

Table 2. Extraction of *S. mahagoni* seeds at different conditions (pressure, P and temperature, T)

Condition (s)	Oil yield (%)
P= 20MPa, T= 40°C	6.556
P= 20MPa, T= 50°C	3.676
P= 20MPa, T= 60°C	1.490
P= 25MPa, T= 40°C	6.640
P= 25MPa, T= 50°C	4.954
P= 25MPa, T= 60°C	4.558
P= 30MPa, T= 40°C	7.024
P= 30MPa, T= 50°C	8.612
P= 30MPa, T= 60°C	14.45

Figure 5 shows the effect of pressure and temperature on oil extraction. At constant temperature, oil yield increases as pressure increases (20-30 MPa). This is due to the increase of density and thus increase the solvating power of SC-CO_2. Effect of pressure influenced both solvating power and intermolecular interaction strength [21,22]. Similar result was reported in SC-CO_2 extraction of green coffee oil [23], wheat bran [24] and *S. mahagoni* seeds [20].

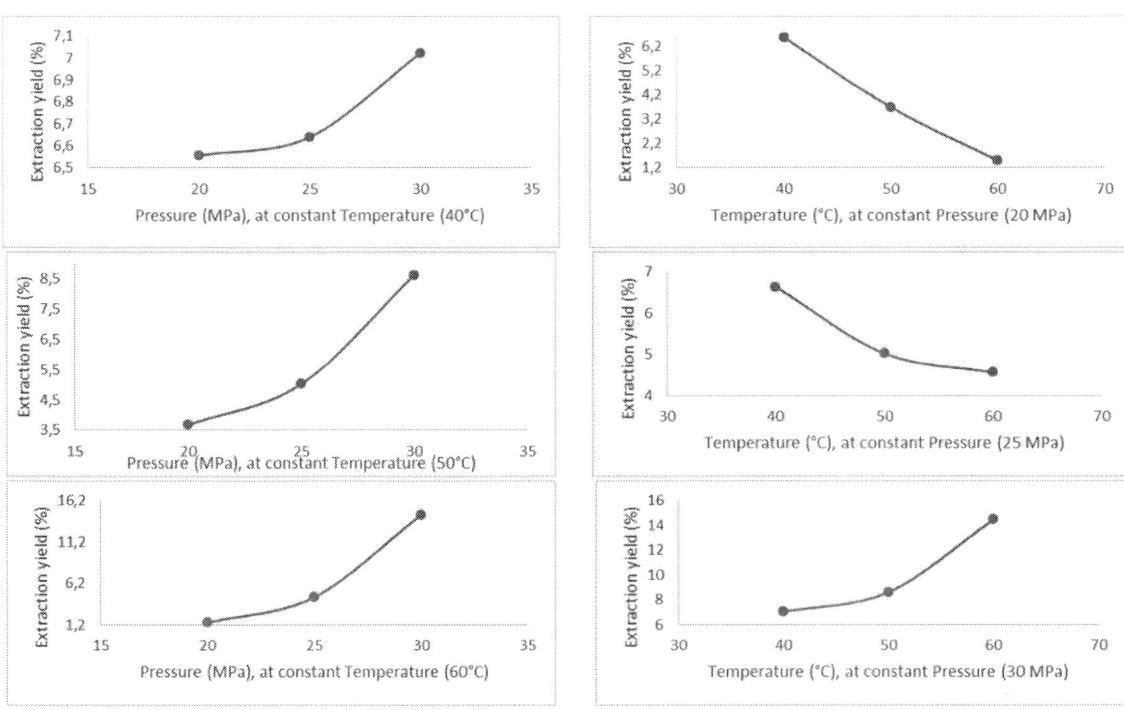

Figure 5. The effect of pressure and temperature on extraction yield

Next, at constant pressure (20 and 25 MPa), oil yield decreases as temperature increases (40-60 °C). But at constant pressure (30 MPa), oil yield increases as temperature increases. Effect of temperature in extraction process is much more difficult to evaluate unlike pressure. This is due to the dual effect of temperature toward extraction process depending on the domination of solute vapour pressure or solvent density [19,25,26]. Decreasing solvent density with the increase in temperature resulted in higher extracted yield because of the increase in vapour pressure. Similar effect were reported on extraction green coffee oil [23] and silkworm pupal oil [26].

4. Conclusion

Supercritical carbon dioxide (SC-CO_2) extraction applied to extract *Swietenia mahagoni*. From the experimental data, the extraction time of 120 minutes, particle size of 0.5 mm, the flowrate of CO_2 of 4 ml/min, at a pressure of 30 MPa and the temperature of 60°C were the best conditions to obtain the highest yield of essential oil.

Acknowledgements

The authors gratefully acknowledge the Universiti Teknologi Malaysia for awarding with the Research University Grant Scheme, GUP (Q.J130000.2546.10H50).

References

[1] Goh B H, Kadir H A, Malek S N A, Ng S W 2010 Swietenolide diacetate from the seeds of Swietenia macrophylla *Acta Crystallogr. Sect. E Struct. Reports Online* 66
[2] Eid A M M, Elmarzugi N A, El-Enshasy H A 2013 A review on the phytopharmacological effect of Swietenia macrophylla *Int. J. Pharm. Pharm. Sci.* **5** 47–53
[3] Falah S, Suzuki T, Katayama T 2008 Chemical constituents from Swietenia macrophylla bark and their antioxidant activity *Pakistan J. Biol. Sci.* **11** 2007–2012
[4] Haque M A, Khan G M A, Razzaque S M A, Khatun K, et al. 2013 Extraction of rubiadin dye from Swietenia mahagoni and its dyeing characteristics onto silk fabric using metallic mordants *Indian J. Fibre Text. Res.* **38** 280–284
[5] Blundell A G, Gullison R E 2003 *For. Policy Econ.* **5** 395–405
[6] Balijepalli M K, Suppaiah V, Chin A-M, Buru A S, et al. 2014 Acute oral toxicity studies of swietenia macrophylla seeds in sprague dawley rats *Pharmacognosy Res.* **7** 38–44
[7] Goh B H, Kadir H A 2011 In vitro cytotoxic potential of Swietenia macrophylla King seeds against human carcinoma cell lines *J. Med. Plants Res.* **5** 1395–1404
[8] Maiti A, Dewanjee S, Kundu M, Mandal S C 2009 Evaluation of antidiabetic activity of the seeds of Swietenia macrophylla in diabetic rats *Pharm. Biol.* **47** 132–136
[9] Liza M S, Rahman, R A, Mandana B, Jinap S, et al. Supercritical carbon dioxide extraction of bioactive flavonoid from Strobilanthes crispus (Pecah Kaca) *Food Bioprod. Process* **88** 319–326
[10] Aionicesei E, Škerget M, Knez Ž 2008 Measurement of CO2 solubility and diffusivity in poly(l-lactide) and poly(d,l-lactide-co-glycolide) by magnetic suspension balance *J. Supercrit. Fluids* **47** 296–301
[11] Pereira C G, Meireles M A A 2009 Supercritical Fluid Extraction of Bioactive Compounds: Fundamentals, Applications and Economic Perspectives *Food Bioprocess Technol.* **3** 340–372
[12] Salleh L M 2010 Extraction of bioactive flavonoid compounds from Pecah Kaca (Strobilanthes Crispus) using supercritical carbon dioxide *Ph. D. Thesis* Universiti Putra Malaysia
[13] Yian L N 2015 Supercritical carbon dioxide extraction of rubber seed oil rich in alpha-linolenic acid *Ph.D. Thesis* Universiti Teknologi Malaysia
[14] Lina S, Fei R, Xudong Z, Yangong D, Fa H 2010 Supercritical carbon dioxide extraction of microula sikkimensis seed oil. *J. Am. Oil Chem. Soc* **87** 1221–1226
[15] Lozano P, Delgado D, Gómez D, Rubio M, Iborra J L 2000 A non-destructive method to determine the safranal content of saffron (Crocus sativus L.) by supercritical carbon dioxide extraction combined with high-performance liquid chromatography and gas chromatography *J. Biochem. Biophys. Methods* **43** 367–378
[16] Leo L, Rescio L, Ciurlia L, Zacheo G 2005 Supercritical carbon dioxide extraction of oil and α-tocopherol from almond seeds *J. Sci. Food Agric.* **85** 2167–2174
[17] Gómez A M, de la Ossa E M 2002 Quality of borage seed oil extracted by liquid and supercritical

carbon dioxide *Chem. Eng. J.* **88** 103–109

[18] Bozan B, Temelli F 2002 Supercritical CO2 extraction of flaxseed *J. Am. Oil Chem. Soc.* **79** 231–235

[19] Machmudah S, Kawahito Y, Sasaki M, Goto M 2007 Supercritical CO2 extraction of rosehip seed oil: Fatty acids composition and process optimization *J. Supercrit. Fluids* **41** 421–428

[20] Hartati, Salleh L M, Yunus, A C M, Aziz A A 2014 Optimization of supercritical CO2 extraction of Swietenia mahagoni seed by response surface methodology *J. Teknol. Sciences Eng.* **67** 15–20

[21] Bulgarevich D S, Sako T, Sugeta T, Otake K, et al. 2002 The role of general and hydrogen-bonding interactions in the solvation processes of organic compounds by supercritical CO2 / n-Alcohol Mixtures *Ind. Eng. Chem. Res* **41** 2074–2081

[22] Morita A, Kajimoto O 1990 Solute-solvent interaction in nonpolar supercritical fluid: a clustering model and size distribution *J. Phys. Chem.* **94** 6420–6425

[23] de Azevedo A B A, Kieckbush T G, Tashima A K, Mohamed R S, et al. 2008 Extraction of green coffee oil using supercritical carbon dioxide *J. Supercrit. Fluids* **44** 186–192

[24] Kwon K, Uddin S, Jung G, Sim J, Chun B 2010 Supercritical carbon dioxide extraction of phenolics and tocopherols enriched oil from wheat bran *Int. J. Biol. Life Sci.* **6** 117–122

[25] Thana P, Machmudah S, Goto M, Sasaki M, et al. 2008 Response surface methodology to supercritical carbon dioxide extraction of astaxanthin from Haematococcus pluvialis *Bioresour. Technol.* **99** 3110–3115

[26] Wei Z J, Liao A M, Zhang H X, Liu J, Jiang S T 2009 Optimization of supercritical carbon dioxide extraction of silkworm pupal oil applying the response surface methodology *Bioresour. Technol.* **100** 4214–4219

IOP Conf. Series: Materials Science and Engineering **162** (2017) 012031 doi:10.1088/1757-899X/162/1/012031

Optimization of supercritical carbon dioxide extraction of Piper Betel Linn leaves oil and total phenolic content

A H A Aziz[1], M A C Yunus[2*], N H Arsad, N Y Lee, Z Idham, A Q A Razak

[1] Centre of Lipids Engineering & Applied Research (CLEAR), Ibnu Sina Institute for Scientific and Industrial research, Universiti Teknologi Malaysia, 81310 UTM Johor Bahru, Johor

[2] Faculty of Chemical and Energy Engineering, Universiti Teknologi Malaysia, 81310 UTM Johor Bahru, Johor

E-mail : azizi@cheme.utm.my

Abstract. Supercritical Carbon Dioxide (SC-CO_2) Extraction was applied to extract piper betel linn leaves. The piper betel leaves oil was used antioxidant, anti-diabetic, anticancer and anti-stroke. The aim of this study was to optimize the conditions of pressure, temperature and flowrate for oil yield and total phenolic content. The operational conditions of SC-CO_2 studied were pressure (10, 20, 30 MPa), temperature (40, 60, 80 °C) and flowrate carbon dioxide (4, 6, 8 mL/min). The constant parameters were average particle size and extraction regime, 355μm and 3.5 hours respectively. First order polynomial expression was used to express the extracted oil while second order polynomial expression was used to express the total phenolic content and the both results were satisfactory. The best conditions to maximize the total extraction oil yields and total phenolic content were 30 MPa, 80 °C and 4.42 mL/min leading to 7.32% of oil and 29.72 MPa, 67.53 °C and 7.98 mL/min leading to 845.085 mg GAE/g sample. In terms of optimum condition with high extraction yield and high total phenolic content in the extracts, the best operating conditions were 30 MPa, 78 °C and 8 mL/min with 7.05% yield and 791.709 mg gallic acid equivalent (GAE)/g sample. The most dominant condition for extraction of oil yield and phenolic content were pressure and CO_2 flowrate. The results show a good fit to the proposed model and the optimal conditions obtained were within the experimental range with the value of R^2 was 96.13% for percentage yield and 98.52% for total phenolic content.

1. Introduction

Piper betel Linn belongs to family of *Piperaceae* which is originated from South East Asia like Malaysia, Indonesia, India, Thailand and China. In Malaysia, this plant known as 'DaunSirih' which is a perennial dioecious creeper and grows about one metre height. This plant has very unique love shaped leave that usually chewed together with areca nut and lime by old folks [1,2]. The *Piper betel* leaves are used by local to treat cold, headache, asthma, pimples, cut, improve digestion and many more. The leaves juice was used in eyedrops and as an inedible ink for marking and labelling garments [3] and to enhance the intoxicating effect in alcoholic beverages [4].

The leaves contain vitamins, mineral, protein, fibre, carbohydrate, fat and essential oil that medicinally important for body [5]. Previous researchers have proved that *piper betle* leave contain of phenolic compound which is medicinal importance or biological properties that can possess as antioxidant [6], antimicrobial [7], antidiabetic [8], antifertility [9], antihemolytic [10], antibacterial [11], [12], antihepatotoxic [13], antidepressant [14], anti-inflammatory [15] and many more. Phenolic compounds of *piper betel linn* has been reported by some researchers which are chavicol by [16], chavibetol and chavibetol acetate by [17] and eugenol by [18]. Hydroxychavicol and Eugenol are the main bioactive compound in the extract of *piper betel*leaves [19]. Folin-Ciocalteu method was used to access the total phenolic content in the *piper betel* leaves.

Extraction and product recovery from various plant parts were considered as the most crucial steps in the evaluation of target molecules [20]. Conventional extraction processes are time consuming, laborious and involve long operational techniques, bulk amount of solvents required and can cause thermal degradation of the target molecules at continuously high temperature. In addition, traces of toxic solvents are rarely removed from the extracts hence low quality of the yield of product and large quantity of solvent waste. Therefore, supercritical fluid extraction is an alternative extraction technique with better selectivity and efficiency. The most common solvent used in supercritical fluid extraction is carbon dioxide (SC-CO_2) because CO_2 is inexpensive, non-toxic, non-flammable and considered as a GRAS (Generally Recognized as Safe) solvent. The low polarity is the only major drawback of CO_2 hence co-solvent can be used to increase the polarity of CO_2.

The aim of this study to optimize SC-CO_2 variables such as temperature, pressure and flowrate of the solvent for the maximum extract yield and total phenolic content from *piper betel* leaves by using response surface methodology (RSM) with Box-Behnken design via Design Expert software version 6.0. RSM was used to develop a model on the percentage yield and total phenolic content (TPC) where the model could be used to optimize the extraction parameters.

2. Materials and methods

2.1. Materials

The green *piper betel linn* leaves were used and obtained from a market place. The leaves were rinsed using distilled water 2 to 3 times to remove any dust and precipitate stick on the leaves. The clean betel leaves were cut into small pieces and placed in the aluminium tray. Then, the tray was placed into the oven-dryer at 40°C for few hours. Next, the dry leaves were ground using grinder and sieve by using sieving tray for particle size of 355µm, 500µm and 710µm. Lastly, the ground betel leaves were put inside the sealed plastic bag and kept in the freezer at -20°C to maintain the freshness of the sample.

2.2. Supercritical Carbon Dioxide (SC-CO_2)

10g of ground *Piper betel* leaves were weighed and placed into the extraction cell. The extraction cell was fitted and placed in an oven to maintain the desired temperature. CO_2 supplied from a cylinder was chilled and pump using HPLC pump (Lab Alliance, Series II Pump). The pressure in the extractor cell was controlled by adjusting back pressure regulator to achieve desired pressure. The extracts were collected in the collection vial attached at the outlet stream. The extraction takes places for 5hours extraction regime with 30minutes fractionation to determine the optimum time of extraction. Experiment regime was reduced to 3.5hours with 30minutes fraction after obtaining the optimum extraction time. The average particle size of the ground *Piper betel* leaves was constant at 355µm since during preliminary study of the average particle size of 500µm and 710µm produced the lowest yield compared to 355µm. Three parameters were tested in this study which are temperature (40, 60, 80°C), pressure (10, 20, 30MPa) and flowrate of CO_2 (4, 6, 8mL/min) in order to obtain high yield of the extracts and phenolic content simultaneously.

2.3. Total Phenolic Content

Folin-Ciocalteu method (FC method) was used to determine the total phenolic content of *piper betel linn*leaves. The *piper betel* extract was diluted in methanol and 1mg/ml of the solution was added into 6ml of distilled water, 0.5ml Folin-Ciocalteu reagent and 1.5ml sodium carbonate. Next, the solution was diluted in 7.9 ml distilled water and incubate for 90 minutes. The absorbance of the mixture was measured at 725nm using a UV-VIS spectrophotometer. The total phenolic content is expressed as gallic acid equivalents (GAE) in mg/g dry weight of the plant.

2.4. Experimental Design

In this study,an experimental design was performed in order to optimize three important parameter of SC-CO_2 extraction which are pressure (X_1), temperature (X_2) and flowrate of CO_2 (X_3) at constant average particle size and extraction regime in order to achieve high yields of *Piper betel* extracts and phenolic content simultaneously. Box Behnken was applied for experimental design of *piper betel* leaves SC-CO_2 extraction. The parameters were assigned at three coded levels of -1, 0 and +1 for temperature of 40, 60 and 80°C, pressure of 10, 20 and 30Mpa, and flowrate of CO_2 of 4, 6 and 8mL/min. The result obtained were analysed using Design Expert Software version 6. Table 1 shows the experimental design for the SC-CO_2 of *piper betel* leaves.

Table 1. Design of the experiment

Pressure, X_1	Temperature, X_2	Flowrate of CO_2, X_3	Percentage Yield [%]	Total Phenolic Content [mg GAE/g]
-1	-1	0	0.671	181.951
1	-1	0	4.930	120.815
-1	1	0	2.451	276.565
1	1	0	8.012	475.983
-1	0	-1	1.777	208.327
1	0	-1	6.723	374.090
-1	0	1	2.073	596.798
1	0	1	6.155	625.910
0	-1	-1	3.779	413.943
0	1	-1	4.294	406.114
0	-1	1	3.307	470.160
0	1	1	4.882	804.854
0	0	0	3.925	480.350

Response surface methodology (RSM) was applied to optimize the operating parameters of SC-CO_2 for the extraction of *piper betel* leaves. RSM model or regression equation which include linear and quadratic variables as well as interaction terms were used to fit the first and second-order polynomial equation based on the experimental data as follows:

$$Y = B_0 + \sum_{i=1}^{k} B_i X_i + \sum_{i=1}^{k} B_{ii} X_i^2 + \sum_i \sum_j B_{ij} X_i X_j \quad (1)$$

Y is the predicted response, B_0 is a constant, B_i, B_{ii}, B_{ij} are the coefficient for linearity, X_i and X_j are independent variables. The statistical testing of the model was assessed through F-test to obtain the mathematical relationship between input and output parameters. F value calculated must be higher than F value tabulated so that the relationship was significant. While, the significance of the model was performed by analysis of variance (ANOVA) through calculation of the correlation factor R^2.

3. Result and Discussion

3.1. Percentage Yield

Table 1 shows that the response of percentage yield of the *piper betel* leaves extraction by using SC-CO_2. The highest percentage yield of *piper betel* extracts is 8.012% at high temperature and pressure with medium flow of carbon dioxide which are 80°C, 30MPa and 6mL/min. More than 8.00% of the yield was consider good enough according to [21] if using SC-CO_2.The regression model from Eq.1 for percentage yield of extract was first order polynomial as shown in Eq. 2 below;

$$Y = -4.672 + 0.087X_1 + 0.118X_2 - 9.75 \times 10^{-3} X_3 \quad (2)$$

From the Eq. 2, the 3D response surface graph as a function of pressure and temperature on percentage yield of piper betel extractat the constant flowrate (6 ml/min) was obtain as shown in Figure 1. At constant pressure, the increase in temperature increases the vapour pressure of the extract contributein enhanced the extraction yield. Therefore, SC-CO_2 density increases with increasing pressure hence improved solubility of the extract and increased the extraction yield.As can be seen in Eq. 2, the temperature shows the most dominant parameter which contributes positive effect with the coefficient of 0.188 compared pressure with the coefficient of 0.087. The slope of the increases temperature shows steeper compared to the increases pressure proves that temperature was the most dominant factor on the yield recovery of piper betel.

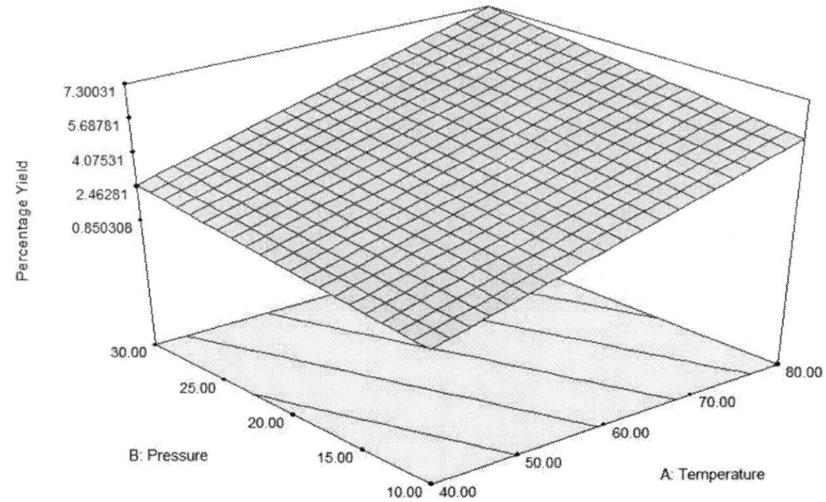

Figure 1. 3D Response Surface for Percentage Yield

ANOVA was performed as in Table 2. F value calculated was 73.12 while F value tabulated at 95% confidence level was 3.86 ($F_c > F_t$). Hence, there is significant relationship between the parameters. The calculated coefficient of correlation (R^2) and adjusted coefficient of correlation (R^2a) were 96.13% and 94.74% respectively. These values indicate that the model adequately represented the experimental data and 94.74% of the variations could be covered by the fitted model Eq. 2.The best operating conditions where the percentage yield at maximum were 80°C, 30MPa and 4.42mL/min with the percentage yield was 7.32%.

Table 2. ANOVA Table for Percentage Yield

Source	df	Sum of Square	Mean Square	F_c value	F_t value	R^2	R^2a
Regression	3	50.45	16.82	73.12	3.86	0.9613	0.9474
Residual	9	2.03	2.3E-01				
Total	12	52.48					

3.2. Total Phenolic Content

The amount of total phenolic content of the extracts of *piper betel* leaves using the FC method was shown in Table 1. The highest TPC from the *piper betel* extracts was 804.854 mg GAE/g at experimental conditions of 60°C, 30MPa and 8mL/min. The RSM model for TPC was second order polynomial as in Eq. 3:

$$Y = -202 + 44.04X_1 - 6.72X_2 - 312.24X_3 + 0.33X_1X_2 - 0.85X_1X_3 + 4.28X_2X_3 - 0.36X_1^2 - 0.72X_2^2 + 28.86X_3^2 \quad (3)$$

From Eq. 3, the 3D response surface graph was generated by the software and showed in Figure 2a-c as a function of pressure, temperature and flowrate parameters. From the Eq. 3, flowrate of carbon dioxide was the most dominant factor towards total phenolic content response. From Figure 2a, asthe temperature increase from 40°C to 60°C, it leads to increases the phenolic compound vapour pressure resulting in enhanced recovery of phenolic content. On the other hand, increasing the temperature from 60°C to 80°C leads to lower SC-CO_2 density and reduces the solubility of phenolic compound thus causes lower recovery of phenolic compound. As well as, there is possible when operate at high temperature, the phenolic content may degrade. Figure 2b and 2c indicate that increases of the flowrate at constant pressure and temperature lead to high value of phenolic content.

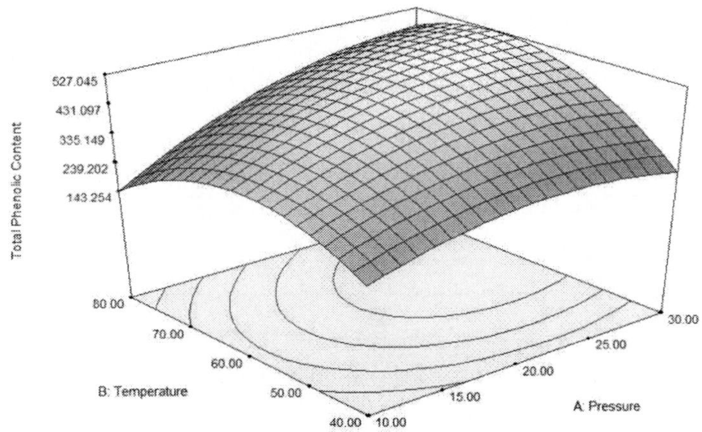

(a) Pressure-Temperature Parameter

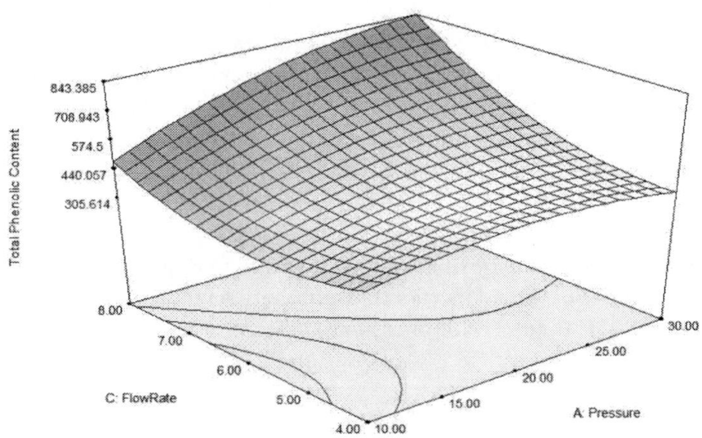

(b) Pressure-Flowrate Parameter

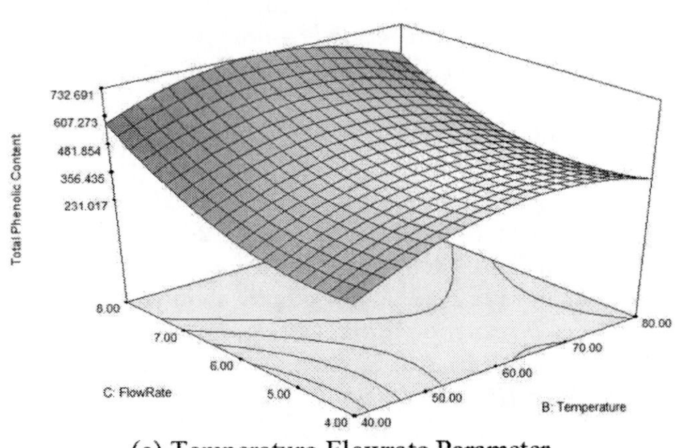

(c) Temperature-Flowrate Parameter

Figure 2. 3D Response Surface for Total Phenolic Content

Table 3 represents ANOVA table for TPC. F value calculated was 22.2 while F value tabulated at 95% confidence level was 8.81 ($F_c>F_t$). Hence, there is significant relationship between parameters. The calculated coefficient of correlation (R^2) and adjusted coefficient of correlation (R^2a) were 98.52% and 94.08% respectively. These values indicate that the model is good agreement to represent the experimental data and 94.74% of the variations could be covered by the fitted model Eq. 3.The optimum operating conditions where the extracts rich with phenolic compound were 67.53°C, 29.72MPa and 7.98mL/min with the value of TPC was 845.085mg GAE/g.

Table 3. ANOVA Table for Total Phenolic Content

Source	df	Sum of Square	Mean Square	F_c value	F_t value	R^2	R^2a
Regression	9	4.38E05	48702	22.2	8.81	0.9852	0.9408
Residual	3	6582.45	2194.2				
Total	12	4.45E05					

4. Conclusion

In this work the optimization of the supercritical fluid extractionof *piper betel* leaves was carried out for three processconditions, namely temperature (40-80°C), pressure (10-30MPa) and flowrate of CO-CO_2 (4-8mL/min), usinga Box–Behnken design of experiments and upon application of theresponse surface methodology. The main objective of this research was to determine the optimum operating conditions of SC-CO_2 for the extraction of *piper betel* leaves. The best conditions to maximize percentage of the extraction yield are 80°C/30MPa/4.42mL/min, which lead to 7.32%. In terms of total phenolic content in the extracts, the optimized operating conditions are 67.53°C/29.72MPa/7.98mL/min, providing 845.085mg GAE/g. The optimum condition for two responses simultaneously where maximum percentage yield as well as high value of TPC are 78°C/30MPa/8mL/min, which lead to 7.05% yield and 791.709mg GAE/g.

Acknowledgement

The financial support from the Ministry of Higher Education (MOHE) for fundamental research grant scheme (RJ130000.7809.4F766) and Universiti Teknologi Malaysia for professional development research university grant (QJ130000.21A2.02E85) are gratefully acknowledged.

References

[1] Foo L W, Salleh E and Mamat S N H 2015 Extraction and qualitative analysis of piper betle leaves for antimicrobial Activities *Int. J. Eng. Tech. Sci. Res.* **2(2)** 1–8

[2] Norton S A 1998 Betel : consumption and consequences *J. Am. Acad. Dermatol.* **38** 81–88

[3] Kharkongar P, Joseph J 1984 *Folklore medico-botany of rural Khasi and Jaintia tribes in Meghalaya. In Glimpses of Indian Ethnobotany, Jain SK (ed.)* (New Delhi : Oxford and IBH Publishing Co.) 133

[4] Balasubramanyam V R, Johri J K, Wawat A K S, Tripathi R D and Chaurasia R S 1994 *Betelvin (Piper Batle L.). Nation Botanical Research Institute (CSIR)* (India: Lucknow)

[5] Guha P 2006 Betel leaf : the neglected green gold of India. *J. Hum. Ecol.* ***19*(2)** 87–93

[6] Arambewela L, Arawwawala M, and Rajapaksa D 2006 Piper betle : A potential natural antioxidant*Int. J. Food Sci. Tech.* **41(SUPPL. 1)** 10–14.

[7] Shameem P M D, Thirumal M B 2013 A preliminary antimicrobial screening on leaves of *Piper betle* Linn. *Contemporary Investigations and Observations in Pharmacy* **2**(1) 22-26

[8] Sujatha H, Menuka A, Rajapaksha A L S, Ariyawansa H S 2011 Piper betle linn : as a remedy for Diabetes mellitus *IJRAP* **2(5)** 601-1603

[9] Pradhan D, Suri K A, Pradhan D K, Biswasroy P 2013 Golden heart of the nature : *Piper betle* L. *J. Pharmacogn. Phytochem.* **1(6)** 147-167

[10] Devjani C, Barkha S 2011 Antimicrobial, antioxidative and antihemolytic activity of *Piper betle* leaf extracts *Int. J. Pharm. Pharm. Sci.* **3(3)** 192-199

[11] Kaveti B and Tan L Antibacterial activity of *Piper betle* leaves *IJPTP* **2(3)** 129-132
[12] Amallia H, Sitompul R 2009 Effectiveness of *Piper betle* leaf infusion as a palpebral skin antiseptic *Univ. Med.* **28(2)**
[13] Young S C, Wang C J, Lin J J, Peng P L, Hsu J L, Chou F P 2007 Protection effect of piper betel leaf extract against carbon tetrachloride-induced liver fibrosis in rats *Arch.Toxicol.* **81(1)** 45-55
[14] Vinayak M, Ruckmani A, Chandrashekar K, Gopala R K V, Madhavi E, Swati B 2012 Antidepressant activity of ethanolic extract of *Piper betle*leaves in mice *Curr. Res. Neurosci.* **2** 11-16
[15] Badrul A M, Fahima A, Nahida P, Rashna S P, Akter S, Chowdhury J 2012 Antioxidant, analgesic and anti-inflammatory activities of the methanolic extract of *Piper betle* leaves *Avicenna J. Phytomed.*1-14
[16] Amonkar A J, et al. 1986 Hydroxychavicol : A new phenolic antimutagen from Betel Leaf. *Food Chem. Toxicol.* **24, 12** 1321-1324
[17] Rimando A M, *et al.* 1986 Studies on the Constituents of Philippine Piper Betel leaves *Arch. Pharm. Res.* **9, 2** 93-9
[18] Nagabhushan M, Amonkar A J, Nair U J, D'Souza A V, Bhide S V 1989 Hydroxychavicol : A new anti-nitrosating phenolic compound from betel leaf *Mutagenesis* **4** 200–204
[19] Pin K Y, Luqman C A, Abdull R A, Mazura M P, Fadzureena J, Vimala S, Rasadah M A 2010 Antioxidant and anti-inflammatory activities of extracts of betel eaves (Piper betle) from solvents with different polarities. *J. Trop. For. Sci.***22** 448–455
[20] Rayaguru K W, Routray and Mohanty S N 2011 Mathemathical modeling and quality parameters of air-dried Betal leaf (*Piper betle*L.) *J. Food Process Pres.* **35** 394–401
[21] Del ValleJ M, Jiménez M and De la Fuente J C 2003 Extraction kinetics of pre-pelletized jalapeño peppers with supercritical CO2 *J. Supercrit Fluids* **25(1)** 33–44

IOP Conf. Series: Materials Science and Engineering **162** (2017) 012032 doi:10.1088/1757-899X/162/1/012032

The effect of filler addition and oven temperature to the antioxidant quality in the drying of *Physalis angulata* leaf extract obtained by subcritical water extraction

R F Susanti[1,2] and Desy Natalia[1]
[1]Chemical Engineering Department, Industrial Engineering Faculty, Parahyangan Catholic University, Bandung, West Java, Indonesia
[2] Korea Institute of Science and Technology, Hwarangno 14-gil 5, Seongbuk-gu, Seoul, 136-791, Republic of Korea

E-mail: santi@unpar.ac.id

Abstract. In traditional medicine, *Physalis angulata* which is well known as ceplukan in Indonesia, has been utilized to cure several diseases by conventional extraction in hot water. The investigation of the *Physalis angulata* extract activity in modern medicine typically utilized organic solvents such as ethanol, methanol, chloroform and hexane in extraction. In this research, subcritical water was used as a solvent instead of organic solvent to extract the *Pysalis angulata* leaf part. **The focus of this research** was the investigation of extract drying condition in the presence of filler to preserve the quality of antioxidant in *Physalis angulata* extract. Filler, which is inert, was added to the extract during drying to help absorb the water while protect the extract from exposure in heat during drying. The effects of filler types, concentrations and oven drying temperatures were investigated to the antioxidant quality covering total phenol and antioxidant activity. Aerosil and microcrystalline cellulose (MCC) were utilized as fillers with concentration was varied from 0-30 wt% for MCC and 0-15 wt% for aerosil. The oven drying temperature was varied from 40-60 °C. The results showed that compare to extract dried without filler, total phenol and antioxidant activity were improved upon addition of filler. The higher the concentration of filler, the better the antioxidant; however it was limited by the homogeneity of filler in the extract. Both of the variables (oven temperature and concentration) played an important role in the improvement of extract quality of *Physalis angulata* leaf. It was related to the drying time which can be minimized to protect the deterioration of extract from heat. In addition, filler help to provide the powder form of extract instead of the typical extract form which is sticky and oily.

1. Introduction

Physalis angulata is a widespread pristine herb found in tropical and subtropical areas of Asia, Afrika and America.[1] It is belongs to Solanaceae family and well known as *ciplukan* or *ceplukan* in Indonesia. It has been utilized in traditional medicine however it is not cultivated well. It grows wildly in yard or rice field without special treatment. Sometimes farmer exterminates those plants before planting season because they considered as weeds. In such a traditional way, *Physalis angulata* is utilized by boiling all plant part to obtain the extract while in modern medicine typical organic solvents such as methanol, ethanol and even hexane and chloroform were used.[2-4] However, those organic solvents are classified as Volatile Organic Solvents (VOC) which the exposure to human body should be limited strictly. Water is considered as a safer solvent, however at normal temperature until around its boiling point, it still behaves as polar solvent. Therefore it is not suitable to draw the less polar substances from *Physalis angulata*. In this research, water at subcritical state was utilized as a solvent. It is because water polarity under pressure and temperature can be adjusted to mimic the organic solvents commonly used in extraction. Subcritical water is water at elevated temperature

[1] To whom any correspondence should be addressed.

(between its boiling point to supercritical state) under pressure to maintain its liquid state [5]. The fixed extraction condition was used to extract *Physalis angulata* leaf (250°C, 100 bar and 15 minutes).

The *Physalis angulata* extract itself has been proven to have activities as antimycrobacterial [3], antinociceptive [6], antihepatoma [4], immunosuppression[7], anti-inflammatory[1] etc. Our group was the first who investigate the antioxidant activities in *Physalis angulata* extract based on each plant part individually [8-11] Antioxidant is any substance which if present in low concentration can inhibit or delay the oxidation of the cell. It can prevent damage of cell in our body arising from chemical reaction involving free radical reaction. Free radical is very dangerous which can weaken our immune system, damage and kill our cells, also generate diseases[12]. In this research, the antioxidant potential of *Physalis angulata* was investigated in related with extract drying condition.

Crude extract which is commonly obtained from extraction of plant material contains oleoresin (oily and solid form at once) often causes difficulty in the determination of drug dosage. The treatment of crude extract into dry extract can be obtained by drying in dryer. However, the bioactive compound in crude extract which contains mostly antioxidant may damage during exposure in heat. In addition just simply drying resulted in a sticky form of extract. The addition of filler was intended to make the drying process faster, to keep the flavor, to add the quantity, and to make the powder form of extract [11]. It is inert and typically added to add the consistency of drugs. The fillers commonly used in pharmacy are sucrose, lactose, dextrose, cellulose, sorbitol etc [13, 14] Therefore the best filler type to preserve the antioxidant in *Physalis angulata* extract and the proper concentration of filler should be investigated. In this research, extract qualities were determined based on total phenol and antioxidant activity. The effects of filler types (Microcrystalline Cellulose (MCC) and aerosil), filler concentrations (0-30 wt% for MCC and 1-15 wt% for aerosil) and drying temperatures (40-60 °C) were investigated to the extract obtained from *Physalis angulata* leaf which were dried in vacuum drying oven.

2. Experimental

The experimental procedure consists of pretreatment, extraction by subcritical water, drying in vacuum drying oven and extract characterization.

2.1 Pretreatment

Physalis angulata plants were collected from several rice fields and yards in Yogyakarta. After identified at Faculty of Pharmacy, Gadjah Mada University, each plant part was separated and investigated individually. In this experiment, leaf part was investigated separately while fruit part has been reported previously[11] . After collection, leaf were washed and dried in tray dryer for 24 hours for then ground and classified into size of -20+30 mesh. The water content was analyzed using moisture analyzer (Mettler Toledo, HR83, USA)

2.2 Subcritical Water Extraction

The reactor system was fabricated from SUS316 with volume of 150 ml, equipped with electric heating for heating the reaction to the desired temperature and cooling jacket for decreasing the reactor temperature once the reaction finished. The temperature and pressure of the system is controlled using control box. The extraction procedure has been described elsewhere [11].

2.3 Drying Experiment

The drying experiments were conducted in vacuum drying oven (Jeio Tech, Korea). The extract with volume about 30 mL was dried upon an addition of filler. The mixture of extract and filler was stirred carefully to obtain a homogenous solution. The drying temperature was set and the extract was dried and weighted every hour. The drying was stopped if the constant weight of extract was observed. The water content then was checked by moisture analyzer (Mettler Toledo, HR83, USA). The dry extracts were stored at refrigerator prior used.

2.4 Analysis

Prior used in analysis, extract powder was dissolved in methanol p.a to separate the extract from filler. The solution then was filtered using filter paper Sartorius 388 (Sartorius Stedim Biotech S.A.), the filtrate was used for analysis while the filler was dried and weighted for calculation purpose. Total phenol and antioxidant activity were quantified according to the method which has been reported in previous publication[11].

2.5 Total phenol

The total phenols of the extracts were analyzed by Follin-Ciocalteu method, which used gallic acid as the standard solution. Accurately, 0.5 mL extracts 500 ppm (5 mg in 10 ml DDI water) were mixed with 2.5 mg Follin Ciocalteu reagent (1:10) and 2 mL of Na_2CO_3 solution (75 g/L). After added with reagent and Na_2CO_3, the mixture solution was incubated at 40°C for 30 minutes, and the absorbance was measured at 735 nm.

2.6 Antioxidant activity

The stock DPPH solution was prepared by dissolving 7.9 mg DPPH in 100 ml methanol p.a. (0.2 mM). The solution was obtained by diluting a specific amount of dry extracts with 100 ml methanol p.a. The different concentration of solutions (20-400 ppm) were made by diluting the extract mother liquor. Then, 2 ml of this solution was allowed to react with 2 ml of the DPPH solution for 24 h and the condition was kept away from any sources of light. Then the absorbance was measured at 517 nm. The IC50 value was calculated by making a plot of % inhibition (ordinat) versus extract concentration (absis) at different extract concentration, then calculating the linear trendline (y=ax+b). From the trend line equation, IC50 is determined by replacing y value with 50 and calculating the x value, which later considered as IC50.
% inhibition was calculated by formula:

$$\% \text{ Inhibition} = A_0\text{-}Aa/A_0 \times 100\% \quad (2.1)$$

With A_0 = absorbance in the absence of extract (blanko)
Aa = absorbance in the presence of extract

3. Result and discussion

The extraction was conducted by subcritical water extraction at temperature of 250°C, 100 bar and 15 min. The extraction condition was not varied since it is not the focus of this research. Two kinds of filler were used in drying, microcrystalline cellulose (MCC) and aerosil. Both of fillers are typical filler used in pharmaceutical. The affectivity of drying then characterized by measuring total phenol and antioxidant activity since they are representative variables to characterize antioxidant.

1.1. The effect of drying temperature and filler concentration on the total phenol

The content of total phenolic in the extract was quantified based on absorbance values of *Physalis angulata* leaf extract reacted with Folin-Ciocalteu reagent and compared with standard solution of gallic acid equivalent as describe in the procedure. The total phenol as a function of filler concentration and drying temperature is illustrated in Figure 1. An increase in the filler concentration leads to an increase in total phenol in the extract in the increment of drying temperature. It can be explained in relation with drying time which presented in table 1. An increase in the filler concentration showed shortening the drying time, therefore decreasing the exposure of extract in heat. For instance, in the absence of filler at 40°C, it is needed 6030 minute to dry 30 mL of extract, however in the presence of aerosil 15 wt%, it only need around 2340 minutes. A significant increase in total phenol at drying temperature of 60°C was observed when concentration of filler was above 10 wt%. That tendency was observed at both of filler type, MCC and aerosil, even though the investigation of aerosil concentration was only observed till 15 wt%. While the trend of total phenol at 40-50°C was similar. It might be because a significant decrease of drying time was observed when drying temperature increased from 50 to 60°C compared to 40 to 50°C. In MCC case, an increase in drying temperature from 50 to 60 °C was decrease the drying time around 40-60% while an increase in the drying temperature from 40 to 50°C only decreased drying time in the range of 10-20%. At the drying temperature of 50°C, the total phenol in all samples was lower compare to that of dried at 40°C and 60°C. It is not clear what causes the trend; however the combination between drying temperature and length of drying time might be one of the reason.

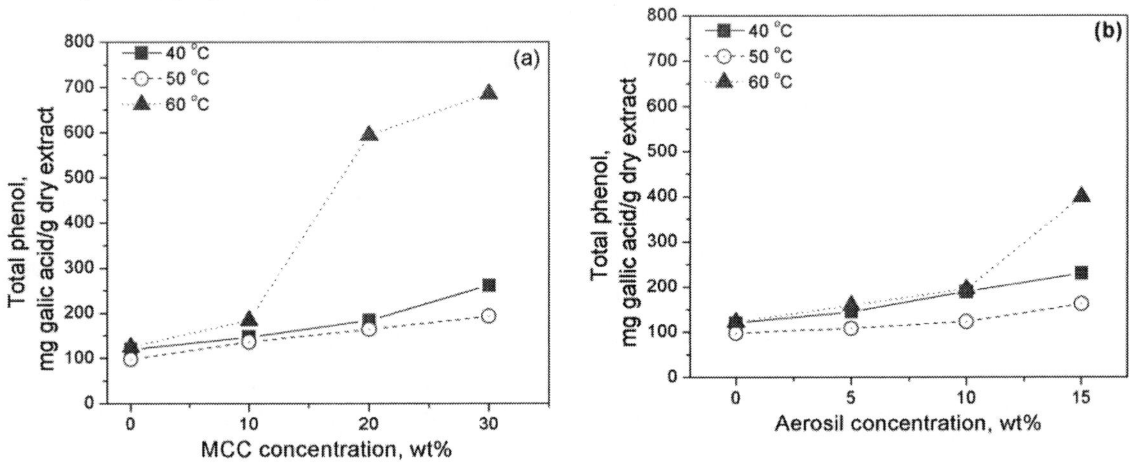

Figure 1. Total phenol as a function of filler (MCC (a) and aerosil (b)) concentration under different oven drying temperature

Table 1. The length of drying time and water content in the drying of *Physalis angulata* leaf extract in addition of MCC filler

Drying temperature (°C)	MCC Concentration (wt%)	Drying time (min)	Water content (wt%)
	0	6820	2.39
40	10	5710	6.56
	20	5455	2.66
	30	5215	3.01
	0	5505	2.37
50	10	5160	1.80
	20	4620	3.28
	30	4335	2.84
	0	3250	2.42
60	10	2535	5.36
	20	2050	4.27
	30	1700	2.98

Table 2. The length of drying time and water content in the drying of *Physalis angulata* leaf extract in addition of Aerosil filler

Drying temperature (°C)	Aerosil Concentration (wt%)	Drying time (min)	Water content (wt%)
	0	6030	2.66
40	5	3090	2.23
	10	2610	1.17
	15	2340	0.98
	0	4980	2.56
50	5	2340	3.12
	10	2130	1.34
	15	1800	2.84
	0	3080	2.39
60	5	2220	1.34
	10	1470	1.12
	15	1320	9.76

3.2 *The effect of drying temperature and filler concentration on antioxidant activity*

Antioxidant activity which is presented as Inhibition Concentration 50 (IC50) is defined as the concentration of extract which is required for 50% inhibition of oxidation. It means the lower the IC50 value, the smaller extract needed for inhibiting the oxidation, means the stronger the antioxidant activity. The dependence of IC50 to the filler concentration and drying temperature can be seen in figure 2. An increase in filler concentration decreases the IC50 value that means higher the antioxidant activity. It is also in agreement with total phenol value which also showed the similar trend. The temperature of 60°C gave the best results among all with IC50 value of 50, which then considered as strong antioxidant.[15, 16] Total phenol value and antioxidant activity trend is similar in both of variables, concentration and oven drying temperature.

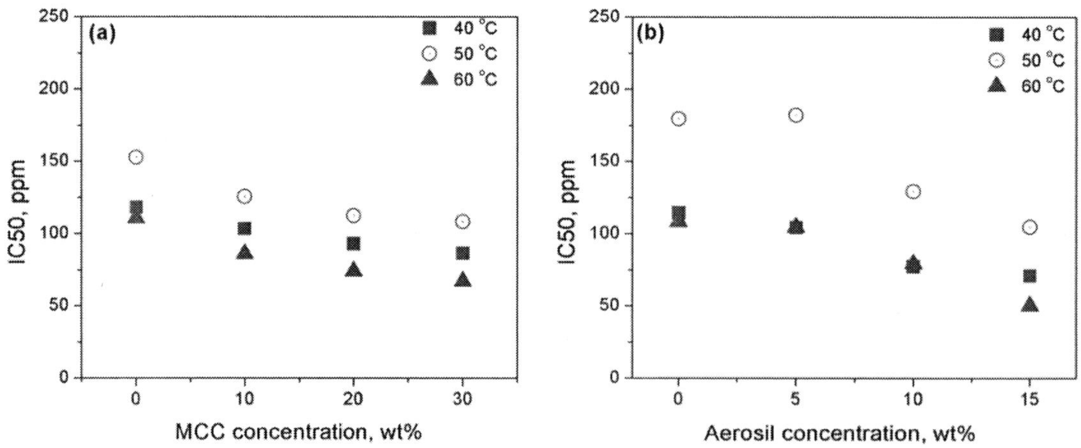

Figure 2. IC50 as a function of filler (MCC (a) and aerosil (b)) concentration under different oven drying temperature

3.3 Effect of filler type on the total phenol and antioxidant activity

Due to the different order of filler density, we investigated the different concentration range for MCC and aerosil. Aerosil has much lower density compare to MCC, therefore for the same volume of extract, aerosil volume is much more compare to MCC. In this experiment, the maximum concentration for homogenous solution of extract, aerosil was 15 wt% while MCC was 30 wt%. Therefore, we can only compare the performance of 10 wt% of MCC and aerosil in the extract under different oven temperature as can be seen in figure 3.

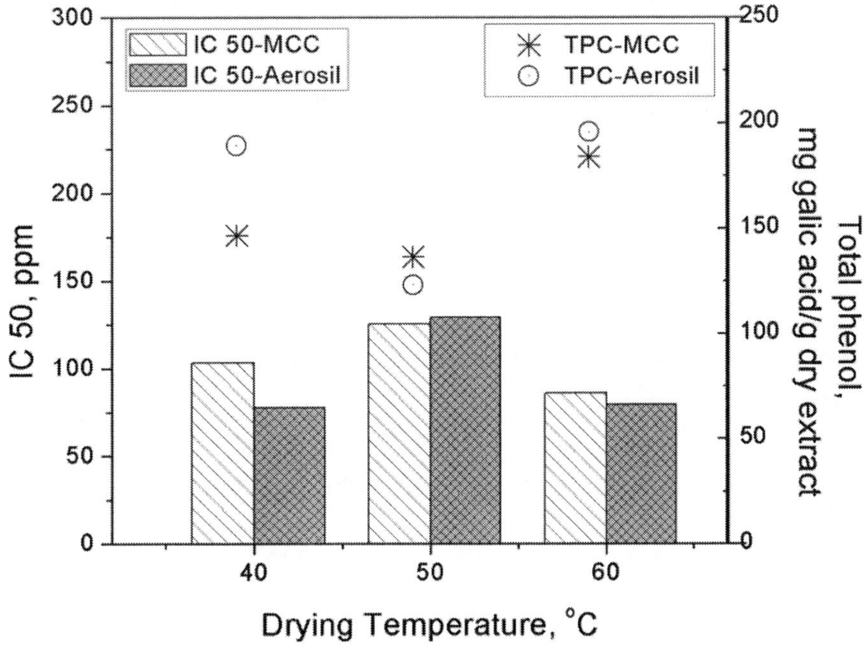

Figure 3. IC50 and total phenol as a function of filler type and oven drying temperature

Aerosil showed better total phenol and antioxidant activity results in all temperatures observed. It may be because aerosil (50 g/L) has much lower density compared to MCC (320 g/L) [14]so that for the same weight, the volume of aerosil was more than MCC. The impact would be in a better ability of aerosil to absorb water compared to MCC for the same quantity, giving faster drying time.

4. Conclusion

Total phenol and antioxidant activity can be improved in addition of filler. Increasing of filler concentration for both of microcrystalline cellulose (MCC) and aerosil as well as increasing drying temperature were proven to give positive effect in the total phenol and antioxidant activity. It is because prolong exposure in heat could deteriorate the extract, however addition of filler could shorten the drying time so the antioxidant in the extract could be maintained. An increase in filler concentration from 0 wt% to 30 wt% for MCC and 0 wt% to 15 wt% for aerosil and oven drying temperature from 40-60°C has been proven to improve the antioxidant quality because of reduction in drying time. Extract was protected from long exposure to heat. At 10 wt%, aerosil showed better performance compare to MCC.

Acknowledgments

The author gratefully appreciate the financial support from Korea Institute of Science and Technology (KIST), Korea through KIST Collaboration Alumni Project and Lembaga Penelitian dan Pengabdian Masyarakat (LPPM) UNPAR through Internal Funding.

IOP Conf. Series: Materials Science and Engineering **162** (2017) 012032 doi:10.1088/1757-899X/162/1/012032

References

[1] Bastos G N T, Silveira A J A, Salgado C G, Picanço-Diniz D L W& do Nascimento J L M 2008 *J. Ethnopharmacol* **118** 246

[2] Pinto N B, Morais T C, Carvalho, Silva C R, Andrade G M, Brito G A C, Veras M L, Pessoa O D L, Rao V S & Santos F A 2010 *Phytomedicine* **17** 740

[3] Pietro R C L R, Kashima S, Sato D N, Januârio A H & Franca S C 2000 *Phytomedicine* **7** 335

[4] Wu S-J, Ng L-T, Chen C-H, Lin D-L,Wang S-S & Lin C-C 2004 *Life Sci.* **74** 2061

[5] Teo C C, Tan S N, Yong J W H, Hew C S & Ong E S 2010 *J. Chromatogr. A* **1217** 2484

[6] Bastos G N T, Santos A R S, Ferreira V M M, Costa A M R, Bispo C I, Silveira A J A & Do Nascimento J L M 2006 *J. Ethnopharmacol* **103** 241

[7] Sun L, Liu J, Liu P, Yu Y, Ma L& Hu L 2011 *Process Biochem.* **46** 482

[8] Susanti R F, Kennedy and Winata A S 2014 *Proc. Int. Sem. of Chem. Eng. in Minerals & Materials Processing (ITB Bandung, Indonesia)*

[9] Susanti R F, Kurnia K, Vania A & Renaldo I J 2015 *Modern Applied Science* **9** 190

[10] Susanti R F, Renaldo I J and Garmi S 2013 *Proc. Reg. Symp. on Chem. Eng.(Bohol Philippines)*

[11] Susanti R F&Christianto G 2016 *IOP Conference Series: Materials Science and Engineering* **105** 012012

[12] Young I S & Woodside J V 2001 *J Clin Pathol* **54** 176

[13] Banker G S&Anderson N R 1986 *The Theory and Practice of Industrial Pharmacy* 3rd ed (Philadelphia: Lea and Febiger)

[14] *Handbook of pharmaceutical excipients* 6th ed.; ed. Rowe RC, Sheskey P J and Quinn M E (London: Pharmaceutical Press) 2009

[15] Fidrianny I, Fikayuniar L & Insanu M 2015 Asian J. Pharm. Clin. Res. **8** 215

[16] Blois M S 1958 *Nature* **181**1199

IOP Conf. Series: Materials Science and Engineering **162** (2017) 012033 doi:10.1088/1757-899X/162/1/012033

Preparation of silver nanoparticles loaded graphene oxide nanosheets for antibacterial activity

T T T Vi and S J Lue
Chang Gung University, 259 Wenhua First Road, Guishan District, Taoyuan City, 33302, Taiwan (R.O.C.)

E-mail : jessie@mail.cgu.edu.tw

Abstract. A simple, facile method to fabricate successfully silver nanoparticle (AgNPs) decorated on graphene oxide (GO) layers via grafted thiol groups. Samples were prepared with different concentrations of $AgNO_3$. Resulting AgNPs were quasi-spherical in shape and attached on the layers of GO. Physical properties were confirmed by X-ray diffraction (XRD), zeta potential, dynamic light scattering (DLS), Fourier transform infrared (FTIR) spectra, thermogravimetric analyzer (TGA), transmission electron microscopy (TEM) and field emission scanning electron microscopy (FE-SEM). Antimicrobial test was effectively showed using MRSA (Staphylococcus areus). The GO-Ag NPs with appropriate Ag NPs content of 0.2M $AgNO_3$ exhibited the strongest antibacterial activity at 48.77% inhibition after 4 hours incubation.

1. Introduction

Antibiotic resistance is one of the biggest threats to modern society. The medicines used to treat dangerous bacterial infections are rapidly losing effectiveness that creating side effects. Nowadays, scientists have found that bacteria are less prone to develop resistance against silver nanoparticles (Ag NPs) than conventional antibiotics[1]. Recently, graphene oxide (GO) is also the potential materials which becomes popular with many applications, one of them is antibacterial treatment. The principle of antimicrobial may come from GO traps bacteria while Ag kills bacteria[2, 3]. Hence, the combination of GO and Ag NPs may suggest to be an efficient material due to biocompatibility, simple synthesis and low cost. In this work, Ag NPs was decorated onto GO sheets via grafted thiol groups. Despite of myriad studies are carried inliterature but there are less report and quest about the difference between graphene oxide and graphene oxide – silver composite bacteria treatment. Herein, to clarify this matter, our main focus of this study is to evaluate the characterization of these two kind materials and analyze their antibacterial properties.

2. Materials and methods

2.1. Chemicals

Graphite powder, NaSH, Propidium iodide were purchased from Sigma-Aldrich, St. Louis, Missouri, USA. Potassium permanganate ($KMnO_4$) powder was purchased from Nihon Shiyaku Industries .Ltd, Osaka, Japan. Sulfuric Acid (H_2SO_4) (95-98%) solution was purchased from Scharlab S.L., Barcelona, Spain. $AgNO_3$ was purchased from Mallinckrodt Baker Inc., Paris.

S. aureus (ATCC 25178) was obtained from Bioresource Collection and Research Center, Taiwan. Bacteria were cultivated in agar disc and keep in fridge. Before using was growth overnight in Difco™ Nutrient Broth under aerobic conditions at 37◦C using a FIRSTEK S300R orbital shaker incubator for 12 hours.

2.2. Preparation of GO-Ag nanocomposites

Graphene oxide was prepared from graphite powder by modified Hummer's method[4]. GO (0.5g) was dispersed in 30mL deionized water (DI) in 50mL beaker while ultrasonic for 20 minutes. NaSH (8g) was gradually added by stirring and then was ultrasonic at 40^0C. The mixture was then maintained under stirring for 20 hour at 55^{0}C to produce thiol groups on the GO surface. The product was filtered and washed with DI water and dried in the vacuum at 50^0 for 3h. The thiolate GO powder 0.1g was dispersed in DI water 30mL by sonication for 30 minutes. Then series of aqueous solution of 0.1M and 0.2M, 0.25M $AgNO_3$ was added to thiolate GO solution under stirring, respectively. Subsequently, 50mL 0.1M NaOH was added and stirred for 20 hours. The GO-Ag powder was obtained by centrifugation at 10000rpm for several times,and then dried in vacuum in 40^{0}C. Power products was then using dialysis tubing in order to remove unreacted Ag and loose bound of Ag NPs. Finally, we get GO-Ag with concentration named GO-Ag 0.1M and GO-Ag 0.2M, GO-Ag 0.25M.

2.3. *Characterization of GO and GO-Agnanocomposites*

The latex of composite supension were determined for transmission electron microscopy (TEM, JEM 2000EXII, JEOL, Tokyo, Japan) and Energy-dispersive X-ray spectroscopy (EDX, model JSM-7500F, Hitachi High- Technologies Corp., Tokyo, Japan), laser light scattering (Zetasier, 2000 HAS, Mavelvern, Worcestershire, UK), UV-visible spectrophotometer (V-650, Jasco, Tokyo, Japan). The latex of composite were dried and the powders measured by Fourier Transform Infrared Spectroscopy (FTIR), X-ray diffraction (XRD, model D5005D, Siemens AG, Munich, Germany).

2.4. *Growth curves of bacteria strains after incubators with samples*

Bacteria was centrifuged in 10.000rpm for 5 min by using Hitachi, the supernatant was discarded, precipitate was collected and diluted with 12Ml PBS, 450μL of the above bacterial suspension was then transfer into 48 well-plate. Following this addition, 50 μL of each sample was added (PBS is used as a control sample) and kept in incubator for 1 hour. Consequently, 10μl of sample mixture was taken out and changed into 96 well-plates, 90μL NB was added afterwards. All treatments were performed two times in triplicate, measured by BioTek microplate reader with OD 600nm after each an hour during 4 hour. The inhibition efficiency was calculated by using following equation[5]:

$$\text{Inhibition (\%)} = \left(1 - \frac{OD\ sample\ \ treatment}{OD\ control}\right) \times 100 \qquad (1)$$

Results were expressed as mean ± standard deviation. Comparative studies of means were performed using excel software with n=3.

3. Results and Discussion

3.1. *Physicochemical characteristics of GO and GO-Ag nanocomposites*

The Ag NPs attached on the graphene was confirmed by UV-visible spectra. Figure 1(a) shifted the peak at 410nm which was consistent with the surface plasmon resonance phenomena of Ag NPs formulation[6] while pristine GO exhibited a typical peak around 247nm corresponding to the C=C aromatic bonding[7]. The XRD was used in order to examine the GO and GO-Ag crystalline structure. As shown on Figure 1(b), the graphene oxide powder exhibited a sharp diffraction peak (001) at 2θ = 11.7°. This results indicated the successful synthesis and matched with references data for GO sythesis [3, 8-12]. GO-Ag composite shifted the peak values at about 38.1°, 44.3°, 64.5°and 77.5°which are assigned to the (111), (200), (220) and (311) crystal lattice planes of face-centered cubic (fcc) Ag nanoparticles, respectively while the peak of GO disappeared, that was reported because the metal-nanoparticles was attached onto the inlayers may lead to the sheeting the signals of graphene oxide peaks [13].Prominently, the sharp peak of 38.1°confirmed that the nanoparticles are composed of pure crystalline Ag [14, 15].In Figure 1(c), FTIR illustrated the functional groups that contains on GO and GO-Ag sample. GO spectrum were observed at 3404, 1718, 1625 and 1055 cm^{-1}. Those spectra associated with the stretching vibration of C-OH (hydroxyl), the stretching vibration of C=O, the vibration of C=C (possibly due to the skeletal vibration of un-oxidized graphite domains), and C-O stretch located, respectively. Based on the presence of those oxygen-containing groups such as carboxyl, hydroxyl and epoxy, it confirmed that the result of oxidation process is GO. This FTIR results are also similar with previous studies [8]. Besides, after decoration with Ag NPs, compare with the spectrum of graphene oxide, the –OH groups became stretched out while the intensity related to the bonding signal slightly decreased due tothe interactions between Ag^+ ions and the oxygen-containing groups on the GO sheets, which may suggest the nucleation sites for anchoring and growth of AgNPs [16]. Remarkably, there has a bonding of thiol group (C-SH) at 910 appeared in Ag-GO sample.

Figure 2(a), (b), (c), (d) shows the TEM images of GO, GO-Ag 0.1M, GO-Ag 0.2M and GO-Ag 0.25M samples, respectively. Fig 2. (a) shows the typical flaky structure like winkle, wavy, transparent puregraphene oxide. At Figure 2(b) and (c) have displayed obviously small dots stick on the layer of graphene layers while at Figure 2(d) the dots tend to aggregate each other which form high ratio volume phenomena. Energy-dispersive X-rays spectroscopy was used to detect the element composition of sample. At Figure 3, the signal of Ag peak was clearly observe by EDS spectrum which indicates the presence of Ag NPs therein GO-Ag samples.

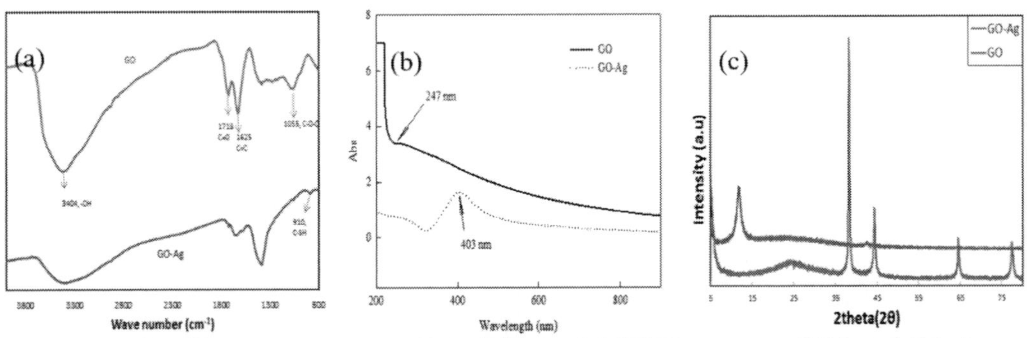

Figure 1(a) illustrates FT-IR, **(b)**UV-visible and,**(c)**XRD patterns of GO and GO-Ag

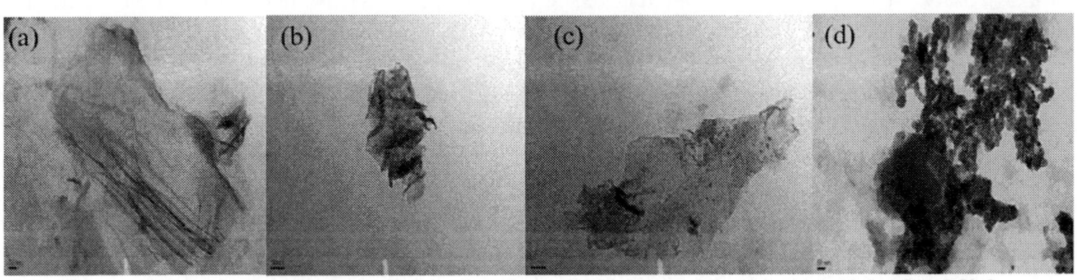

Figure 2(a) TEM images of pure-GO, **(b)**GO-Ag 0.1M, **(c)**0.2M, **(d)**0.25M

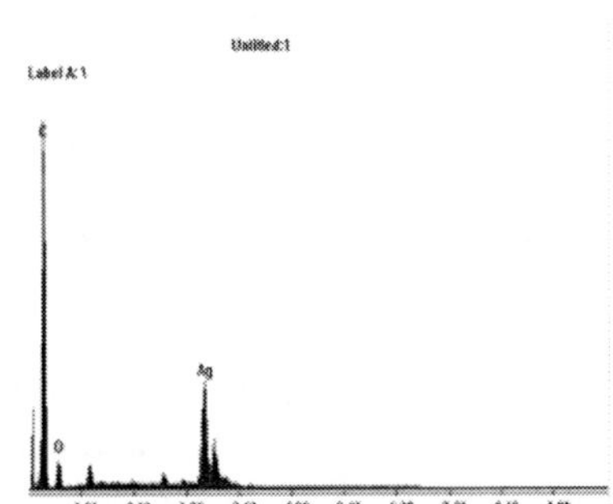

Figure 3. EDS spectrum of GO-Ag nanocomposite confirming the presence of Ag, C and O

3.2. Antibacterial of GO and GO-Ag nanocomposites

Elisa reading with 600nm absorbance was performed by imbedded of each sample (PBS was used as a control) after each hour reading during 4 hours. Time curves growth was described in Figure3 and inhibition results in details has shown in Table 1. Obviously, GO-Ag 0.2M which was the most significant decreased after time incubation. For more details, table 1 showed GO had no inhibition at all while GO-SH still had antibacterial activity even lower than other GO-Ag and gradually increased from GO-Ag 0.1M to GO-Ag 0.2M from 10.8%, 16.7% and 48.77%, respectively.Interestingly, we find that the inhibition of GO-Ag 0.25M was 19.21% lower than GO-Ag 0.2M 48.77% even though the $AgNO_3$ amount is higher. This is because Ag NPs in GO-Ag 0.25M tends to agglomerate reducing the contact surface area of Ag NPs with bacteria. This suits with TEM images about the distribution of Ag NPs on GO sheets.

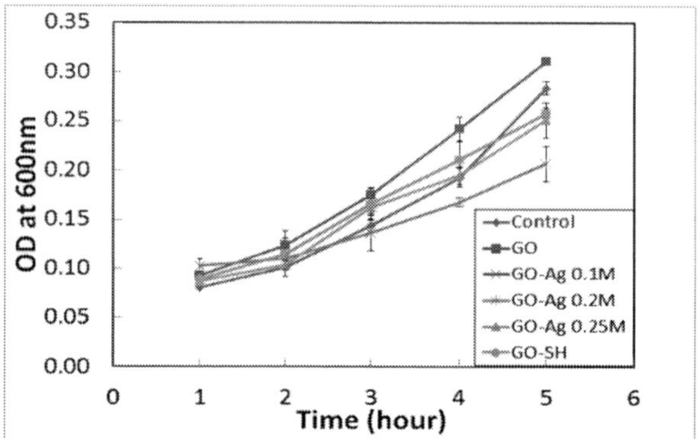

Figure 4. Illustrates growth curves of optical density measurements of bacteria at a wavelength of 600nm

4. Conclusion

The graphene oxide loading silver nanoparticle based on thiol groups was successfully fabricated. The XRD, FTIR, UV-visible spectra results were confirmed the property of sample. The silver was decorated well-distributed to the layer of graphene oxide by TEM observation. The difference of silver ratio contributed to the antibacterial activity at concentration. At 0.2M $AgNO_3$ was obtained the effectively antimicrobial at 48.77% higher than that one in other concentrations. That is reasonable results because many studies consider that too much silver may lead to agglomeration[17]. The synthesized GO-Ag NPs may have potential in antibacterial in particular and in biomedical in general.

Acknowledgment

This study is supported by Chang Gung University (BMRP 326) and Chang Gung Memorial Hospital (CMRPD2F0051).

References

[1] Beyth N, et al. 2015 Alternative Antimicrobial Approach: Nano-Antimicrobial Materials *Evidence-Based Complementary and Alternative Medicine*. **2015** 16

[2] Fernando K A S, et al. 2014 Migration of Silver Nanoparticles from Silver Decorated Graphene Oxide to Other Carbon Nanostructures *Langmuir* **30**(39) 11776-11784

[3] Zou X, et al. 138 Mechanisms of the Antimicrobial Activities of Graphene Materials *Journal of the American Chemical Society* **138**(7) 2064-2077

[4] Dimiev A M and Tour J M 2014 Mechanism of Graphene Oxide Formation *ACS Nano* **8**(3) 3060-3068

[5] Li Q, et al 2008 Antimicrobial nanomaterials for water disinfection and microbial control: Potential applications and implications *Water Research* **42**(18) 4591-4602

[6] Lukman A I, et al. 2011 Facile synthesis, stabilization, and anti-bacterial performance of discrete Ag nanoparticles using Medicago sativa seed exudates *Journal of Colloid and Interface Science* **353**(2) 433-444

[7] Cushing S K, et al. 2014 Origin of Strong Excitation Wavelength Dependent Fluorescence of Graphene Oxide *ACS Nano* **8**(1) 1002-1013

[8] He J, et al. 2015 Killing Dental Pathogens Using Antibacterial Graphene Oxide *ACS Applied Materials & Interfaces* 7(9) 5605-5611

[9] Krishnamoorthy K, et al. 2012 Antibacterial Activity of Graphene Oxide Nanosheets *Science of Advanced Materials* **4**(11) 1111-1117

[10] Liu S, et al. 2011 Antibacterial Activity of Graphite, Graphite Oxide, Graphene Oxide, and Reduced Graphene Oxide: Membrane and Oxidative Stress *ACS Nano* **5**(9) 6971-6980

[11] Notley S M, Crawford R J and Ivanova E P 2013 Bacterial Interaction with Graphene Particles and Surfaces *Advances in Graphene Science*

[12] Stobinski L, et al. 2014 Graphene oxide and reduced graphene oxide studied by the XRD, TEM and electron spectroscopy methods *Journal of Electron Spectroscopy and Related Phenomena* **195** 145-154

[13] Xu C and Wang X 2009 Fabrication of Flexible Metal-Nanoparticle Films Using Graphene Oxide Sheets as Substrates *Small* **5**(19) 2212-2217

[14] Das M R, et al. 2011 Synthesis of silver nanoparticles in an aqueous suspension of graphene oxide sheets and its antimicrobial activity *Colloids and Surfaces B : Biointerfaces* **83**(1) 16-22

[15] El-Dessouky R, Georges M and Azzazy H M E 2012 Silver Nanostructures: Properties, Synthesis, and Biosensor Applications, in *Functional Nanoparticles for Bioanalysis, Nanomedicine, and Bioelectronic Devices Volume 1*. *American Chemical Society* 359-404

[16] Yuan L, et al. 2014 Facile synthesis of silver nanoparticles supported on three dimensional graphene oxide/carbon black composite and its application for oxygen reduction reaction *Electrochimica Acta* **135** 168-174

[17] Tang J, et al. 2013 Graphene Oxide–Silver Nanocomposite As a Highly Effective Antibacterial Agent with Species-Specific Mechanisms *ACS Applied Materials & Interfaces* **5**(9) 3867-3874

The effect of activated carbon support surface modification on characteristics of carbon nanospheres prepared by deposition precipitation of Fe-catalyst

H Kristianto[1, a], A A Arie[1, b *], R F Susanti[1, c], M Halim[2, d] and J K Lee[2, e]

[1] Department of Chemical Engineering, Faculty of Industrial Technology, Parahyangan Catholic University, Ciumbuleuit 94 Bandung 40141 Indonesia

[2] Center of Energy Convergence, Korea Institute of Science and Technology, Hwarangno 14-gil 5, Seongbuk-gu, Seoul 136-791, Republic of Korea

E-mail : [a]hans.kristianto@unpar.ac.id, [b]arenst@yahoo.com, [c]santi@unpar.ac.id, [d]612514@kist.re.kr, [e]leejk@kist.re.kr

Abstract. In this study the effect of activated carbon support modification to synthesis of CNSs was observed. Modification of activated carbon was done by using nitric acid. The effect of modification was analyzed from its FTIR spectra. The Fe catalysts were deposited on to the support by using urea deposition precipitation method at various initial catalysts concentration. CNSs was synthesized by utilizing cooking palm oil as renewable carbon source, and pyrolized at 700°C for 1 hour under nitrogen atmosphere. The products obtained then analyzed using SEM-EDS, TEM, XRD, and Raman spectroscopy. The modification of activated carbon support had increased the oxygen functional group. This increase resulted on increase of metal catalysts deposited on activated carbon surface. Peak of C (100) was observed, while ID/IG of samples were obtained around 0.9, which is commonly obtained for CNSs. High catalysts loading on modified activated carbon support caused decomposition of CNSs and formation carbon onion.

1. Introduction

Metal catalysts play important role on carbon nanomaterial synthesis. Transition metal catalysts are usually used in carbon nanomaterial synthesis, both in metallocene forms (Fe, Co, Ni, etc.) [1], and in salt forms [2]. The loading of metal catalysts, and its size should be considered carefully in synthesis of carbon nanomaterials [3]. Generally, the increase of oxygen functional groups on activated carbon could increase the adsorption of metal ions [4, 5]. This characteristics could be used to increase the metal catalyst loadings on activated carbon support. Xiong, et al. (2014) obtained increase of Niobia catalyst loading on activated carbon support by modifying activated carbon's functional groups using nitric acid [6]. Similar results were also obtained for different metal catalyst on activated carbon support [7, 8]. It is known that the surface modification of activated carbon increase its hidrophilicity.

There are several methods used in catalyst deposition on the surface of supports, namely impregnation and deposition precipitation. In this study, we used urea deposition precipitation method for catalyst preparation. This method has some advantages, such as smaller metal particle size, more homogenous catalyst distribution, higher catalyst loading, reproducible, and strong interaction of catalyst and support, compared with impregnation method [6, 9]. Tripathi, et al. (2012) synthesized CNSs and CNTs from castor oil by utilizing this method to obtained nano-sized Ni and Co catalyst[10].

In this study, the effect of activated carbon support's modification to the synthesis of CNSs was studied. The modification was done using simple wet oxidation method of nitric acid. Fe-catalyst loading was varied at 10%, 20%, and 30% of initial catalyst concentration. The CNSs synthesis was done using simple batch pyrolisis method, utilizing cooking palm oil as carbon source. The effect of modification to activated carbon's functional group was studied using Fourrier Transform Infrared Spectroscopy (FTIR), while the product of CNSs synthesis was observed using Scanning Electron

Microscope- Energy Dispersive X-ray Spectroscopy (SEM-EDS), Transmission Electron Microscopy (TEM), X-ray Diffraction (XRD) and Raman spectroscopy.

2. Experimental

Commercial activated carbon (AC),obtained from Merck (p.a.), was used as support for the growth of CNSs. The modified activated carbon used in this experiment was obtained by using nitric acid modification (HNO_3 65%– Merck p.a.). Activated carbon was mixed with nitric acid at mass ratio 1:1. This mixture was shaken for 4 hours at 60°C, and then repeatedly washed by aquadest until pH 6.5-7.0. The modified activated carbon then oven dried at 110°C for 24 hours (AC-NA). Fe-catalyst was prepared by urea (technical grade) deposition precipitation of iron (III) nitrate nonahydrate ($Fe(NO_3)_3.9H_2O$ – Merck p.a.). Activated carbon, $Fe(NO_3)_3.9H_2O$ (10%, 20%, and 30% mass ratio to activated carbon), and urea (1,5 moles per mole of iron [11]) was homogenously mixed in 50 mL distilled water. These mixtures was shaken for 4 hours at 90°C, and let cooled down for overnight. The activated carbon containing Fe-catalyst was separated from the liquid. The resulting solid then oven dried at 110°C for 24 hours. The dried powder was mixed with palm oil at 1: 3 mass ratios. This mixture was carbonized using electrical furnace at 700°C for 1 hour then let cooled down under N_2 atmosphere. The sample was labelled AC-xx or AC-NA-xx, where xx is the initial Fe-catalyst concentration.

The effect of surface modification to the surface functional group of activated carbon was studied using FTIR (Shimadzu FTIR 8400, KBr pellet method). Morphology and structural characteristics of carbon product was investigated by SEM-EDS (FE-SEM, S4100, HITACHI), TEM (FEI Tecnai TEM), Raman spectroscopy (Renishaw inVia Spectrometer System) and XRD (Bruker D8 Advance, Cu-K α radiation).

3. Results and Discussions

The effect of activated carbon to the surface functional groups was studied by FTIR spectra, presented in Figure 1. It could be observed some increase of stretching vibration of AC-NA at wavelength ranges of 1060-1200, 1530-1570, 1710, 3300-3400cm^{-1}. Based on the FTIR assigments of functional groups on carbon surfaces, these bands were assigned to stretching of C-OH (phenolic, ethers), quinones, carboxylic acid, and alcohols respectively [12, 13]. The increase of these bands indicates that surface modification of activated carbon have developed surface oxygen functional groups on the activated carbon. This result is confirmed by EDS atomic composition analysis presented in Table 1.

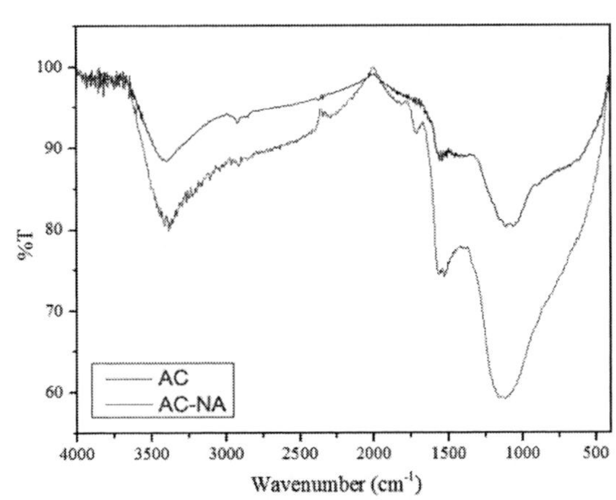

Figure 1. FTIR spectra of activated carbon before and after modification

The effect of activated carbon support modification could be observed from the EDS atomic composition analysis, presented in Table 1. The AC-NA samples had more Fe composition than the unmodified AC samples. The similar effect of activated carbon modification to deposition of various metal catalysts, such as Nb [6], Fe [7] and Pt [8] has been reported before. This increase was possible due to the increase of functional groups, thus making the carbon surface more hydrophilic [6], thus resulting on an increase of interaction between metal catalysts with the support.

The SEM images of AC and AC-NA samples at various samples are presented in Figure 2. It could be observed that CNSs are present as conglomeration of spheres, as stated by Nieto-Marquez (2010) [14]. The effect of catalyst concentration could be observed from the SEM images. There are increasing amount of spheres with higher catalysts concentration, as the catalysts promote the growth of carbon nanomaterial. It could be visually observed that smaller CNSs was obtained with higher catalysts concentration. This result is confirmed by observation of TEM analysis. Similar result of catalyst concentration to the formation of CNSs has been reported before [15]. However, there are no spheres observed in AC-NA 30 samples.

The result observed from TEM analysis is in accordance with SEM analysis, which the higher catalyst concentration, the more CNSs are obtained, and its size became smaller. It also could be observed that the CNSs presented as dense spheres decorated with iron catalysts. For AC-NA 30 samples, it could be observed onion-like spheres (carbon nano onion) were formed. We speculate carbon onion were formed from decompostion of CNSs at high catalysts concentration. It is known that carbon onion is formed by several graphitic layers with few nanometers in thickness, with or without catalysts trapped within its structure [16]. This results is supported by XRD and Raman analysis of the samples.

Table 1. %mass composition of AC and AC-NA samples at various catalyst concentration

Sample	%mass			Sample	%mass		
	C	O	Fe		C	O	Fe
AC	98,31	1,69	0	AC-NA	90,46	9,5	0,04
AC 10	84,27	11,76	3,96	AC-NA 10	83,15	3,75	13,1
AC 20	86,14	9,62	4,24	AC-NA 20	78,38	11,92	9,7
AC 30	67,87	16,51	15,62	AC-NA 30	51,5	18,78	29,72

XRD spectra of standard activated carbon, AC, and NA 10, 20, and 30 are presented in Figure 4. It could be observed that standard activated carbon had wide broad peak at 2θ of 20-30°, and 40°, indicating amorphous structure of activated carbon [17]. The spectra of AC 10, 20, 30, and NA 10, 20 show no significant difference, where C (100) and Fe peaks are observed. The C (100) peak is observed at 43°, indicating weak graphitization of carbon structure that is normally observed in CNSs [14]. Iron peaks are also observed at 30° (220), 35.4° (311), 57.09° (422), and 62.4° (440), indicating magnetite (Fe_3O_4) structure [18]. Strong peak of C (002) at 26.5° is observed at NA 30 sample, indicating sp^2 carbon structure of graphite, while sharp peaks at 45° and 65° are body centered cubic α-Fe catalyst [19]. The α-Fe is known as graphitization catalysts, thus influencing formation of graphitic C (002) structure.

IOP Conf. Series: Materials Science and Engineering **162** (2017) 012034 doi:10.1088/1757-899X/162/1/012034

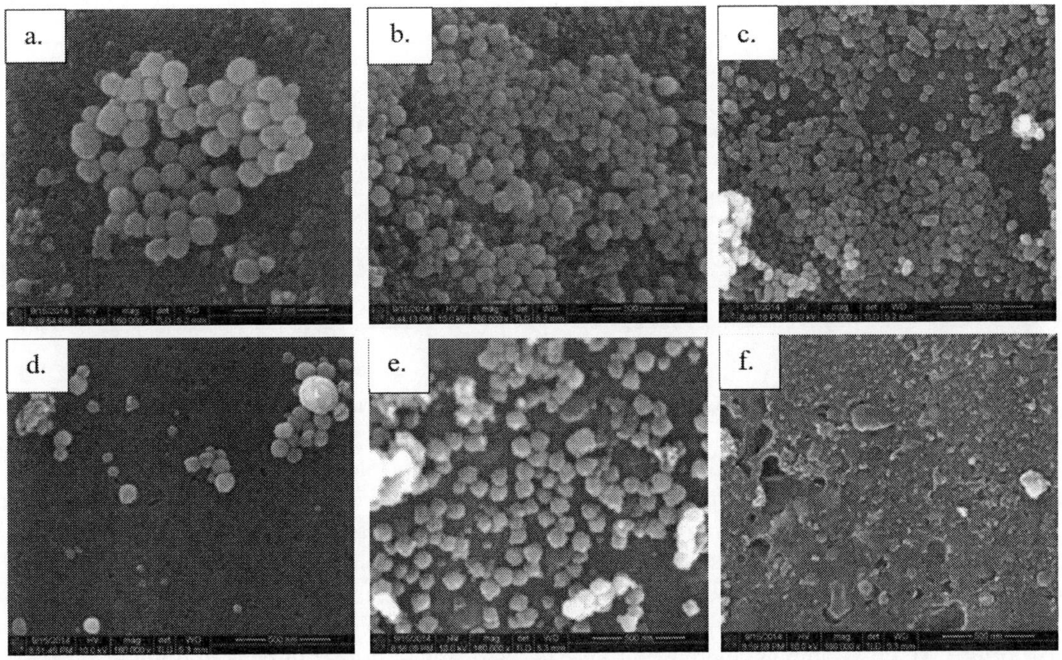

Figure 2. SEM images of AC 10 (a), AC 20 (b), AC 30 (c), and AC-NA 10 (d), AC-NA 20 (e), and AC-NA 30 (f) at 160,000× magnification

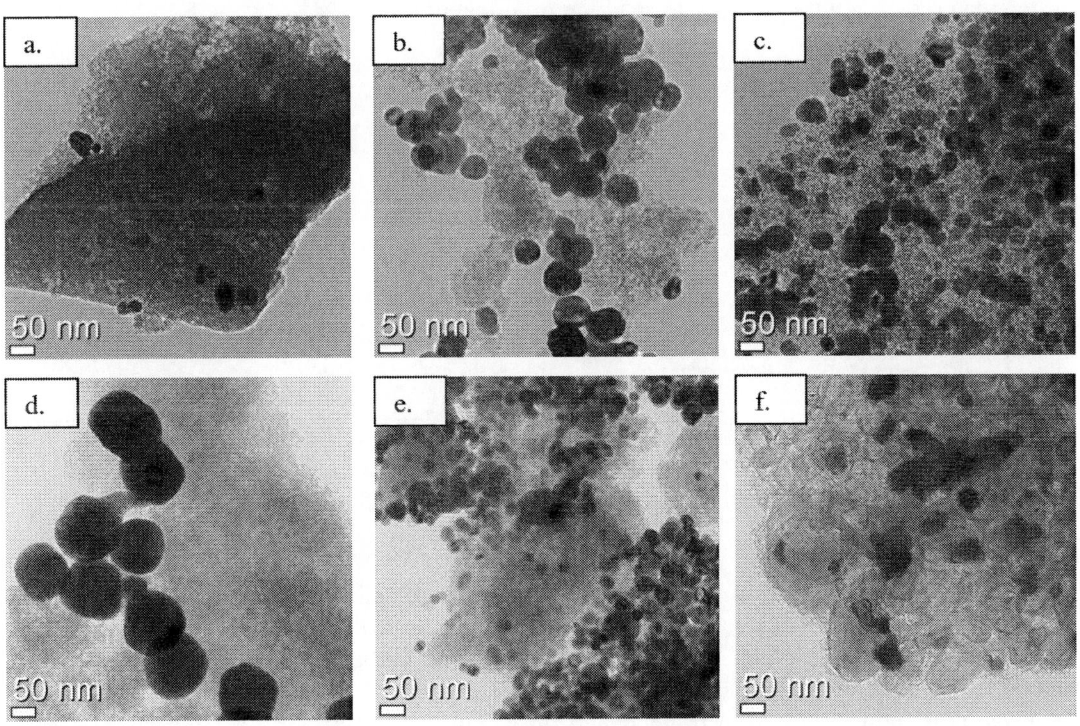

Figure 3. TEM images of AC 10 (a), AC 20 (b), AC 30 (c), and AC-NA 10 (d), AC-NA 20 (e) , AC-NA 30 (f)

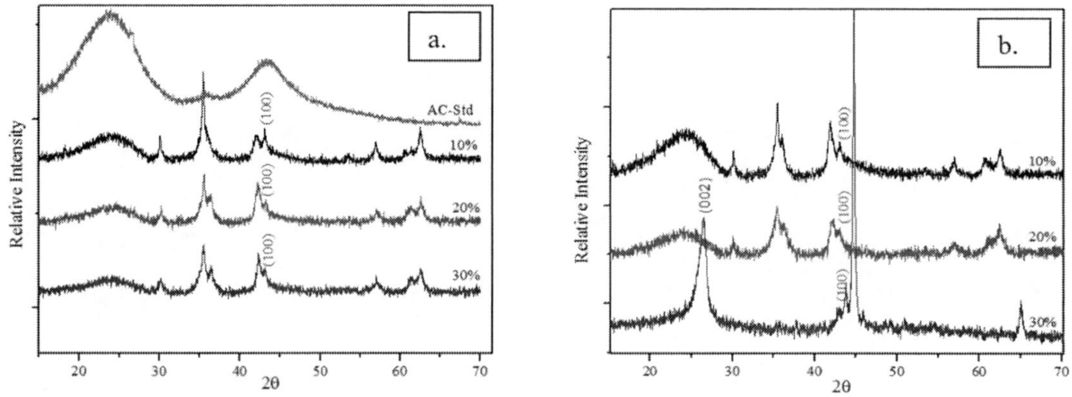

Figure 4. XRD spectra of AC Standard, AC 10, 20, and 30 (a), and AC-NA 10, 20, and 30 (b)

Raman spectra of AC and AC-NA are presented in Figure 5. There are two peaks observed on all AC and AC-NA samples. The D peak of disordered graphite structure is observed at 1344cm^{-1}, and the G peak of sp^2 bonded carbon atoms bonded in hexagonal lattice at 1593cm^{-1}. Although more CNSs are observed in SEM and TEM images, the ID/IG ratio show similar number around 0.9, indicating similar graphitic structure are obtained, with formation of graphitized carbon (CNSs), and smaller number of structural defects [20]. ID/IG ratio of CNSs have been reported to be aroud 0.86 to 1.20 [21-23]. For AC-NA samples, increase of Fe-catalyst leads to higher G peak and lower ID/IG ratio of 0.67. This result indicates formation of more graphitized carbon at higher catalyst concentration. The 2D peak is also observed in AC-NA 30 at 2700cm^{-1}. It is known that the 2D peak of Raman spectra of carbon materials is highly sensitive to stacking of graphene sheets[24].

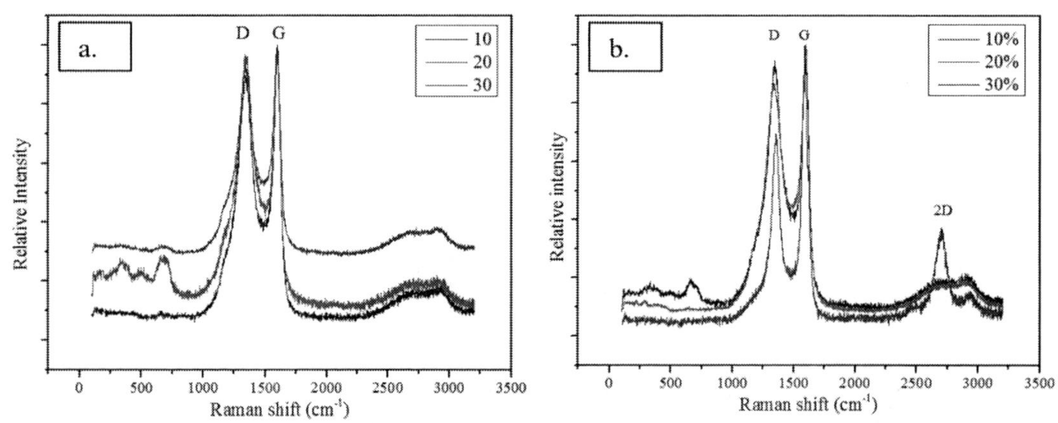

Figure 5. Raman spectra of AC 10, 20, 30 (a) and AC-NA 10, 20, 30 (b)

4. Conclusions

In summary, in this paper we have presented the effect of activated carbon support modification by utilizing simple wet oxidation method. The modification had successfully increased the oxygen functional groups and oxygen contents on the activated carbon support. The increase of oxygen functional groups, increase the activated carbon's hydrophilicity and its interaction with metal catalysts, thus more catalysts were deposited on its surface, influencing the formation of CNSs. At high metal catalysts concentration, carbon onion was obtained as product. We speculate the carbon onion was formed from decompostion of CNSs due to high metal catalysts content in the support.

Acknowledgements
This study was supported by The Indonesian Ministry of Education (Higher Education Commission) through DIKTI project: International Research Collaboration and Scientific Publication and Parahyangan Catholic University Centre of Research and Community Service contract No. III/LPPM/2016-02/66-P. The authors are immensely grateful for the financial supports.

References

[1] Mahanandia P, Vishwakarma P N, Nanda K K, Prasad V, Barai K and Mondal A K 2008 *Solid State Commun.* **145** 143-148
[2] Deshmukh A A, Mhlanga S D and Coville N J 2010 *Mater. Sci. Eng.R* **70** 1-28
[3] Yu Z, Chen D, Tøtdal B and Holmen A 2005 *Catal. Today* **100** 261-267
[4] Chen J P and Wu S 2004 *Langmuir* **20** 2233-2242
[5] Bhatnagar A, Hogland W, Marques M and Sillanpää M 2013 *Chem. Eng. J* **219** 499-511
[6] Xiong H, Nolan M, Shanks B H and Datye A K 2014 *Appl. Catal. A-General* **471** 165- 174
[7] Quintanilla A, Casas J A and Rodriguez J J 2007 *Appl. Catal. B-Environmental* **76** 135-145
[8] Burguete C P, Solano A L, Reinoso F R and d Lecea C S –M 1989 *J Catal* **115** 98-106
[9] Richardson J T and Dubus R J 1978 *J. Catal.* **54** 207-218
[10] Tripathi S, Sharon M, Maldar N N, Shukla J and Sharon M 2012 *Arch. Appl. Sci. Res.* **4(4)** 1788-1795
[11] Steen E V and Prinsloo F F 2002 *Catal. Today* **71** 327-334
[12] Shen W, Li Z and Liu Y 2008 *Recent Pat. Chem. Eng.* **1** 27-40
[13] Figueiredo J L, Pereira M F R, Freitas M M A and Orfao J J M 1999 *Carbon* **37** 1379-1389
[14] Nieto-Marquez A, Romero R, Romero A and Valverde J L 2010 *J. Mater. Chem.* **21** 1664-1672
[15] Kristianto H, Putra C D, Arie A A, Halim M and Lee J K 2015 *Procedia Chem.* **16** 328-333
[16] He C, Zhao N, Shi C, Du X, Li J and Cui L 2006 *J. Alloy Compd* **425** 329-333
[17] Senthilkumar S T, Senthilkumar B, Balaji S, Sanjeeviraja C and Selvan R K 2011 *Mater. Res. Bull.* **46** 413-419
[18] Kwon J H, Wilson L D and Sammynaiken R 2014 *Synthetic Met.* **197** 8-17
[19] Sun L, Tian C, Li M, Meng X, Wang L, Wang R, Yin J and Fu H 2013 *J. Mater. Chem.* A **1** 6462-6470
[20] Peng H, Ma G, Sun K, Mu J, Zhang Z and Lei Z 2014 *ACS Appl. Mater. Interfaces* **6** 20795-20803
[21] Wang Y, Su F, Wood C D, Lee J Y and Zhao X S 2008 *Ind. Eng. Chem. Res.* **47 (7)** 2294-2300
[22] Yuan D, Chen J, Zeng J and Tan S 2008 *Electrochem. Commun.* **10** 1067-1070
[23] Zhao N, Wang J, Shi C, Liu E, Li J and C He 2014 *J. Alloy Compd.* 816-822
[24] Malarda L M, Pimenta M A, Dresselhaus G and Dresselhaus M S 2009 *Phys. Rep.* **473** 51-87

IOP Conf. Series: Materials Science and Engineering **162** (2017) 012035 doi:10.1088/1757-899X/162/1/012035

Separating xylose from glucose using spiral wound nanofiltration membrane: Effect of cross-flow parameters on sugar rejection

N F M Roli[1,a], H W Yussof[1,b*], M N A Seman[1,c] , S M Saufi[1,d] and A W Mohammad[2,e]

[1] Faculty of Chemical & Natural Resources Engineering, Universiti Malaysia Pahang, Lebuhraya Tun Razak, 26300 Kuantan, Pahang, Malaysia

[2] Department of Chemical and Process Engineering, Faculty of Engineering and Built Environment, Universiti Kebangsaan Malaysia, 43600 UKM Bangi, Selangor, Malaysia

E-mail : [a]nurulfatihahmahamadroli@gmail.com, [b]hafizuddin@ump.edu.my, [c]mazrul@ump.edu.my, [d]smsaufi@ump.edu.my, [e]wahabm@eng.ukm.my

Abstract. A solution model consisted of two different monosaccharides namely xylose and glucose were separated using a pilot scale spiral wound cross-flow system. This system was equipped by a commercial spiral wound nanofiltration (NF) membrane, Desal-5 DK, having a molecular weight cut off (MWCO) of 150-300 g mol^{-1}. The aim of this present work is to investigate the effect of the cross-flow parameters: the trans-membrane pressure (TMP) and the feed concentration (C_0) on the xylose separation from glucose. The filtration experiments were carried out in total reflux mode with different feed concentration of 2, 5, and 10 g/L at different TMP of 5,8 and 10 bar. The performances of the NF membrane were evaluated by measuring the permeate flux and sugar rejection for each experiment. All the samples were quantified using a high performance liquid chromatography equipped by a fractive index detector. The experimental results indicated an increase in pressure from 5 to 10 bar which was a notable increase to the permeate fluxes from 2.66 x 10^{-3} to 4.14 x 10^{-3}L $m^{-2}s^{-1}$. Meanwhile, an increase in the C_0 increases the xylose rejection. At TMP of 10 bar and C_0 of 5 g/L, the observed xylose rejection and glucose rejection were measured at 67.19% and 91.82%, respectively. The lower rejection in xylose than glucose suggested that larger glucose molecule were not able to easily pass through the membrane compared to the smaller xylose molecule. The results of this phenomena proved that NF with spiral wound configuration has the potential to separate xylose from glucose, which is valuable to the purification of xylose in xylose production as an alternative to chromatographic processes.

1. Introduction

Chromatography separation method becomes vital in the field of biorefining industry. The separation process involved in this industry was included various sugar separation for example glucose-fructose separation, desugarisation of molasses, and xylose recovery from spent sulphite liquor for production of xylitol. Xylitol becomes a sweetener alternative in the food industry that has equal sweetness as sucrose [1]. Xylose mainly comes from hydrolysis of hemicellulose of agriculture waste, which consists around 55 % of total sugar. Another monosaccharide of interest, which is glucose, also results from the hydrolysis of hemicellulose covering around 25% of the total sugar [2]. Form previous study, the estimated cost of production of xylitol using chemical process and cost of hydrogenation of xylose to xylitol was about $ 350/ton (RM 754/ton) is less than 20% of the total cost of xylitol production. More than 80% of the total cost of xylitol production came from the cost of xylose crystal production which is about $ 2,300 - 2,500/ton (RM 7541 - 8,197/ton). There are few reasons for the high cost of xylose crystal production [3].

Thus, chromatographic separation would be the most suitable technique in sugar biorefinery [4]. However, the performance cost of this technique is relatively higher than that of other separation methods, and selecting an appropriate stationary phase is usually difficult [5] . Among various method of separation currently in study, nanofiltration (NF) offers cost-effective and easy-maintenance alternative separation of xylose from glucose [6,7]. Nanofiltration (NF) membrane separation process is a filtration process employing size exclusion membranes with pore size of 1-10 nm. NF membrane has advantage of low operation pressure, and high permeates flux. Nanofiltration membrane is capable of separation of charged and uncharged substance [8]. Sjoman et al. (2007) observed rejections of xylose were from 0 to 80% and glucose rejections were from 10 to 90% by using NF membrane. The observed rejection was highly dependent on the permeate flux of the membrane affected by pressure. Higher pressure will result higher permeate flux, hence the high rejection of monosaccharide and also few separation parameters that have an effect on nanofiltration beside pressure such as temperature, and total solution composition and concentration have been studied [9].

In the present study, the separation of xylose form glucose was conducted in a pilot scale NF system with Desal-5 DK spiral wound membrane. Experimental variables such as xylose-glucose concentration and transmembrane pressure were studied. In addition, streaming potential of the membrane was measured at various sugar concentration to provide evidence of sugar rejection on separation performance.

2. Materials and Methods

2.1 Model solutions

D-(+)-Xylose (>99%) and D-(+)-Glucose (>99.5%) were purchased from Sigma-Aldrich, Inc. The xylose to glucose concentrations was kept at the ratio 1:1. The feed concentrations were 2,5 and 10 g/L. The transmembrane pressure was tested at 5,8 and 10 bar.

2.2 Membrane and module

Table 1 shows the properties of the DK membrane used in this study. A Desal-5 DK NF membrane module was purchased from GE Water & Process Technologies, USA. This membrane is a three-layer thin film composite membrane with a polyamide top layer and a polysulfone support layer. The molecular weight cut-off of the membrane is 150-300 Da. Since the molecular weight of xylose and glucose are ~ 150 - 180 g mol^{-1} respectively [6], it was expected that concentration of xylose could be achieved by this membrane.

Table 1. Properties of Desal-5 DK membrane

Parameter	Characteristics
Manufacturer	GE Osmonics
Support material	Polysulfone
Surface material	Polyamide
Filtration area [m^2]	1.6
Average pore diameter [nm]	0.42 [10],~1 [11]
Maximum Temperature [^{0}C]	50 [12], [13]
Maximum Pressure [bar]	4.6-26.6 [11]
Molecular weight cut-off [g/mol]	150-300 [6], 200 [13]

2.3 Filtration experiment

The Desal-5 DK membrane was flushed with demineralized water at pressure of 5 bar for 30 minutes. Then the pressure was gradually raised to 10 bar and the membrane compaction was done at high pressure for 1 h and 30 minute. Next, the pure water flux of the virgin membranes was monitored every 15 minutes in 90 minutes until the volume of water was constant. This test was done to check the stability of the membrane flow. Before each filtration the pure water flux was measured at 35 ^{0}C to

IOP Conf. Series: Materials Science and Engineering **162** (2017) 012035 doi:10.1088/1757-899X/162/1/012035

control the cleanliness of the membranes. Pure water flux (J_w) was calculated by using Eq. (1) [14]. The sugar filtration was done in total recycled mode, i.e. permeate and retentate were circulated back to the feed tank to avoid any concentration change in the feed solution. The filtrations pressure used were 5,8 and10 bar. The temperature of the feed solution was monitored and maintained below 35°C to prevent damage to the membrane. The experiment variable was varies into two variables which transmembrane pressure (TMP) and feed concentration (C_0) to determine sugar rejection.

$$J_p = \frac{V}{A \leftrightarrow t} \tag{1}$$

Eq. (1) showed the calculation for permeate flux where V is permeate volume as a function of time, t in hour and A is the area of commercial membrane.

2.4 Chemical analyses
Samples were analyzed by high performance liquid chromatography (HPLC) to determine the amount of xylose and glucose. All samples were filtered through a 0.22-μm filter first. The chromatography system (Agilent 1200 LC) used in this study consisted of a pump with degasser, auto-sampler, chromatography column, column thermal controller and refractive index (RI) detector. The chromatography analysis was performed on a Rezex RHM-monosaccharide H^+, 300 x 7.8 mm column at 85 °C and water was used as mobile phase at flow rate of 0.6 mL min^{-1}. For each analysis, 20 μL of sample solution was injected into the system. The observed retention (R_i^{ob}) was calculated from Eq. (2), where C_f and C_p are the concentration of feed sugar in the tank and permeate, respectively. During the experiments, samples from the tank and permeate stream were taken and concentrations were determined.

$$R_i^{ob} = \left(1 - \frac{C_p}{C_f}\right) \times 100\% \tag{2}$$

Alternatively, the separation factor, X_{xyl} can be used to evaluate the performance of the membrane. In Eq. (3) where C_p (xyl) and C_p (glu) are the concentrations of xylose and glucose in the permeate and C_f (xyl) and C_f (glu) are the concentrations of xylose and glucose in the feed.

$$X_{xyl} = \frac{\left(C_p(xyl) / C_p(glu) \right)}{\left(C_f(xyl) / C_f(glu) \right)} = \frac{1 - R_{xyl}}{1 - R_{glu}} \tag{3}$$

3. Result and Discussion

3.1 Effect of transmembrane pressure (TMP)
The concentration used in this study was 2g/L, 5g/L and 10g/L. The flux for each concentration was increased as the pressure increased. Increasing concentration of xylose in the feed solution with time, increases the concentration polarization of these molecules on the NF membrane surface causing an increase in the osmotic pressure and a corresponding decrease in transmembrane pressure difference across the barrier. This results in a reduction in flux [16]. As shown in Figure 1, the permeation flux at 2g/L, 5g/L and 10 g/L was quite similar for each TMP. The effect of high permeation flux does not affect the sugar rejection and also the separation factor of xylose to glucose.

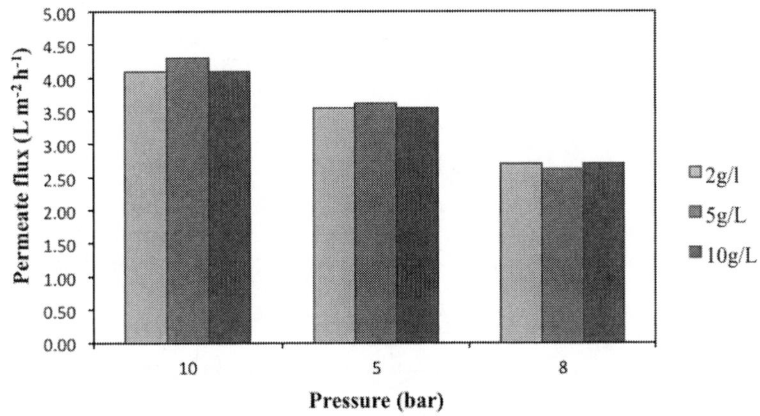

Figure 1. Permeate flux,(L/m^2h) ,J_w vs Pressure (bar)

3.2 Effect of feed concentration and TMP on xylose and glucose retentions

Separation factor is used to determine the performance of the membrane based on sugar rejection. Theoretically, the glucose will retain on the membrane while xylose passes through, according to the properties of xylose-glucose shown in Table 2. The high separation factor with average at 3.25 with 52.57 % xylose retention and and 85.64 % glucose retention was achieved when the feed concentration at 5g/L and at low pressure (5 bar) as shown in Figure 2. However, the results indicate that at 10 bar, the xylose retention was the highest which is 67.19 % but the xylose separation factor for 10g/L and 5g/L was declined. Previous study has been reported that the xylose separation factor will slightly decreased when the pressure increased. The possible causes were membrane compression and concentration polarization occurs at higher applied pressure with concentrated solution [10]. The result in this study was in contradicted to [6], which is the higher separation factor of xylose achieve at higher feed concentration. However, at total feed concentration 2g/L the separation factor was increased when the pressure increases. This might be because of at low concentration with high pressure, less resistance for smaller molecule concentration to pass through the membrane.

Table 2. Properties of Xylose and Glucose [15,10]

Properties	Xylose	Glucose
Molar mass (g/mol)	150.13	180.16
Stokes diameter (nm)	0.64	0.73
Equivalent molar radius (nm)	0.34	0.36

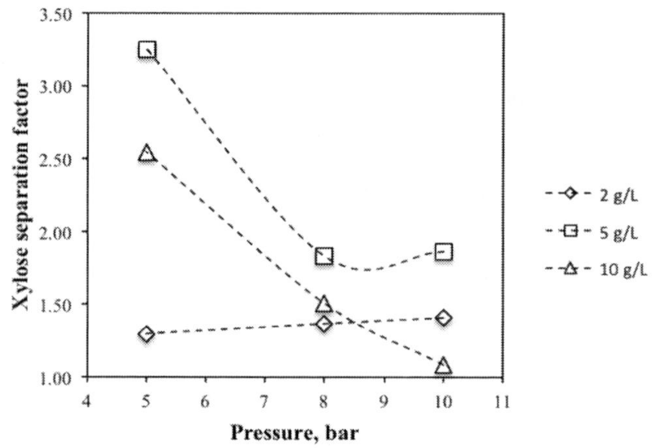

Figure 2. Separation factor xylose to glucose vs pressure at 5,8 and 10 bar

4. Conclusion

Based on the commercial process, sugar was usually separated by chromatography. In this paper, the separation of glucose and xylose was studied by using commercial nanofiltration membranes. Xylose, a pentose sugar, and glucose, a hexose sugar, were chosen as model compounds because xylose purification is an important step in commercial xylitol production. According to the results, nanofiltration has possibilities to enhance the yield and partially replace chromatographic methods in xylose production. There were several factors that will affect the separation of the performance of nanofiltration membranes like pressure, and concentration of feed that will lead to the changes in membrane structure, like compaction and swelling. The performance of Desal-5 DK membranes show that the effects of permeate flux in this study did not led to better xylose separation factor. The high separation factor with average at 3.25 achieved when the feed concentration at 5g/L and at low pressure (5 bar). The decreasing of xylose separation factor from this study indicates that concentration polarization had built a high resistance barrier restricting xylose from passing through the membrane. However, according to the results, nanofiltration showed a promising potential for xylose recovery.

References

[1] Rangaswamy S 2003 *Xylitol Production From D-Xylose by Facultative Anaerobic Bacteria*
[2] Flickinger M C and Drew S W 1999 Encyclopedia of Bioprocess Technology: Fermentation, Biocatalysis, and Bioseparation, vol 4 John Wiley
[3] da Silva A K, Chande 2012 D-Xylitol: Fermentative Production, Application and Commercialization *Springer*
[4] Hellstén S 2013 *Recovery of Biomass-derived Valuable Compounds Using Chromatographic* Finland
[5] Liu H, Zhao L, Fan L, Jiang L, Qiu Y and Xia Q 2016 Establishment of a nanofiltration rejection sequence and calculated rejections of available monosaccharides *J.Sep Purif Technol* **163** 319–330
[6] Sjoman E, Manttari M, Nystrom M, Koivikko H and Heikkila H 2007 Separation of xylose from glucose by nanofiltration from concentrated monosaccharide solutions *J. Memb. Sci.* **292** 106–115
[7] Goulas A K, Kapasakalidis P G, Sinclair H R, Rastall R A and Grandison A S 2002 Purification of oligosaccharides by nanofiltration *J. Memb. Sci.***209** 321–335
[8] Boussu K, Vandecasteele C and der Bruggen B 2008 Relation between membrane characteristics and performance in nanofiltration *J. Memb. Sci.* **310** 51–65

[9] Boussu K, der Bruggen B V, Volodin A, Haesendonck C V, Delcour J A, der Meeren P Vand Vandecasteele C 2006 Characterization of commercial nanofiltration membranes and comparison with self-made polyethersulfone membranes *J. Desalination.* **191** 245–253

[10] Mah K H, Yussof H W, Jalanni N A, Seman M N A and Zainol N 2014 Separation of Xylose from Glucose Using Thin Film Composite (TFC) Nanofiltration Membrane : Effect of Pressure, Total Sugar Concentration and Xylose/Glucose Ratio *Jurnal Teknologi.* **1** 93–98

[11] Weng Y H, Wei H J, Tsai T Y, Chen W H, Wei T Y, Hwang W S, Wang C P and Huang C P 2009 Separation of acetic acid from xylose by nanofiltration *J. Sep Purif Technol* **67** 95–102

[12] Balannec B, Vourch M, Rabiller-Baudry M and Chaufer B 2005 Comparative study of different nanofiltration and reverse osmosis membranes for dairy effluent treatment by dead-end filtration *J.Sep Purif Technol* **42** 195–200

[13] Mohammad A W, Teow Y H, Ang W L, Chung Y T, Oatley-Radcliffe D L and Hilal N 2005 Nanofiltration Membranes Review: Recent Advances and Future Prospects *J.Desalination* **356** 226–254

[14] Road B 2007 *Novel Fouling-Reducing Coatings for Ultrafiltration, Nanofiltration, and Reverse Osmosis Membranes* Austin

[15] Qi B, Luo J, Chen X, Hang X and Wan Y 2011 Separation of Furfural from monosaccharides by nanofiltration *J. Bioresour. Technol.* **102** 7111

[16] Murthy G, Sridhar S, Shyam M, Shankaraiah B and Ramakrishna M 2005 Concentration of xylose reaction liquor by nanofiltration for the production of xylitol sugar alcohol *J.Sep Purif Technol* **44** 221–228

Optimization of process condition of nanosilica production by hydrothermal method

N Qisti[1,a], N S Indrasti[2,b*], and Suprihatin[2,c]
[1] Student of Agroindustrial Technology, Bogor Agricultural Institute, 16680 Dramaga, Bogor, Indonesia
[2] Supervisor of Agroindustrial Technology, Bogor Agricultural Institute, 16680 Dramaga, Bogor, Indonesia

E-mail : [a]nuqyss@gmail.com, *[b] nastiti.indrasti@gmail.com, [c]suprihatin167@gmail.com

Abstract. Bagasse ashes have high silica content thus it can be used in nanosilicaproduction to increase its benefit value. This study aimed to get the best time for synthesis and to determine the optimum synthesis time and temperature. This study used the hydrothermal method, a simple method with relatively low reaction temperature and provide a good chemical homogeneity. Time varieties in synthesizing silica were 8,10 and 12 hours, at the temperature of 150 ^{O}C. But the results were not as expected. Moreover, optimization of synthesis temperature and time used 4 hours at the temperature of 150 ^{O}C based on previous studies. Optimization was conducted using the Response Surface Methodology (RSM). Later, a testusing PSA (Particle Size Analyzer) was performed to obtain particle sizes and PDI values (Polydispersity Index). The results showed that the prediction model of temperature synthesis was 152.67 ^{O}C, synthesis time of 6 hours, particle size of 276.288 nm and PDI value of 0.189642. The tests showed that the size of particle obtained was 330.39 nm and PDI value at 0.3580. Actual results and predicted results were not significant different.

1. Introduction

Bagasse ashes is the solid waste of burned bagasses. Its improved presence in the industrial area caused environmental pollution, especially air pollution caused by dust. Indonesia's sugar industry is still operating in various capacities which generates residue of bagasse combustion on boilers as bagasse ash in huge amount. Total production of bagasse ash is about 0.3% of the bagasse weight, so that sugar industry with production capacity of 5 000 tons a day results in bagasse ashes for about 15 tons a day [1]. Bagasse ashes utilization is currently only limited as the manufacture of organic fertilizers and burial material [2], the rest is disposed as a solid waste. The most dominant inorganic mineral elements of bagasse ash is silica (SiO_2) with maximum concentration up to 70.97% [3]. Thus, idea to utilize bagasse ash by making nanosilica as alternative product is generated. Nanosilica is a modified silica by using some treatments to produce nano-size particle.

Nanosilica used in science and industrial applications, such as catalysts, pigments, pharmaceuticals, [4], medicines, cosmetics, and food [5]. One material to be a deep concern of researchers is nanoparticles of silica (SiO_2) due to its good stability, chemically inert, biocompatible character that able to work in harmony with the body's work systems, and ability to form a single spherical [6].It is also revealed that the silica nanoparticles can be used as support material that is ideal for magnetic nanoparticlesas it is able to experience functionalization with ease, prevent anisotropic dipolar magnetic attraction when get an external magnetic field, and increase resistance to corrosion of magnetic nanoparticles. Silica particles have different roles for each resulted products as well as the quality of the product is determined by the size and particle distribution of silica itself within the system [7].

Sizing silica to the nanoscale requires special treatment by several methods such as sol - gel process, gas phase process, coprecipitation method, emulsion techniques, and plasma spraying and foging process (polymerization of dissolved silica into organo silica) [8]. In this study, the method used was the hydrothermal method in synthesizing SiO_2 to produce nanosilica that was uniform and homogeneous.

Hydrothermal method has many advantages such as simple in preparation, the reaction temperature is relatively low, uniform dispersion of the doping metal ions, as well as stoichiometry control and provide a good chemical homogeneity [9-10] stated that the hydrothermal method can improve the crystallinity, thermal stability, surface area and photocatalytic activity. In addition, the hydrothermal method used in this study because this method can be used in large-scale industry due to the equipment used is quite simple.

Based on statement [11], it is suggested touse higher temperature variations or longer variation of hydrothermal time in expectingto get crystalline silica without calcination. This study was reviewed by results of previous studies [12] which had already obtained best temperature in the synthesis process of nanosilica. Thus, this study only performed synthesis time modification by extending the synthesis time to get the best time fornanosilica synthesis so that it canproceed to the next stage, the optimization stage. Synthesis temperature and time optimization of hydrothermal methods in this research used the Respone Surface Methodology (RSM) to obtain information about the relationship between the response and several influenced factors [13]. In this study, responses to be included are nanosilica particle size and yield linked with the process factors of temperature and time. Expected results to be obtained are the relationship between the response and factors of nanosilica production process using the hydrothermal method as well as the optimum condition of temperature and time that will proceed to the next stage that is the scale up stage.

2. Methodology

This study was divided into two steps: Nanosilica Synthesis and Optimization of Synthesis temperature and time.

2.1 First Step: Nanosilica Synthesis

This step was conducted to get the best treatment of synthesis time.

2.1.1 Preparation of materials

Bagasse ashes were washed with distilled water then dried on oven at 105 °C for 5 hours. Dry ashes were heated at 700 °C for 6 hours to remove minerals and other impurities compounds. Results of incineration produced ash powder. Incinerated bagasse ashes then were stored in a desiccator to maintain the water contain[14].

2.1.2 Production of nanosilica

About 10 grams of boiler ashes were extracted in 80 ml of NaOH for 3.5 hours. The solution was filtered and washed using 20 ml of hot water. filtrate then was cooled down to the room temperature and was added with 5 N H_2SO_4 to get pH 2 and was added with NH_4OH until pH 7. Sol subsequently formed through the aging process for 3.5 hours at room temperature and then was dried at 105 °C for 12 hours [15].

Pure silica which was obtained then was put in a hydrothermal reactor to be processed. Temperature used product was flushed with sulphate was 150 °C while the time applied were 8, 10 and 12 hours. After that, the product of hydrothermal was titrated with 5 N $H_2SO_4$to get pH 8.5. Furthermore, it was rinsed for 7 times using warm distilled water to remove impurities contained in the product. Products which had been free of impurities then was aginned for 3 hours at temperature of 60 °C and was dried on oven with temperature of 105 °C for 12 hours. Product of drying results then was mortared to get nanosilica with fine powder. The weighing then was done to obtain the yield.

2.1.3 Characterization of nanosilica

The particle size distribution was observed with Vasco Particle Size Analyzer. A total of 0.1 grams of nanosilicapowder was dispersed in distilled water, was rotated with a magnetic stirrer for 10 minutes and was sonicated for 1 to 2 minutes. Scanning of nanosilica particles was carried by PSA for 2 to 5 minutes (Ismayana 2014).

2.2 Second Step: Optimization of synthesis temperature and time

This step was conducted to obtain the optimum temperature and time of synthesis.

2.2.1 Experimental design

The experimental design used to determine the optimization of synthesis temperature and time was Respone Surface Methodology (RSM) with ANOVA ($\alpha = 5\%$). This method was expected to obtain the optimum time and temperature of nanosilica synthesis by observing the relationship between responses (particle size, PDI and yield) and factors (temperature and time) of nanosilica production. Before establishing the experimental design, a trial was performed to determine the optimum conditions based on variable factors used. After that, median (center point) was determined in accordance with the optimum conditions obtained. The experimental design was the Central Composite Design (CCD) with two variable factors: X1 (temperature) and X2 (time).

Data of study results further were processed by regression analysis using software Design Expert 7.0.0 to generate a polynomial equation and contours of relationship between variables (factors) and responses. Validation of the Final Optimum Conditions in this study was the validation of optimal conditions in response to the particle size, PDI and yield recommended by the program. The validation phase was aimed to prove the response value of solution of factor combination recommended. After the trial phase, actual response result obtained was compared with the value of response prediction which was generated by the program.

The results of the optimum time and temperature of nanosilica synthesis then analyzed with theanalysis of characterization as described below.

The particle size distribution was observed by Vasco Particle Size Analyzer. A total of 0.1 grams of nanosilicapowder was dispersed in distilled water, was rotated with a magnetic stirrer for 10 minutes and sonicated for 1 to 2 minutes. Scanning nanosilica particles was carried by PSA for 2 to 5 minutes [16].

3 Results and Discussion

3.1 Nanosilica Synthesis

3.1.1 Results of particle size test and particle size distribution

The best treatment of particle size test by [12] became the basic reference in this study which functioned to determine the treatment of first step which is able to be seen in Figure 1.

Results in Figure 1 showed that there was decline of particle size at temperatures of 150 ^{o}C which was the best temperature in the previous study [12] in sixth hours. Based on statements of [11], it was suggested to use higher temperature variations or longer variation of hydrothermal time with expectation to getcrystalline silica without calcination. This suggestion led to hypothesis that longer extraction time will reduce the particle size, thus treatment time of nanosilicafor 8, 10 and 12 hours were conducted in step one.

Particle size measurement in this step showed that there was increase in the particle size for each increase time of synthesis reaction when reaction time was added from 8 hours to 10 hours or from 10 hours to 12 hours. This was due to the agglomeration process between particles which formed larger particles along with the increasing time of synthesis. It was also associated with Ostwald's Law in which particles with a big size will be formed by dissolving the particles with a smaller size [17].

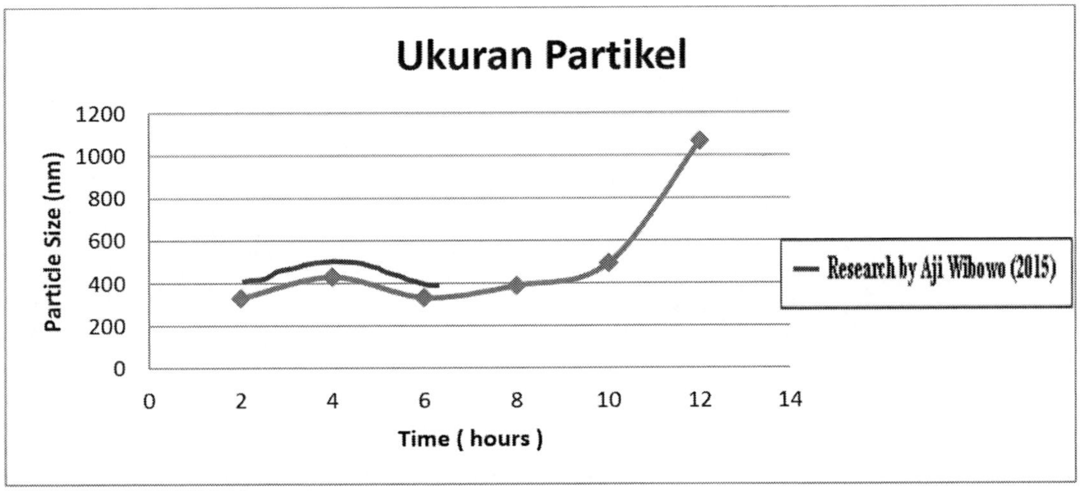

Figure 1. The test results of particle size

After comparing the two results, it was obtained that the best time that was 4 hours. Furthermore, the best treatment obtained was 150 ^{0}C for 4 hours which will used to determine the optimum temperature and time of this treatment.

This step can be seen PDI value as homogeneous indicator for nanosilica product. Polydispersity Index (PDI) can be seen on figure 2.

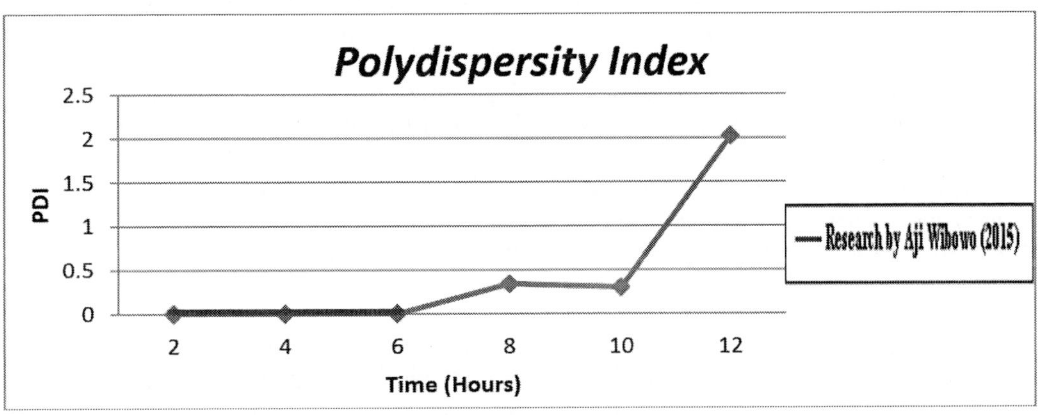

Figure 2 The test results of PDI

Figure 2 showed that results of PDI value were high enough. It showed in synthesis time on 8, 10 and 12 hours. It indicated that homogeneous level of nanosilica was bad when synthesis time increased.

3.2 Optimization of synthesis temperature and time

3.2.1 Analysis of factor combination and response optimization of particle size

Result of particel size test resulted in the particle size of 193.7 nm - 879.7 nm with average of 397.67 nm. Appropriate models to optimize the process conditions with particle size response was quadratic polinominal model.

The graph of contour and 3D surface of particle size value are presented in Figure 3.

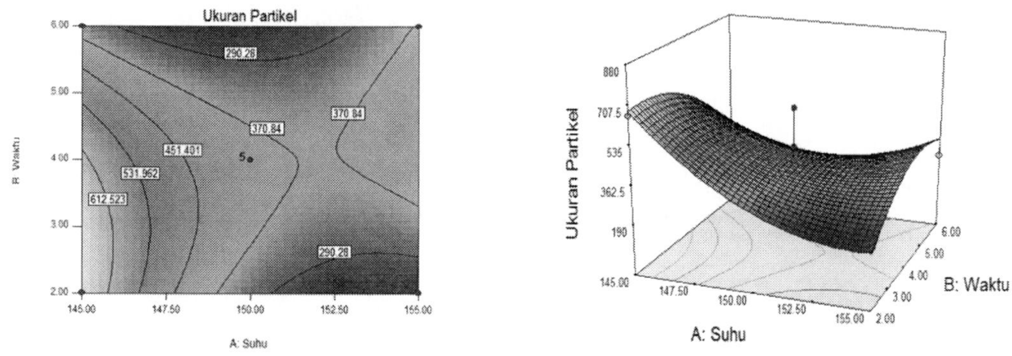

(a) **(b)**

Figure 3 Effect of synthesis temperature and time ofnanosilica production on particle size response of nanosilica: graph (a) contour and (b) three-dimensional response

Based on Figure 3(a), it is seen that there are circular contour lines with a red point places between the third and the fourth inner circle. Five red points on the contour are the central points of the design made. In particle size response, the value to be determined isnanosilica with minimum particle size. Circular contour line which points out directionand has blue area showed the best response value as it resulted in smaller size of particle. Five central points on the contour are not exactly at the center point of the circle. These conditions shows that the best response will be obtained by conditioning factors not at the center, but by shifting them to the top and the bottom right toward the blue area. Response of particle size will be optimum in an increased concentration of synthesis temperature and synthesis time in certain time range.

Figure 3 (b) is a 3D surface response of nanosilica particle size that shows that synthesis time and temperature was significantly affected the particel sizeof produced nanosilica. It is also seen that the relationship between temperature and time was very influential. When the synthesis temperature was increased and the synthesis time was lowered, particle size of nanosilica produced decreased. This result occured since the rising temperature (to some extent) impacted on an increase of solubility level of nanosilica produced. The Increase of solubility level generated a high level of ionic mobility, low viscosity and higher ion concentration that resulted in agglomerates to be separated and formed smaller size of particle [18].

Mean while, factor of particle size also became smaller when time was decreased (to some extent). This is due to the reason that agglomeration between particles did not occur in relatively short time. Therefore, longer reaction time of agglomeration process between particles impacted on larger particle formed when time was increased (up to certain extent). It is also associated with Ostwald's Law in which big size particles will be formed by dissolving smaller size particles [17]. However, within therange of 5 - 6 hours, particle size decreased as a result of other dissolving process of large size particle. In addition, the separation between the particles also occured thus the particle size tended to decrease [19].

3.2.2 Analysis of factor combination and response optimization of PDI (Polydispersity Index)

Result of PDI test showed PDI value of 0.0001 – 0.56 with average value of 0.32. The average value of PDI indicated that particle distribution was quite good because the value was less than 0.7. Based on statement [20], polydipersity index (PDI) is a parameter defines particle size distribution. PDI value between 0.01 to 0.5 - 0.7 is classified as monodispers (homogeneous) particle. Appropriate model to optimize the process conditions with PDI response is the quadratic polinominal model.

Contour graph and 3D surface of particle size value are presented in Figure 4.

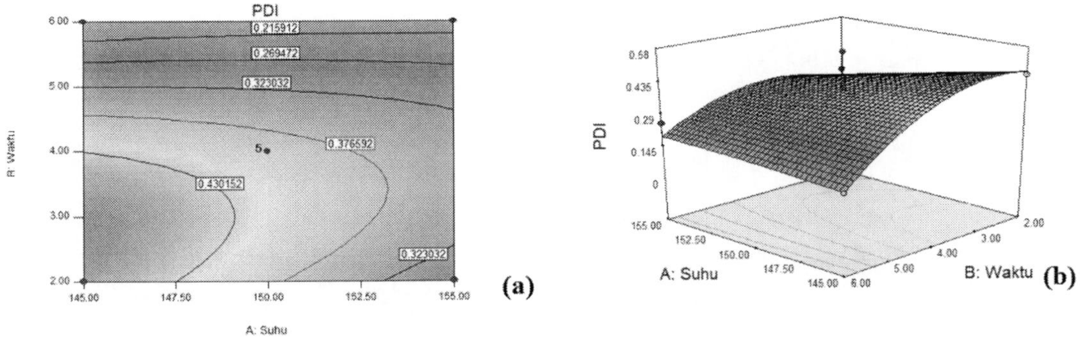

Figure 4 Effect of synthesis temperature and time ofnanosilica production on responses of nanosilicaPDI value: graph (a) contour and (b) three-dimensional response

Based on Figure 4 (a), circular contour lines with a red point places between the fourth and the fifth outer circle is seen. Five red points on the contour are central points of the design made. In response of particle size, the value to be determined is nanosilica with minimum value of PDI. Circular contour line which points out directionand has blue area shows the best response value where the PDI value generated was getting smaller. Five central points on the contour are not exactly located at the center point of the circle. This condition showed that the best response will be obtained not by conditioning factors at the center, but by shifting them upward towards the blue area. Response of particle size will be optimum in an increased concentration of synthesis temperature and a certain range of synthesis time.

Figure 4 (b) is the 3D of nanosilica PDI value surface response which indicates that temperature and time of synthesis significantly affected the PDI value produced by nanosilica. It is also seen that the relationship between temperature and time was very influential. When synthesis temperature and synthesis time were increased, PDI value of nanosilica produced decreased. The declineof PDI value indicated that uniformityof particle size which resulted in stronger bond between particles. According to [21], when synthesis temperature and time were raised, value of PDI will decrease. Thus, the longer the reaction time, the more controlled the process of formation and growth of particles. Similarly, increase on synthesis temperature up to certain point will result inlower PDI value. This is because the higher the synthesis temperature, the faster the formation rate and the growth of such particles [22]. In addition, it can also be influenced by the tendency of particles joining one another which forms particles with larger size and high stability.

3.2.3 Analysis of factor combination and yield response optimization

Based on the result of particle size test, particle size value of 15.7% - 32.8% with average yield of 26.86% was obtained. Appropriate model to optimize the process conditions with yield response was the 2FI polinominal model.

The contour graph and 3D surface of yield valueare presented in Figure 5.

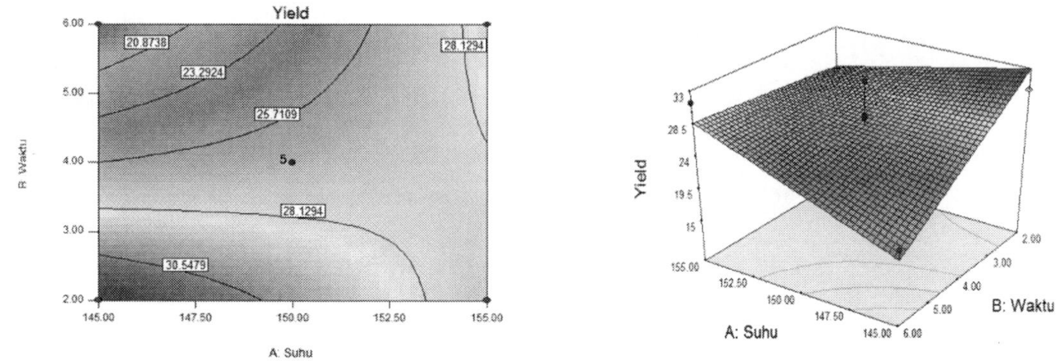

(a) **(b)**

Figure 5 Effect of synthesis temperature and time of nanosilica production on responses of nanosilica yield: graph (a) contour and (b) three-dimensional response

Figure 5 (a) and 5 (b) above were results which were not accepted by the model as synthesis temperature and time did not effect the yield response. Large or small influence on yield generated affects the washing process of materialsusedto remove the sodium compounds, thus pure silica and nanosilica can be obtained.

3.2.4 Optimization and validation of optimum condition

Particle size and PDI value are extremely important variable innanosilica production as they significantly affect the characteristics of nanosilica produced and are also used as quality standard of good nanosilica. After all variables and response were given the importance weight, solution of optimization results was further obtained to be 152.67 ^{0}C for 6 hours synthesis process. In this case, the model will be considered as good model if the value of response prediction is close to the actual verification value [23].

After the optimization solution wasobtained, validation on the prediction of given response variable was conducted. Result of Design Expert 7.0.0 program showed that there was value of responseprediction, followed by 95% of prediction interval whichwas the confidence value of individual observations. Prediction Interval (PI) was divided into two, namely 95% PI low and 95% PIhigh. As the 95% PI low was the lowest value of the predicted interval, 95% PI high was the highest value of the predicted interval. Comparison between response prediction value of optimization solutions and the actual value can be seen in Table 1.

Table 1 Comparison between the value of response prediction and actual value optimization solution

Response	Actual	Prediction	95% PI	
			Low	High
Particle size	330.39	276,288	346.79	639.03
PDI	0.3580	0.189642	0.137	0.42

Based on the result of validation test of response variable, it was revealed that nanosilica obtained from the synthesis temperature treatment of 152.67 ^{o}C and synthesis time of 6 hours produced particle with size of 330.39 nm and PDI value of 0.3580. Compared with the value of response prediction given by the program, response value of validation result was not significantly different.

4 Conclusions

Result of preliminary study showed that the particle size increasedwhen the synthesistime was increased. Thus, the best result used was from the previous study that was synthesis at temperature of 150 ^{o}C for 4 hours.

Data processing of RSM resulted in prediction modelshowing synthesis temperature of 152.67^{o}C, synthesis timeof 6 hours, particle size of 276.288 nm and PDI value of 0.189642.

Validation results were in the form of particle size of 330.39 nm and PDI value of 0.3580. Overall, actual resultand predictive model resultwere not significantly different.

Acknowledgement

This study is supported by Nano Technology Project of Agroindustrial Technology of Bogor Agricultural Institute.

5 References

[1] Huda M H 2012 Adsorption-desorpsi compound paraquatdiklorida with silica gel from the waste dregs of sugar cane (Sachharumofficinarum) *Thesis* (Yogyakarta :Sunan Kalijaga State Islamic University)

[2] Paramita W 2002 Influence of addition of Starter EM 4 against anaerobic composting process scales household (housing case study of PT Sarana Pensions Permai) *Thesis* (Surabaya :Environmental and Engineering FTSP-ITS)

[3] Hernawati I G 2010 A hybrid maize cultivation (Yogyakarta : State Islamic University Sunan Kalijaga)

[4] Zawrah M F, El Kheshen AA, Abd-El-All H 2009 Facile and economic synthesis of silicananoparticles *J. Ovonic Res.* **5** (5) 29-133

[5] Nabeshi H, Yoshikawa T, Arimori A, Yoshida T, Tochigi S, Hirai T, Akase T, Nagano Y, Abe K, Kamada H, Tsunoda, Shin-ichi, Itoh N, Yoshioka Y, Tsutsumi Y E Effect of surface properties of silica nanoparticles on their cytoxicity and cellular distribution inmurine macrophages Nanoscale Res. Lett.**6** (23)100-103

[6] Yuan Go H, Zhang F Z, Miao L, Yu R, Zhao H, Ian M 2010 Study of controllable preparation of silica nanoparticles with multi-sized and their size-dependent cytoxicity in pheochromocytoma cells and human embryonic kidney cells J. Health Sci. **56** (6) 632-640

[7] Fernandez R B 2012 The synthesis of nanoparticles SiO_2 using the sol-gel method and Application against the cytotoxic activity *Dissertation* (Padang : The University of Andalas)

[8] Jayanti N D 2014 The Optimalisasi Parameter PH pada Sintesis Nanosilika dari Pasir Besi Merapi dengan Ekstraksi Magnet Permanen Menggunakan Metode Kopresipitasi *skripsi* (Yogyakarta : State Islamic University SunanKalijaga)

[9] Zhang H F, MH, Yu LE 2012 Hydrothermal synthesis and photocatalytic performance of metal-ions doped TiO_2*App. Cat. A.***General 414** (12) 238-244

[10] Kim D S, Louie Y S 2007 The hydrothermal syntehis of mesoporus TiO_2 with high crystallinity, thermal stability, 1 wasarage surface area, and enhanced photocatalytic activity *App. Catal. A.* **General 323** (1)110-115

[11] Manusir M 2015 Pengaruh waktu tahan pada proses hidrotermal dan temperatur kalsinasi terhadap kekristalan silika dari bahan alam pasir kuarsa (Surabaya :Universitas Negeri Surabaya)

[12] Wibowo A 2015 Sintesis nanosilikadengan perlakuan variasi waktu dan suhu hidrotermal berbahan baku abuketel *Skripsi* (Bogor :InstitutPertanian Bogor)

IOP Conf. Series: Materials Science and Engineering **162** (2017) 012036 doi:10.1088/1757-899X/162/1/012036

[13] Nogales J M R, Roura E, Contreas E 20015 Biosynthesis of ethyl butyrate using immobilized lipase : a statistic approach *Process biochem* **40** (3) 63-68
[14] Thuadaij N Nuntiya A 2008 Preparation of nanosilica powder from rice husk ash by precipitation method. Chiang Mai J. Sci. **35** (1)206-211
[15] Ismayana A 2014 Perancangan proses co-composting dan nanoteknologi untuk penanganan limbah padat industri gula *Dissertation* (Bogor : InstitutPertanian Bogor)
[16] Setiawan W K 2015 Preparasi Nanosilika dari Abu Ketel dengan Metode Kopresipitasi sebagai Aditif Membran Elektrolit Berbasis Kitosan *Thesis* (Bogor : The Institute Pertaninan Bogor)
[17] Yao J H, Elder K R, Guo H, Grant M 1992 Theory and simulation of Ostwald ripening *Phys. Rev. B* **47** (21)
[18] Byrappa K, Masahiro Y 2001 Handbook of hydrothermal technology (New Jersey : Noyes Publications)
[19] Cao X, Yong C S, Yong N H,Guang P L, Liu C 2013 Integrated process of large-scale and size-controlled SnO_2 nanoparticles by hydrothermal method (China : Transactions of nonferrous metlas society of china) **23**725-730
[20] Nidhin M, Indumathy R, Sreeram K J, Nair B U 2008 Synthesis of iron oxide nanoparticles of narrow size distribution on polysaccharide templates *Bull. Matter. Sci.* **31**93-96
[21] Babu K, Dhamodaran R 2009 Synthesis of polymer grafted magnetic nanoparticle with the highest grafting density via controlled radical polymerization (India : Indian Institute of Technology)
[22] Nabifar A 2007 Investigation of kinetic aspect in nitroxide-mediated radical polymerization of styrene *Thesis* (Canada : University of Waterloo)
[23] Madamba P S 2005 Determination of optimum intermittent drying conditions for rough rice (*Oryza sativa, L.) Lebensm.-Wiss. U.-Technol.* **38** 157-156

IOP Conf. Series: Materials Science and Engineering **162** (2017) 012037 doi:10.1088/1757-899X/162/1/012037

Synthesis and characterization of polyester thin film composite membrane via interfacial polymerization: Fouling behaviour of uncharged solute

K H Mah[1,a], H W Yussof[1,b*], M N A Seman[1,c] and A W Mohammad[2,d]

[1] Faculty of Chemical & Natural Resources Engineering, Universiti Malaysia Pahang, Lebuhraya Tun Razak, 26300 Kuantan, Pahang, Malaysia

[2] Department of Chemical and Process Engineering, Faculty of Engineering and Built Environment, Universiti Kebangsaan Malaysia, 43600 UKM Bangi, Selangor, Malaysia

E-mail : [a]mahkahhong@gmail.com, [b]hafizuddin@ump.edu.my, [c]mazrul@ump.edu.my, [d]wahabm@eng.ukm.my

Abstract. Most hydrolysis studies on biomass in Malaysia produce high amount of xylose and glucose compared to other monosaccharides and most of them are acidic. Thin film composite (TFC) membrane developed via interfacial polymerization using triethanolamine (TEOA) and trimesoyl chloride (TMC) as monomers allows separation at low pH to occur without damaging its performance. Comparative studies were carried out on membranes with and without the thin film layer formed via interfacial polymerization on the polyethersulfone (PES) support. The surfaces of the membranes were characterized by field emission scanning electronic microscopy (FESEM), attenuated total reflectance Fourier transform infrared (ATR-FTIR) spectroscopy, and hydrophilicity via contact angle measurement. In addition, the performance and uncharged solute fouling behaviour of TFC membrane were also investigated. The TFC membrane used for characterization purposes was prepared at TEOA concentration of 4 % w/v in 1 x 10^{-6} M sodium hydroxide solution, TMC concentration of 0.25 % w/v in pure hexane, reaction time of 45 minutes, and cured at temperature of 60 °C. Characterization results showed a huge different between the synthesized TFC membrane and the un-synthesized PES membrane in term of surface properties and morphology. Nanofiltration results indicate that the formation of thin layer on top of PES support membrane improved the separation performance compared to PES support membrane. The synthesised polyester TFC membrane have irreversible fouling of 11.02 (±5.60) % and reversible fouling of 5.59 % using water as cleaning agent.

1. Introduction

Surface chemistry and morphology of membrane are greatly influence the transmembrane transport of components as well as on the efficiency of the membrane process [1]. Thus, surface modification of previously formed membranes is a promising approach to confer new properties to the existing membranes providing surfaces with tailor-made separation properties, energies and chemical functionalities [2]. The emerging of this asymmetric composite membrane also known as TFC membrane has changed the membrane industry significantly. TFC membranes is synthesized by first fabricating the asymmetric membranes and then coat another ultrathin barrier layer on top of the fabricated membranes. The coating techniques that had been introduced were interfacial polymerization, grafting polymerization, and layer-by-layer deposition [3,4]. Although a lot of effort has been put in developing membrane using grafting polymerization, a relatively small number of grafted polymeric materials found their way in commercial and large-scale applications [5]. Layer-by-layer deposition membranes have superiority over the other two fabrication techniques, but not practical in term of real applications where almost all studies are still conducted in lab scale [6]. Besides, most commercial TFC RO and NF membranes were fabricated through IP. IP allows the properties of top and bottom layer to be controlled independently, has pave ways to a wide variety of TFC membranes [3].

IOP Conf. Series: Materials Science and Engineering **162** (2017) 012037 doi:10.1088/1757-899X/162/1/012037

Interfacial polymerization techniques gained popularity over the other coating techniques when a variety of TFC membrane was successfully developed by many companies, allowing wide application in the separation processes industry. Interfacial polymerization technique allows properties of both bottom support and top barrier film to be individually tailored and optimized to achieve desired separation selectivity and rate [4]. Only few works have carried out interfacial polymerization technique to prepare other polymeric thin-film polyester membranes in comparison to thin-film polyamide membranes [2]. Polyester RO membrane synthesized by Kwak et al. [7] had water permeate flux eleven times higher than that of polyamide membrane. In addition, the incorporation of ester linkage increased the oxidation resistance of the membrane which has significantly increased the membrane tolerance on chlorine attack [2]. In recent years, many studies focused on the preparation of TFC membranes using interfacial polymerization technique with monomers having special functional groups [8].

Tang et al. [9] had utilized TEOA, an economical monomer in enhancing the performance of the thin-film composite membrane. TEOA belong to the tertiary amino group, where its molecules can be flexibly transferred into quaternary ammonium group through variation of feed pH. The changes on the membrane surface also increases membrane hydrophilicity at lower pH feed [9,10].

TFC membranes are susceptible to fouling. After a long period of academic and industrial researches, it is generally accepted that hydrophilicity, surface charge, and surface roughness of the TFC membrane surface are the major factors which affect the membrane susceptibility to fouling [1]. Membrane fouling is the accumulation of substances on membrane surface and/or within membrane pores, which causes a reduction in permeate flux (i.e., flow of water through membrane) with time for constant pressure filtration, or an increase in trans-membrane pressure with time for constant permeate flux filtration [11]. Membrane fouling classified into reversible fouling and irreversible fouling. The former can be cancelled by physical means of cleaning, whereas the latter needs chemical cleaning to be cancelled [12].

This work aims to prepare, characterize, and compare the characteristics of synthesized polyester TFC membrane via interfacial polymerization with its initial state (PES membrane). In addition, this works also add information on the xylose and glucose fouling behaviour of polyester TFC membrane. The synthesis parameters used in this work were based the best condition found in our previous study [13].

2. Experimental

2.1 Materials

The polyethersulfone membrane was purchased from AMFOR Inc. (China) with the commercial name of UF PES50. The membrane has a nominal molecular weight cut-off of 50 kDa, and water flux of 260 LMH at 25 °C and 50 psi. The chemicals used in this study were triethanolamine (R & M Marketing, Essex, UK), tri-mesoyl chloride (Alfa Aesar, UK), sodium hydroxide (Merck, Germany), n-hexane (Merck, Germany), xylose (Sigma Aldrich, USA), glucose (Sigma Aldrich, USA), and acetonenitrile (J.T. Baker, USA). All chemicals were analytical grade with high purity (> 99%) and acetonitrile with High Performance Liquid Chromatography (HPLC) grade.

2.2 Preparation of TFC membrane

The polyester thin film composite membrane was prepared based on previous studies [9,10,14] with slight modification in the preparation of aqueous phase solution. The aqueous phase solution composed of 4 % w/v TEOA in 1 x 10^{-6} M sodium hydroxide aqueous solution. The organic phase solution was prepared by dissolving 0.25 % w/v of TMC in pure hexane. PES support membrane was first immersed in aqueous phase solution for a period of 30 minutes. After that, the membrane was then drained and rolled with a glass rod to remove excess liquid. Then, the membrane was immersed in the organic phase solution for a period of 45 minutes. The polyester thin film composite membrane was cured at a temperature of 60 °C for 30 minutes.

2.3 Membrane characterization s (Field emission scanning electron microscope (FESEM), Attenuated total reflection Fourier transform infrared (ATR-FTIR) spectroscopy, Contact angle measurement)

The morphology of membrane surface was examined by FESEM equipment (Model JSM-6330F, JEOL, Japan). The membrane surface was examined by sticking the membrane on top of the support with carbon tape. All the membranes were coated with platinum under vacuum.

ATR-FTIR spectra were obtained using a Thermo Scientific Nicolet iS-10 FT-IR spectrometer equipped with Attenuated Total Reflection (ATR) element of Smart iTX AR Diamond and an Omnic 5.1 software. The membrane active layers were pressed tightly against the crystal plate. The TFC membrane, PES membrane, TEOA and TMC were analysed, with each spectrum averaged from 256 scans collected from 650 to 4000 cm^{-1} at 2 cm^{-1} resolution. The experiments run with air as the background and no baseline or ATR corrections were applied.

The hydrophilicity of all the membranes were measured using contact angle goniometer (JY-82, Chengde testing machine Co. Ltd., China) using the sessile drop method. A glass slide was immersed in pure ethanol for 10 min, rinsed with ultrapure water and then dried before being used as an inert flat surface. Membrane was then placed on top of the glass slide and clipped making sure the surface of membrane is parallel to the glass flat surface. A droplet of 5 µL of ultrapure water at the end of 'I'-shaped needle was carefully deposited to membrane surface to avoid the effect of falling force by gravity. A static image of the droplet in equilibration with the membrane surface was taken. Later, image analysis and contact angle computation were performed using the analysis software. For ensuring the accuracy, the measurements were performed at five different locations, and then the average value was regard as the final contact angle result.

2.4 Nanofiltration set-up and fouling test

A Millipore stirred cell (Model 8200, Millipore-Amicon Corporation, USA) having a maximum volume uptake of 200 mL and an effective membrane area of 2.87 x 10^{-3} m^2 was used in all experiments. Prepared TFC membrane and virgin PES membrane was fitted into the membrane holder at the bottom of the stirred cell. Other parts are then assembled together and place on top of a magnetic stirrer (Model MS-20D, Daihan Scientific Co. Ltd., South Korea).

The fouling experiment was conducted according to the following procedure: (a) pre-compaction of the membrane with ultrapure water at 4 bar for 10 minute to stabilize the pressure-induced membrane deformation; (b) filtration of the ultrapure water to determine initial water flux, J_0 for fouling index calculation by measuring mass of permeate collected at 4 bar and stirring speed of 300 rpm in 1 hour; (c) filtration of 180 mL of model sugar solution (1 % w/v of both xylose and glucose at ratio 1:1) at 4 bar for 6 hour and stirring speed of 300 rpm, where mass of permeate collected every 20 minute for normalized flux calculation; (d) cleaning of membrane with ultrapure water with stirring speed of 350 rpm for 30 minute without applying pressure to remove weakly adsorbed solutes on membrane surface; (e) filtration of the ultrapure water to determine final water flux, J_f for fouling index calculation by measuring mass of permeate collected at 4 bar for 1 hour and stirring speed of 300 rpm. All the mass observed in this experiment is converted to volume using density of water at 25 °C. The fouling of membrane was measured using relative flux reduction (RFR) and normalized relative flux (NRF). RFR is a measure of irreversible fouling calculated using Eq. 1. NRF demonstrates the fouling trend during filtration calculated using Eq. 2 where J_0 is the pure water flux before filtration ($Lm^{-2}h^{-1}$), J_f the pure water flux after filtration ($Lm^{-2}h^{-1}$), and J_t is the permeate flux at time t ($Lm^{-2}h^{-1}$).

$$RFR=\left(\frac{J_0-J_f}{J_0}\right)\times 100\% \tag{1}$$

$$NRF=\frac{J_t}{J_0} \tag{2}$$

3. Results and Discussion

3.1 Membrane characterizations

The membrane was characterized using ATR-FTIR spectroscopy to analyse the surface chemistry of support PES membrane before and after interfacial polymerization between TEOA and TMC. Figure 1 showed spectra belonging to virgin PES membrane and freshly prepared TFC membrane. A strong peak at wavenumber 1724 cm^{-1} was seen in the spectra of TFC membrane (Figure 1 (a)) beside the typical PES bands (Figure 1 (b)). This peak falls into the band belonging to a functional group of ester ($C = O$ stretch) [2,6]. TFC membranes developed using TEOA and TMC in the past were particularly suitable for treating acidic feeds due to the formation of polyester thin-layer [9,10].

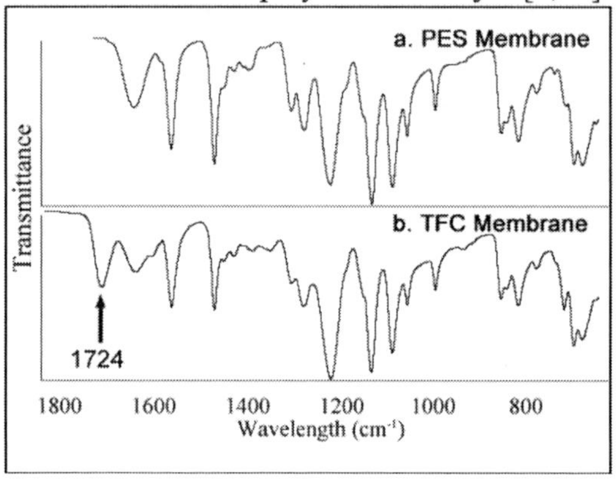

Figure 1. ATR-FTIR Spectra for (a) PES and (b) TFC membrane

FESEM analysis was performed to observe the morphology of the support PES membrane surface after undergo interfacial polymerization to become TFC membrane. Figure 2 shows the FESEM surface images of both PES and TFC membrane at 5,000x magnification. The surface of TFC membrane in Figure 2 (b) spotted a "ridge and valley" structure, which were identical to previous studies [9,13] which also synthesized polyester membrane from TEOA and TMC monomer. A distinct different of surface roughness and morphology can be seen from both images, between Figure 1 (a) and (b) proved that there was the formation of TFC layer on top of the support PES membrane by interfacial polymerization.

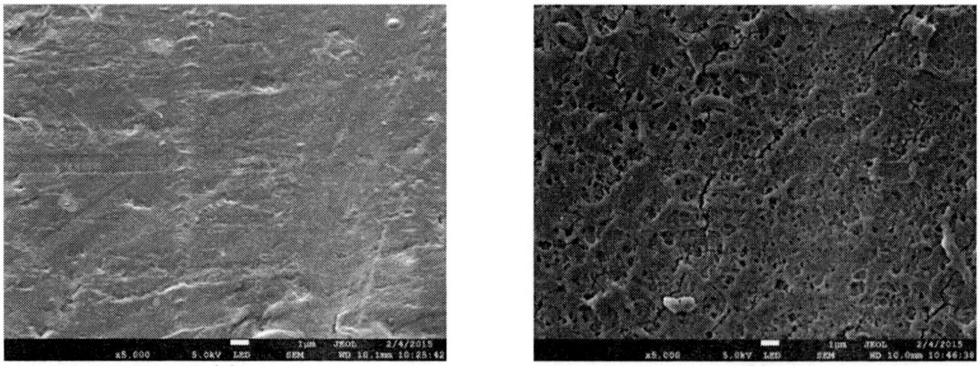

(a) (b)

Figure 2. FESEM for (a) support PES and (b) TFC membrane at 5,000x magnification

The contact angle for support PES membrane was 51.03 ± 3.68 ° and TFC membrane was 30.08 ±1.27 ° as shown in Figure 3. TFC membrane has smaller contact angle than the PES implying that formation of polyester increases the hydrophilicity of the membrane, although the surface of membrane become rougher. Past studies [15,16] point out hydrophilic compounds such as xylose can easily pass through the membrane due to hydrophilic-hydrophobic interaction between membrane and compounds. Commercial membrane with better hydrophilic surface or lower contact angle also have been reported to have higher xylose separation factor [17].

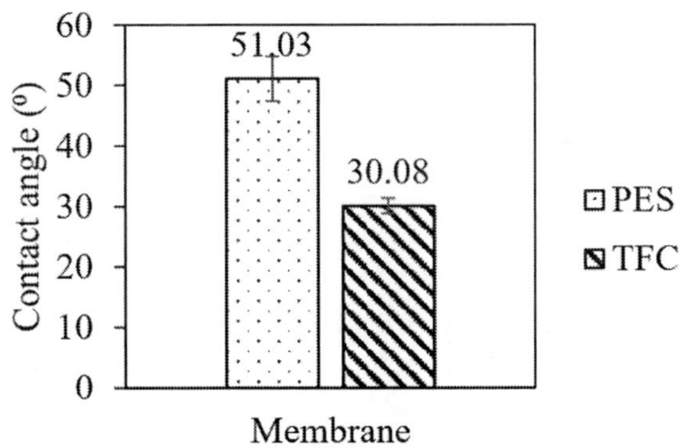

Figure 3. Contact angle for support PES and TFC membrane

3.2 Fouling behaviour

The flux reduction over 6 hours (360 minutes) filtration time was observed as illustrated in Figure 4, where the NRF after a course of this filtration was 16.61 (±3.53) %. The TFC membrane has RFR of 11.02 (±5.60) % which represent the irreversible fouling. A reversible fouling of 5.59 % can be obtained after cleaning with ultrapure water. From Figure 4, it can clearly see that the rate of fouling was less severe compared to fouling of humic acid in previous study [10]. The fouling observed in this study is in good agreement with commercial membrane [18] where fouling of sugar was less susceptible in operations under 2 days.

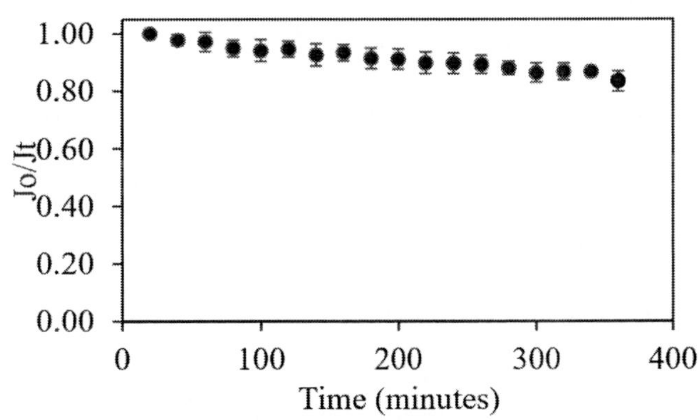

Figure 4. NRF during nanofiltration of model sugar solution

4. Conclusion

The formation of polyester thin layer on top of polyethersulfone membrane were successfully verified using ATR-FTIR and FESEM analysis. The polyester thin layer formed altered the surface of the membrane from smooth and homogeneous to rough surface with "ridge and valley" structure. The formed polyester thin layer not only altered the morphology of the membrane surface but improved the hydrophilicity of the surface itself, which can be verified by contact angle measurement. Besides, the synthesised membrane has less severe fouling rate of xylose and glucose.

Acknowledgement
The authors wish to express their thanks to Ministry of Education through the financial aid from grant LRGS/2013/UKM-UKM/PT/03, RDU140901, and Universiti Malaysia Pahang's postgraduate research grants scheme (GRS1403113). The authors also wish to acknowledge the Ministry of Education Malaysia for sponsoring K.H. Mah postgraduate's study via MyBrain15.

References
[1] Tarboush B J A, Arafat H A, Matsuura T, Rana D 2009 Recent advances in thin film composite (TFC) reverse osmosis and nanofiltration membrances for desalination *J. Appl. Membr. Sci. Technol.* **10** 41-50
[2] Seman A M N, Khayet M, Hilal N 2010 Nanofiltration thin-film composite polyester polyethersulfone-based membranes prepared by interfacial polymerization *J. Memb, Sci* **348** 109-116
[3] Lalia B S, Kochkodan V, Hashaikeh R, Hilal N 2013 A review on membrane fabrication: Structure, properties and performance relationship *Desalination* **326** 77-95
[4] Lau W J, Ismail A F, Misdan N, Kassim M A 2012 A recent progress in thin film composite membrane: A review, desalination **287** 190-199
[5] Nasef M M, Guven O 2012 Radiation-grafted copolymers for separation and purification purposes: Status, challenges and future directions *Prog. Polym. Sci.* **37** 1597–1656
[6] Xu G R, Wang S H, Zhao H L, Wu S B, Xu J M, Li L, Liu X Y 2015 Layer-by-layer (LBL) assembly technology as promising strategy for tailoring pressure-driven desalination membranes *J. Memb. Sci.* **493** 428-443
[7] Kwak S Y, Yeom M O, Roh I J, Kim D Y, Kim J J 1997 Correlations of chemical structure, atomic force microscopy (AFM) morphology, and reverse osmosis (RO) characteristics in aromatic polyester high-flux RO membranes *J. Memb. Sci.* **132** 83-191
[8] Lau W J, Gray S, Matsuura T, Emadzadeh D, Chen J P, Ismail A F 2015 A review on polyamide thin film nanocomposite (TFN) membranes: History, applications, challenges, and approaches *Water Res.* **80** 306-324
[9] Tang B, Huo Z, Wu P 2008 Study on a novel polyester composite nanofiltration membrane by interfacial polymerization of triethanolamine (TEOA) and trimesoyl chloride (TMC) : I. preparation, characterization and nanofiltration properties test of membrane *J. Memb. Sci* **320** 198-205
[10] Jalanni N A, Seman M N A, Faizal C K M 2013 Investigation of new polyester nanofiltration (NF) membrane fouling with humic acid solution *J. Technol.* **65** 69-72
[11] Jamal S, Chang S, Zhou H, Filtration behavior and fouling mechanisms of polysaccharides, Membranes (Basel) **4** 319-332
[12] Peinemann K V, Pereeira-Nunes S, 2010 *Membranes for water treatment* (Wiley-VCH Verlag GmbH & Co)
[13] Mah K H, Yussof H W, Seman M N A, Mohammad A W 2016 Separation of xylose using thin-film composite nanofiltration membrane: Screening of interfacial polymerization factors *RSC Adv.* **6** 69454-69464
[14] Mah K H, Yussof H W, Jalanni N A, Seman M N A, Zainol N 2015 Separation of xylose from glucose using thin film composite (TFC) nanofiltration membrane: Effect of pressure, total sugar concentration and xylose/glucose ration *J. Technol* **70** 93-98
[15] Boussu K, Vandecasteele C, der Bruggen B 2008 Relation between membrane characteristics and performance in nanofiltration *J. Memb. Sci.* **310** 51-65
[16] Braeken L, Ramaekers R, Zhang Y, Maes G, Der Bruggen B, Vandecasteele C 2005 Influence of hydrophobicity on retention in nanofiltration of aqueous solutions containing organic compounds *J. Memb. Sci.* **252** 195-203

[17] Sjoman E, Manttari M, Nystrom M, Koivikko H, Heikkila H 2007 Separation of xylose from glucose by nanofiltration from concentrated monosaccharide solutions *J. Memb. Sci.* **292** 106–115
[18] Murthy G, Sridhar S., Shyamsunder M, Shankaraiah B, Ramakrishna M, 2005 Concentration of xylose reaction liquor by nanofiltration for the production of xylitol sugar alcohol *Sep. Purif. Technol.* **44** 221–228.

Institute of Physics
Dirac House, Temple Back
Bristol BS1 6BE UK

ISSN: 1757-8981
ISBN 978-1-5108-3525-2

9 781510 835252

CURRAN ASSOCIATES INC.
proceedings.com

34th PIANC World Congress 2018

Connecting Maritime Hubs Globally

Dredging

Panama City, Panama
7 - 11 May 2018